Practical Amateur Astronomy

Digital SLR Astrophotography

Second Edition

Digital SLR cameras have made it easier than ever before to photograph the night sky. Whether you're a beginner, nature photographer, or serious astronomer, this is the definitive handbook to capturing the heavens. Starting with simple projects for beginners such as cameras on tripods, it then moves onto more advanced projects including telescope photography and methods of astronomical research. With 80% revised and updated material, this new edition covers nightscapes, eclipses, using cameras with sky trackers and telescopes, and tools for identifying celestial objects and investigating them scientifically. Image processing is discussed in detail, with worked examples from three popular software packages – *Nebulosity*, *MaxIm DL*, *PixInsight*, and *DeepSkyStacker*. Rather than taking a recipe-book approach, Covington explains how your equipment works as well as offering advice on many practical considerations, such as choice of set-up and the testing of lenses, making this a comprehensive guide for anyone involved in astrophotography.

MICHAEL A. COVINGTON is one of America's leading amateur astronomers and the author of the highly acclaimed *Astrophotography for the Amateur* (Cambridge University Press, second edition, 1999). He was a research scientist in computational linguistics and artificial intelligence at the University of Georgia. Now retired from academia, he runs a consulting business in Athens, Georgia, from where he continues to take pictures of the stars.

Practical Amateur Astronomy

Digital SLR Astrophotography

Second Edition

Michael A. Covington

CAMBRIDGE
UNIVERSITY PRESS

CAMBRIDGE
UNIVERSITY PRESS

University Printing House, Cambridge CB2 8BS, United Kingdom

One Liberty Plaza, 20th Floor, New York, NY 10006, USA

477 Williamstown Road, Port Melbourne, VIC 3207, Australia

314-321, 3rd Floor, Plot 3, Splendor Forum, Jasola District Centre, New Delhi – 110025, India

79 Anson Road, #06-04/06, Singapore 079906

Cambridge University Press is part of the University of Cambridge.

It furthers the University's mission by disseminating knowledge in the pursuit of education, learning, and research at the highest international levels of excellence.

www.cambridge.org
Information on this title: www.cambridge.org/9781316639931
DOI: 10.1017/9781316996799

First published 2007
Second edition 2018

A catalog record for this publication is available from the British Library.

ISBN 978-1-316-63993-1 Paperback

Additional resources for this publication available at www.cambridge.org/covington2

Soli Deo gloria

Contents

Preface

When I wrote the first edition of this book, I said that the time was not yet ripe for a comprehensive handbook of DSLR astrophotography. Now it is, and I have rewritten almost the entire book from scratch because so much has changed and so much more knowledge is available.

And the torrent of new developments never stops. Please check this book's web site, www.dslrbook.com, for updates and additional information immediately.

Not everyone will read all the chapters of the book straight through. To cover such a complicated, technical subject, I have had to spiral outward through the subject matter, passing through several regions more than once. Avid daytime DSLR photographers may go straight to Chapter 4, and experienced astrophotographers will find the later chapters more useful. Many readers will skim Chapter 2 on the first pass and then come back to it as needed.

Two notes about pictures:

- Throughout this book, if the caption of a picture specifies only a lens and its *f*-ratio, such as "300-mm *f*/4 lens," you can assume the lens was used wide open, as is usual in astrophotography. If it is stopped down, the caption will say so, such as "300-mm *f*/4 lens at *f*/5."
- You can assume that all the pictures in Chapter 5 and later were calibrated in the normal manner with dark frames, flats, and flat darks or bias frames, unless I say otherwise.

I want to thank my wife Melody and my daughter Sharon for their patience and for help with illustrations and URL checking. I thank several people who contributed images and data; they are acknowledged where their material appears. All the images not otherwise credited are my own work.

Michael Covington

Part I
DSLRs for Astrophotography

Chapter 1
Welcome to DSLR Astrophotography

Digital single-lens reflex (DSLR) cameras are an unusually cost-effective way to photograph the sky. The lower-cost DSLRs marketed to amateur photographers perform almost identically to the high-end professional models.

Unlike compact digital cameras and smartphones, the DSLR is designed for versatility and performance. It has a larger sensor, giving higher sensitivity to light and lower noise; interchangeable lenses, with the ability to couple to telescopes and other optical instruments; and full manual control.

Besides ready-to-view JPEG files, DSLRs also deliver "raw" image files that indicate the number of photons that reached each pixel. With raw files, you can subtract out the brightness of the sky background, correct for measured irregularities in the sensor, and combine multiple exposures of the same subject. These capabilities make the DSLR a powerful tool for photographing faint celestial objects.

By accumulating faint light, DSLRs can capture views of the sky that cannot be seen by human eyes in any telescope. Consider for example Figure 1.1. This is a six-minute exposure of the Andromeda Galaxy, with the contrast and brightness adjusted in *Photoshop* as if it were a daytime photograph; no special processing of raw files was done.

No telescope can give you that view. The reason is that the human eye cannot accumulate light. The outer parts of the galaxy are too dim for the eye to see, and telescopes don't make extended objects (surfaces) brighter. (If they did, sailors would hurt their eyes using telescopes to view distant ships on a sunny day.) Even with a telescope, the eye can see the spiral arms of the galaxy only faintly, under ideal conditions. By accumulating light and subtracting out the background glow of the suburban sky, the DSLR made the spiral arms clearly visible under conditions that were far from ideal.

As you move past the beginner stage, you can do just as much computer control and image enhancement with a DSLR as with an astronomical CCD camera. Some astrophotographers bring a laptop computer into the field and run their

Figure 1.1. The galaxy M31 as the image came from the camera, with no processing except adjustment of brightness and contrast. Canon Digital Rebel (300D); single 6-minute exposure through a 300-mm lens at *f*/5.6, captured as JPEG.

DSLR under continuous computer control. Others, including me, prefer to use the camera without a computer and do all the computer work indoors later.

1.1 What is a DSLR?

1.1.1 Digital Single-Lens Reflex Cameras

A DSLR is a digital camera that is built like a film SLR (single-lens reflex) and has the same ability to interchange lenses. You can attach a DSLR to anything that will form an image, whether it's a modern camera lens, an old lens you have adapted, or a telescope, microscope, or other instrument.

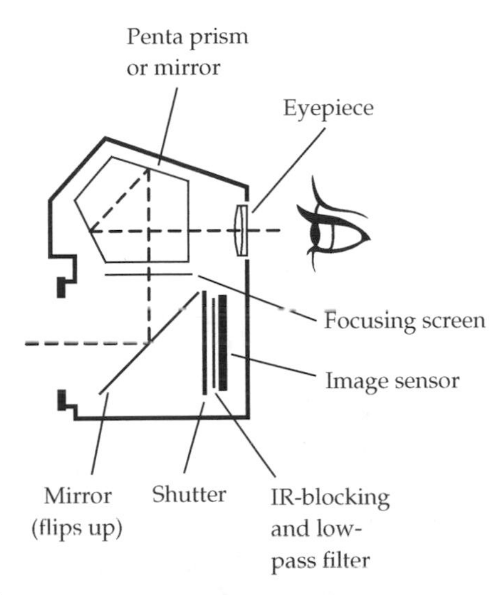

Figure 1.2. A DSLR is a single-lens reflex camera with a digital image sensor. Mirror and eyepiece allow you to view the image that will fall on the sensor when the mirror flips up and the shutter opens.

"Reflex" means that the camera has a mirror that enables you to view the image formed by the lens (Figures 1.2 and 1.3). The mirror directs the image to a focusing screen and eyepiece. When you take the picture, the mirror flips up, the view in the eyepiece goes dark, the image sensor is turned on, and the shutter opens.

With a film SLR and some early DSLRs, focusing through the eyepiece is your only option. Most newer DSLRs, however, offer Live View (live focusing) as an alternative: you can turn on a continous display of the image, and even view it magnified, on the screen on the back of the camera.

For astronomy, Live View is almost indispensable. There is no substitute for viewing a star image, highly magnified, as actually captured by the sensor, while you refine the focus.

1.1.2 DSLRs without Mirrors: MILCs

If you have Live View, do you need a mirror and focusing screen? No, say the makers of mirrorless interchangeable-lens cameras (MILCs), which work like DSLRs but rely on electronic previewing to frame and focus the image.

MILCs are gaining popularity with serious photographers. Figure 1.4 shows one, an Olympus OM-D. Another, the Sony α7S (A7S), has impressed astrophotographers with its high sensitivity to dim light, though it has also reportedly had problems with a "star eater" (Section 3.7.2).

MILCs are a less mature technology than conventional DSLRs, with more difference in performance from model to model. That is partly because some of

Figure 1.3. A more elaborate view of what's inside a DSLR. Note computer circuitry ("DIGIC II") at right. (Canon USA)

Figure 1.4. A mirrorless interchangeable-lens camera (MILC) is like a DSLR without mirror and focusing screen. Eyepiece, if present, shows an internal electronic display. (Grant George Buffett)

them were conceived of as cut-down DSLRs, and others, with smaller sensors, as scaled-up compact digital cameras. If you are contemplating using one for astronomy, I recommend checking online astronomy forums to find out how well it performs; the situation is changing rapidly.

The shallower body design of an MILC allows the lens to be closer to the sensor than in a DSLR, so that the MILC's wide-angle lenses perform better than lenses of comparable focal length for a DSLR. The shallow body also allows room for adapters so that one MILC can take lenses designed for several brands of DSLRs.

A drawback of MILCs for astronomy is that the sensor is normally operating all the time, to allow for image previewing. This tends to heat it up, increasing noise in the image. (The original rationale for the DSLR, in fact, was that the sensor would operate only for brief moments when taking pictures, and this would keep it cool.) The best MILCs compensate for the problem to some extent by using especially good heat sinks on their sensors; besides, you can always turn the camera off and let it rest when you are neither focusing nor exposing a picture.

For the rest of this book, I'll consider MILCs a type of DSLR even though they lack the reflex mirror. Practically everything I say about DSLRs applies to them; all the image processing techniques are exactly the same.

1.2 DSLRs versus Other Cameras

1.2.1 Dedicated Astrocameras

Should you be using a DSLR or a camera designed just for astrophotography, a dedicated astrocamera? The question takes on new urgency now that DSLR-like, multi-megapixel sensors are available in dedicated astrocameras (Figure 1.5) that operate connected to a separate computer and power supply. Astrocameras with large sensors used to be a lot more expensive than DSLRs, but the price gap is narrowing. Adapters are available to use astrocameras with camera lenses as well as telescopes.

Compared to a DSLR, an astrocamera is vibrationless (no shutter or mirror) and better at keeping the sensor cool to avoid thermal noise; many of them have thermoelectric cooling, and all have large heat sinks. Although newer DSLRs have much less thermal noise than those of ten years ago, cooled astrocameras still have an advantage, particularly in warm weather and for longer exposures.

Not all astrocameras produce color images directly. Those that do are called one-shot color (OSC) cameras; like DSLRs, they have alternating red, green, and blue pixels on the sensor. The rest are inherently monochrome but can still produce high-quality color pictures by taking separate exposures through red, green, and blue filters (R, G, and B), often accompanied by an unfiltered luminance (L) exposure to pick up more faint detail.

Figure 1.5. A dedicated astrocamera with thermoelectric cooling (ZWO ASI1600MM-Cool, 16 megapixels, monochrome). (ZWO Company photo.)

The disadvantage of the astrocamera is that it has to be tethered to a computer; you can't use it by itself. If you already use a computer with your telescope for telescope control or autoguiding, that is not a serious drawback. It does mean that the astrocamera cannot be used for daytime snapshots.

There is also a smaller kind of astrocamera used mainly for video and short-exposure imaging (Figure 1.6). These originated as cheap webcams with an eyepiece tube substituted for the lens; indeed, I imaged the 2003 apparition of Mars with a modified webcam. A small video astrocamera can complement a DSLR by serving as an autoguider (Chapter 9) and by enabling you to do planetary work beyond the DSLR's capabilities (Chapter 14). Video astrocameras are inexpensive, and you will probably end up with one.

There are even astrocameras made out of DSLRs and MILCs by adding cooling and other modifications. These are offered by CentralDS (www.centralds.net) and other firms. I classify them as dedicated astrocameras because they require a computer connection and cannot be used handheld for daytime photography.

1.2.2 Fixed-Lens Digital Cameras?

Digital cameras other than DSLRs and MILCs are usually not very suitable for astrophotography because the lenses are not removable, the sensor is small and

Figure 1.6. A video astrocamera (ImagingSource DMK series).

hence less sensitive, the lens is small (a disadvantage where starlight is concerned, regardless of f-ratio), and the camera does not deliver raw files suitable for calibration and stacking.

There are two exceptions to what I just said. One is that some of the best fixed-lens digital cameras are almost like DSLRs whose lenses cannot be removed, and they can be used as such, within the limits of the lens. Even so, DSLRs offer more for the money.

The other exception is that any digital camera, even a smartphone, can be aimed into the eyepiece of a telescope. This technique, called afocal photography (Figure 6.6), is suitable only for single exposures of the moon and possibly bright planets. Because its lens is the same size as the human eye or even smaller, a compact digital camera works well with an eyepiece and can even be handheld; the moon seen through a telescope is like a daytime terrestrial scene. Surprisingly good results have been obtained this way, and pushing a limited camera past its limits can be a satisfying challenge, but it is beyond the scope of this book.

1.2.3 What about Film?

Although film photography lives on as an art form (and I practice it as such), there is, as far as I can determine, no longer any situation in which film outperforms digital imaging for astrophotography.

Film is nonlinear; that is, its response is not proportional to the number of photons, so calibration and subtraction of sky fog are not practical. (Film images can still be digitized and stacked.) What's worse, film suffers reciprocity failure, so that most of the photons in a long exposure are forgotten; that's why the typical deep-sky exposure takes 30 minutes or more, rather than one or two minutes as with a DSLR. Further, my experiments have shown that the smallest points of light that register on fine-grained film are about five times as large as the pixels of a modern DSLR. That is, the DSLR is five times as sharp as film.

1.3 Choosing a DSLR

1.3.1 Canon vs. Nikon vs. Others

New models of DSLRs are introduced every few months, and the major manufacturers compete with each other so closely that their products perform very much alike. See Chapter 16 for the details of how DSLR image sensors are evaluated. Here I will not recommend specific models but only give general guidelines.

Most DSLR astrophotographers use Canon or Nikon cameras, not because other brands don't work, but because there's safety in numbers. Canon and Nikon are the two manufacturers that have made cameras specifically for the astronomy market (the Canon 20Da and 60Da and Nikon D810A). That doesn't mean you have to buy one of those models, but it does show that the makers care about astronomy, as well as other kinds of scientific photography.

Further, Canon and Nikon both have large astrophotographic user communities, and the cameras and their file formats are supported by numerous astronomical software packages and special accessories.

Canon and Nikon differ in design philosophy. The Canon EOS system is very consistent. Canons all work alike, except for documented differences in features, and all take the same lenses (except that some lenses only fit cameras with APS-C sensors, and of course the MILCs have a different line of lenses). Nikon, on the other hand, pursues somewhat different design philosophies in different cameras, and even the set of Nikon lenses that you can use differs slightly from one DSLR to another. The lens mount dates back to the 1959 Nikon F but has undergone many variations (see Section 7.3).

You can use Nikon lenses on Canon bodies, with an adapter, but not vice versa, because the Canon DSLR body is shallower, so the lens can still be the right distance from the sensor even though space is taken up by the adapter. MILC bodies are even shallower and can take a wide range of DSLR lenses. For more about lens adapters, see Section 7.3.3.

One final difference is that with Canon DSLRs, you can use the exposure meter and aperture-priority auto exposure with any lens or telescope. This capability is handy for daytime photography and for taking flat fields for calibration. Nikon DSLRs can only meter and auto-expose with newer Nikon lenses that contain electronics; if you attach other optics, even old Nikon lenses, you can expose manually but cannot use the meter.

Canon's nomenclature can confuse you. The EOS Digital Rebel, EOS Kiss, and EOS 300D are the same camera (one of the first good, affordable DSLRs), but the EOS 300 and EOS Rebel are film cameras from an earlier era. The EOS 200D, on the other hand, is one of the newest DSLRs, much newer than the 300D. The EOS 30D is a good DSLR from about a decade ago, but the EOS D30 is an early DSLR from before Canon developed sensors suitable for astronomy. And so on. If you want to buy a secondhand camera, pay attention to the names.

Pentax, Sony, Olympus, and other DSLR makers are highly respected but have not achieved as large a following among astrophotographers, and I have not tested their products. Sony makes the sensors for Nikon and several other brands (not Canon), although of course the firmware and internal image processing are different. Pentax K-series DSLRs reportedly have better deep-red (hydrogen-alpha) sensitivity than other non-astronomical DSLRs, though I have not tested this; also, Pentax has an interesting feature called the Astrotracer (see below) for fixed-tripod star-field images. Before buying any camera, you should search the Web and get astrophotographers' opinions of it; also make sure its file formats and camera interface are supported by astrophotography software.

1.3.2 Camera Features

Live View: *Practically Essential*

The most important feature to look for in a DSLR is **Live View** (live focusing), the ability to view the image on the screen while you focus the camera. Normally the live view image can be magnified ×5 and ×10, so you can judge it very carefully.

Live View was introduced experimentally on the Canon 20Da (not 20D) in 2005 and soon became standard on almost all DSLRs. It made a lot of other focusing techniques obsolete. Before then, we did our best to focus in the viewfinder, usually with a magnifier such as the Canon Angle Finder. Then, to confirm focus, we took 5-second test exposures and viewed them magnified or downloaded them to a computer. We put various kinds of masks in front of the telescope to exaggerate the effect of focusing error so we could judge it more easily. No more! Viewing an actual star image at pixel resolution is all we need.

Flip-out (Vari-angle) Screen: *Useful*

Although professional DSLRs don't have it because it's not considered rugged, a **flip-out screen** is very handy to have. It enables you to focus and adjust a DSLR that is aimed up at the sky without having to crouch under it. The alternative is to view the screen with a small hand-held mirror (Figure 3.5).

Electronic First-curtain Shutter: *Useful for Lunar and Planetary Work*

Considerably less essential, but useful if you have it, is **electronic first-curtain shutter** (EFCS), a vibration-saving feature that is very helpful with still pictures of the moon, sun, and eclipses (see Figure 3.10) but largely irrelevant to deep-sky work or anything with an exposure longer than half a second or so.

The name refers to the two curtains of the shutter. Normally, the first curtain uncovers the sensor, and then the second curtain, approaching from the same side, covers it again at the end of an exposure. Finally the two curtains retract to their original position; as they do this, they are in contact with each other so that no light gets through.

EFCS means that the first curtain is simulated electronically. That is, when you start an exposure, the real first curtain is already open (for Live View) and the sensor is cleared and turned on electronically to begin the exposure. The second curtain ends the exposure in the normal way (and then, if Live View is active, the shutter opens again). Vibration from the second curtain is not objectionable because nearly all of it occurs after the exposure has ended.

The way EFCS works on particular cameras can be tricky. On most Canon DSLRs, EFCS is used when you take a picture while Live View is active. Autofocus and flash must be off, and if the camera offers a feature called Silent Shooting (on pro and semi-pro models), that feature must be turned on.[1] Nikons generally don't offer EFCS, even when taking a picture in Live View, but the D810 offers EFCS and the D850 offers electronic first and second curtains for totally vibration-free photography. "Quiet shutter release" on low-end Nikons is not EFCS; it merely suppresses the autofocus beep and slows down the mirror a little.

To determine whether a camera has EFCS and how it works, you may have to observe it in action. With the lens removed, make a manual exposure of 2 seconds while Live View is active (2 seconds so that you can distinguish the exposure time from other movements). Watch whether the shutter and mirror move before and after the exposure.

Pentax Astrotracer: *Useful in Some Situations*

The Pentax Astrotracer, built into some DSLR models and available as an accessory for others, enables Pentax DSLRs to track the stars from a fixed tripod by moving the sensor. It takes advantage of movement capability that was already built in for the purpose of counteracting vibration in daytime action photography.

Of course, the total amount of tracking that is possible is limited, but the Astrotracer has been used successfully with lenses as long as 300 mm and exposures as long as 20 seconds (which can of course be stacked). What is impressive is that it "flies blind," using GPS and other sensors to determine the position of the camera, without requiring any alignment on the stars. It is the most portable possible setup for doing deep-sky photography.

[1] I thank Charles Krebs, photomicrographer, for detailed information here.

Bearing that in mind, of course the Astrotracer is no substitute for the rest of astrophotographic guiding technology. Very simple camera trackers will track the stars, with any camera, at least as well as the Astrotracer does, because you can align them on the stars. Still, the Astrotracer is an impressive technical achievement.

Maximum ISO Rating: *Less Important*

Now for features that don't matter so much. The **highest ISO rating** of a DSLR might seem important, but it is not a feature we astrophotographers look at directly. In general, a camera with a higher maximum ISO will indeed have a better sensor, but that is not a direct way to measure it, and regardless of what the camera is capable of, we almost always use ISO ratings no higher than 1600 for greater dynamic range.

Megapixels: *Less Important*

The number of **megapixels** is also not a direct measure of camera quality. Newer sensors do have more megapixels and are generally better in other ways, but given two sensors of the same size and intrinsic quality, the one with larger pixels – fewer megapixels – will pick up more light with less noise. That is how the 12-megapixel Sony α7S achieved outstanding low-light performance compared to 32- and 48-megapixel cameras in the same product line.

Note, however, that a finished, full-page picture, even a very crisp, sharp 8×10-inch print, contains only no more than about 4 megapixels of visible detail. The higher pixel count of the camera is to give you flexibility for cropping or reducing the image.

Full-frame Sensor: *Less Important*

Finally, there is nothing sacrosanct about a **full-frame sensor**, which is simply a sensor the same size as a frame of 35-mm film, 24×36 mm. The common APS-C sensor (about 25×17 mm) is actually a better fit to telescope eyepiece tubes, and when used with a lens designed for a full frame, it uses only the central, sharpest part of the field, which can be advantageous. It could be argued that if we want bigger sensors, they should be square (maybe 30×30 mm), since round lenses fill a round field, but in daytime photography, square pictures went out of fashion with the Kodak Instamatic.

1.3.3 Shopping Strategy

Because of rapid technological progress, you generally want the *newest* DSLR that works well for astrophotography, not the most ruggedly built one. It's better to buy a low-end DSLR today and another one in three years with a new, improved sensor, rather than sink all your money into a professional-grade camera that will commit you to using today's technology for a decade. Another disadvantage of pro-grade DSLRs is that they can be heavy enough to unbalance a medium-sized amateur telescope.

Buying the *very* newest camera has some drawbacks. This week's hot new DSLR may not yet be supported by your software, although updates usually come quickly. Its firmware (the software inside the camera) may also have undiscovered bugs; watch the manufacturer's web site for firmware upgrades.

You can buy DSLRs from the same stores as personal computers and other consumer electronics. The small-town camera stores of the twentieth century have largely disappeared, but some major photographic dealers continue to stand out for their full range of products and reliable service. These include B&H in New York (www.bhphotovideo.com), their neighbor Adorama (www.adorama.com), Samy's in Los Angeles (www.samys.com), KEH in Atlanta (www.keh.com), and Jessops in the UK (www.jessops.com).

1.4 Choosing Software

Picture-editing software such as *Adobe Photoshop* is not sufficient for serious astrophotography. Stacking, dark-frame subtraction, and other kinds of calibration are done on raw (linear) images that have not been converted into viewable pictures, and picture-editing software cannot perform these operations.

Learning to process images can be daunting. My own approach is to keep things simple and straightforward. The guiding principle is, *concept first, procedure second.* That is, first understand what the software is supposed to be doing, then how to make it do it. Skipping the first step will leave you bewildered. Getting too wrapped up in processes that you don't understand will lead you to invent slow, complex ways to do things that could have been done much more simply.

Table 1.1 lists some major image processing packages. They have different design philosophies.

PixInsight is a kit of a thousand tools; it offers you every possible image processing algorithm, including new experimental ones, and leaves all the decisions up to you. It is on the cutting edge of scientific astrophotography, and new features are regularly contributed by users.

DeepSkyStacker (freeware) and *Astro Pixel Processor* are just for the early stages of processing (calibration and stacking); they are especially easy to use, but you have to take their output files into other software (either another astronomical package or *Photoshop*) to produce the finished image. I process most of my pictures with *DeepSkyStacker* followed by *PixInsight*.

The other three packages are full-featured and relatively easy to use because of their menus that emphasize the operations you actually need. *Nebulosity* is pitched at beginners, but you may never outgrow it; it does the processing quickly and efficiently and is easy to use. *MaxIm DL* combines power and versatility with ease of use; it is available in several editions with more or fewer features, one of them targeted to DSLR users. *Images Plus* is another

Table 1.1 *Some major image processing software packages.*

Name	Vendor	OS	Ease of use	Power and versatility
PixInsight	www.pixinsight.com	Windows, Linux, macOS	–	+++
DeepSkyStacker	deepskystacker.free.fr	Windows	+++	See text
Astro Pixel Processor	www.astropixelprocessor.com	Windows, Linux, macOS	++	See text
Nebulosity	www.stark-labs.com	Windows, macOS	++	+
MaxIm DL	www.diffractionlimited.com	Windows	+	++
ImagesPlus	www.mlunsold.com	Windows	+	++

widely-used, well-respected general purpose image processing package. Free trial versions of all of these can be downloaded.

In a book of this length, I can't give you full "press this button, click that box" instructions for every software package. If I did, not only would the book be very thick, it would be obsolete as soon as someone released a minor software update. But I can give you plenty of worked examples so that you can see what the most important operations look like, in detail. Chapter 12 takes you through the basic process with *DeepSkyStacker* (as far as it goes), *Nebulosity, MaxIm DL,* and *PixInsight*. Other chapters use the latter three software packages in their worked examples.

You may still need picture-editing software such as *Photoshop, Lightroom, Photoshop LE,* or *GIMP* to do the final adjustment of pictures to be displayed or printed.

Software for other specific purposes, such as camera control and planetary video imaging, will be mentioned as it comes up.

1.5 Choosing the Computer

You can process images with any computer suitable for general office work. A dual-core Intel or AMD 64-bit processor with 8 megabytes of RAM is fine. In particular, you do not need a special display or graphics card. Having said that, I should add that a faster CPU, more cores, and more RAM do speed things up. If you process a lot of images, you may want the fastest PC you can afford.

You will certainly need plenty of storage. Consider making a regular practice of archiving your images to DVD, and be sure to keep the raw files that came out of the camera, not just the processed result; better processing techniques may become available in the future.

One last thought. If you plan to use a computer by remote access, consider how you will get the files to it. An imaging session consisting of 100 24-megapixel raw image files can easily comprise 3 gigabytes of data.

1.6 Choosing the Telescope or Lens

1.6.1 The Aperture Counterrevolution

DSLRs work well with small telescopes and telephoto lenses. The small size of the telescope you need may come as a surprise. Back in the 1980s, we spoke of an "aperture revolution" when, for the first time, telescopes larger than 8 inches (20 cm) became widely available to amateurs. Today we are in the middle of a counterrevolution. DSLRs are so much more sensitive to light than film or the human eye that nowadays, serious work is often done with apertures more like 3 inches (7.5 cm) or even less. Many of the best deep-sky images are taken with telephoto lenses.

1.6.2 The 500-mm Optimum

A large factor in this counterrevolution is what I call the 500-mm optimum. Several factors – the pixel size of digital sensors, the turbulence of the air, and the accuracy with which the mount can track – all converge, so that we get the best results with a focal length on the order of 500 mm, which implies a small telescope.

The turbulence of the air normally limits the sharpness of the image to about 1 or 2 arc-seconds in a long exposure. That is also the accuracy with which a good mount can reliably autoguide. With typical DSLR sensors, a 500-mm focal length gives an image scale of about 1.5 arc-seconds per pixel, about equal to the resolution limited by other factors. The focal length need not be exactly 500 mm, of course; the point is that 1000 mm or more is probably longer than ideal, while 150 mm is definitely not pushing the limits.

During the film era, the situation was different. The lower resolution of film necessitated longer focal lengths to pick up detail. Reciprocity failure made it impossible to photograph faint objects with small telescopes. Mounts and guiding were not as good either, but the unsharp nature of film concealed their limitations. Those were the days of twelve-inch (30-cm) telescopes in observatory domes and laborious hand-guided 1-hour exposures. Nowadays, much smaller instruments give them serious competition. Of course, larger telescopes are still useful, but getting the most out of them – getting results substantially better than a smaller telescope could give – is an exacting process.

1.6.3 Ease of Use

Another argument for small, simple, portable telescopes is that more complex equipment is always harder to use, regardless of your level of experience. Owners of large telescopes often find themselves buying smaller ones for convenience. Of course, your equipment must not lack anything necessary, but an unduly elaborate setup can hinder rather than help. Beware of buying too much just to impress yourself!

1.7 Choosing the Mount

The only way to photograph faint objects is to make long exposures while tracking the stars, or rather tracking the earth's rotation, with an equatorial mount. The most rapid technological progress in all of astrophotography is currently happening with mounts, drives, and guiders, not with cameras or telescopes, which are, by comparison, relatively mature technology.

Mounts and tracking are covered in detail in Chapters 8 and 9, but let me begin with an exhortation: *Don't skimp.*

By "don't skimp" I mean two things. First, acknowledge that the mount is likely to be the most expensive part of your setup. It is not uncommon or improper for the mount to cost more than the camera and telescope that it carries. There are low-cost solutions of course, and most of the pictures in this book were taken with budget-priced mounts (Celestron AVX and CGEM), but with today's technology, there is a large gain from moving up the price scale, and even the best is not so good that you cannot wish for more. The challenge, for amateur astronomers, is to do for $1000 what can certainly be done for $10 000.

Second, don't skimp on capacity. A handy rule for astrophotographers is to use a mount rated for a telescope twice as heavy as you actually plan to use. That gives you a lot of extra steadiness. It also acknowledges that telescopes tend to gain weight as guidescopes, focusers, and other accessories are added.

There is a lower-cost alternative. If you take wide-field pictures at lower magnification, you can do well with an inexpensive "sky tracker" that only carries a camera and telephoto lens (Figure 8.5). The beauty of the sky is still yours to capture, but the choice of objects to photograph is different.

One word of caution: Be hesitant to buy a new model of mount that has just been introduced. Unlike cameras and telescopes, mounts often need to be "debugged" for a couple of years as manufacturing ramps up. Reliability issues are corrected, minor engineering improvements are made, and microprocessor firmware is revised. Unit-to-unit variation is the weak spot of low- and mid-priced mounts, and by waiting for the manufacturing process to mature, you can avoid much of the risk. If you get a mount and become convinced it has problems, send it for repair before the warranty expires.

1.8 The Craft of Astrophotography

1.8.1 Building your Skill and Judging your Achievements

Astrophotography is a craft requiring development of skill, not just good equipment. Time spent building your skill pays off, including "low-quality time," time under less than ideal conditions, such as in town or under moonlight. You cannot become an expert astrophotographer just by spending a few evenings a year at a dark desert site. Back home, at a mediocre site, you can build familiarity with your equipment, become adept at setting it up in the dark, test and calibrate autoguiders, give mechanisms enough exercise that the lubricants don't stiffen up, and, often, take surprisingly good pictures.

When judging your finished images and those of your colleagues, remember that *every good picture is just below the threshold of showing its faults.* If it were magnified larger, or if the contrast were boosted a little more, defects would be visible – imperfect guiding, uneven sensor response, minor optical flaws, or something. (This is particularly the case now that pixels are smaller than the resolving power of the lens.) The art of image processing is largely the art of bringing out the best without bringing out the worst.

1.8.2 Pushing Limits or Staying within Them

Any equipment configuration, at any site, is good for some things, marginal for others, and incapable of yet others. You can suffer endless frustration by trying to do things that are completely out of reach, or enjoy constant satisfaction by doing what your equipment and site enable you to do well. In between, you can enjoy the challenge of extending your reach.

Accordingly, an important part of building your skill is determining what can be done reliably and reproducibly with your equipment and site. An important threshold is crossed when the question is no longer "Will I get anything?" but rather you have confidence that you will get the picture you have planned. It can take you two or three years of regular observing with a particular set of equipment to reach this stage.

Bear in mind that the award-winning pictures that you see in magazines and on the Web are often the result of dozens or even hundreds of tries, combined with good luck. By contrast, the pictures in this book are not prize-winners. They are examples of what can be done reliably by a moderately experienced astrophotographer with affordable equipment.

1.8.3 Testing as a Means or an End

Astrophotographic equipment is very good at measuring its own imperfections, and many tests are described in the later chapters of this book. Always remember that the real goal is to get good pictures, not good test results. It can be all too easy to convince yourself that any lens or mount is intolerably flawed – and meanwhile, others using the same or worse equipment are getting outstanding pictures. We want to see your image of M33, not your autoguider graph.

1.8.4 Philosophical and Ethical Issues

There have been squabbles between astrophotographers with two different goals. Some astrophotographers' mission is to express their creative artistic vision of the beauty of the cosmos. For them, a picture starts with (or at least quickly becomes guided by) an artist's concept and is modified and retouched as necessary to express it.

Other astrophotographers, probably the majority, take as their starting point the historic role of astrophotography as a scientific record of the universe. To them, every astronomical image is not just a thing of beauty; it is also an observational record. They balk at manipulating images other than through straightforward computational processes. They are so accustomed to the scientific tradition of realism that they have a hard time seeing astrophotography any other way.

I belong to the latter camp, but to both sides I urge tolerance and open-mindedness. I also urge honesty: if an image has been altered so that it is not a scientific record, say so, so that no one will try to use it as such. (Fifty years from now, your picture may be someone's source for the prediscovery brightness of a nova.) And I caution the creative artists that some people are likely to be put off by a picture of anything, no matter how beautiful, if it is astronomically false or impossible, such as a combined image of two separate nebulae or an eclipsed sun in front of clouds.

1.8.5 Amateur or Professional?

One last thought. Most DSLR astrophotographers are individual enthusiasts, but a significant number are educators or researchers. More importantly, the *methods* of amateur and professional astronomy are pulling together.

We are emerging from a period (about 1970–2000) when amateur and professional astrophotography were unusually far apart because of technological advances that only professionals could afford. Professionals had autoguiders, special photographic plates, and CCDs, while amateurs used 35-mm film SLRs, hand-guiding, and consumer-grade film. Professionals got excellent, scientifically valuable images of the universe; amateurs were pleased to get anything recognizable.

The gap has narrowed or even closed. Thanks largely to digital image sensors, but also improvements in telescopes, mounts, computers, software, and Internet information sources, amateurs are much closer to professional capabilities than they used to be. It's almost like the days of Isaac Roberts and Percival Lowell all over again – amateurs again have access to celestial objects not well known to science, together with the tools to investigate them, and there's little to prevent amateurs from doing some kinds of work that professionals do. (See for example the research ideas in Chapter 18.) NASA actually consults amateurs about the processing of planetary images, and professionals are increasingly using high-end amateur equipment when they need portability. I think amateur and professional astronomy will continue to converge for many years to come.

Chapter 2
Digital Image Technology

DSLR astrophotography ties together so many different concepts and principles that you may feel you can't do *anything* until you know *everything*. To work around this problem, this chapter surveys digital imaging as a whole, defining terms and introducing concepts that figure prominently in later chapters.

This chapter focuses on concepts that are not specific to astronomy, though the examples have an astronomical slant; we'll get to specifically astronomical techniques in Chapters 11–14. Experienced daytime photographers may want to skim this chapter and come back to it as needed. For much more detailed coverage, with algorithms, see R. Berry and J. Burnell, *The Handbook of Astronomical Image Processing* (second edition, Willmann–Bell, 2011), and R. C. Gonzalez and R. E. Woods, *Digital Image Processing* (fourth edition, Pearson, 2017).

2.1 What is a Digital Image?

2.1.1 Bit Depth

A digital image is fundamentally an array of numbers that represent levels of brightness (Figure 2.1). Depending on the *bit depth* of the image, the numbers may range from 0 to 255 (8 bits), 0 to 65 535 (16 bits), or some other range. Each position in the array is called a *pixel* (picture element). A million pixels are called a *megapixel*.

The eye cannot distinguish even 256 levels, so 8-bit graphics are sufficient for finished pictures. The reason for wanting more levels during manipulation is that we may not be using the full range at all stages of processing. For instance, a badly underexposed 16-bit image might use only levels 0 to 1000, which are still enough distinct levels to provide smooth tones. An 8-bit image underexposed to the same degree would only go from 0 to 4 and would be useless.

For greatest versatility, some software supports floating-point data, so that levels can be scaled with no loss of precision; in a floating-point system, you can divide 65 535 by 100 and get 655.35. You can also use large numbers without

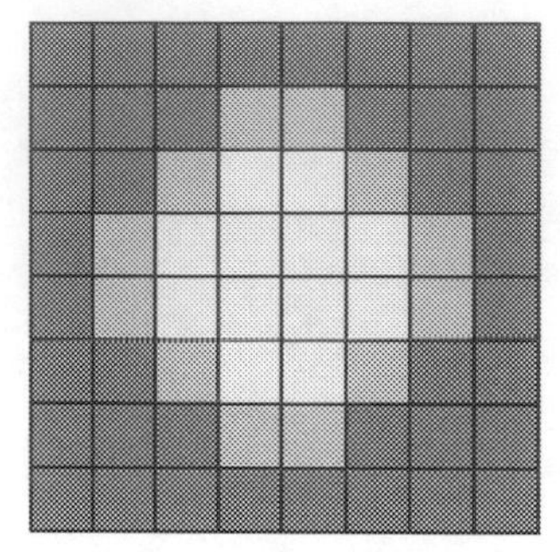

10	10	10	10	10	10	10	10
10	10	10	50	50	10	10	10
10	10	50	90	90	50	10	10
10	50	90	90	90	90	50	10
10	50	90	90	90	90	50	10
10	10	50	90	90	50	10	10
10	10	10	50	50	10	10	10
10	10	10	10	10	10	10	10

Figure 2.1. A digital image is an array of numbers that represent levels of brightness. (From *Astrophotography for the Amateur.*)

going out of range; if you do something that produces a value greater than 65 535, it will not be clipped to maximum white.

Note that *Photoshop* always reports brightness levels on a scale of 0 to 255, regardless of the actual bit depth of the image. This is to help artists match colors.

2.1.2 Linear or Gamma-corrected?

The numbers in an image file may mean either of two things. In raw files they represent the number of photons that reached each pixel on the sensor. In files designed for viewing, they represent brightness levels on the computer screen, which are not proportional to the number of photons; instead they mimic the logarithmic way the human eye perceives light.

The conversion from one to the other is called *gamma correction* and is covered in more detail in Section 2.5.4. If you display a file without gamma correction, its midtones will look very dark even though maximum whites may be correct. What this means is that the essence of gamma correction is to lighten the midtones considerably.

Lightening the midtones is a nonlinear operation. What this means is that the numbers that come out are not proportional to the numbers that went in. Midtones get lightened more than highlights.

Before gamma correction, the image is described as *linear*. After gamma correction, it is nonlinear, and image arithmetic (such as dark-frame subtraction) cannot be performed on it; the calculations would not come out right. Most other kinds of image editing can be performed before or after gamma correction, with slightly different results.

The camera performs its own gamma correction to produce viewable JPEG files. However, astrophotos are seldom "correctly exposed" by the standards of daytime photography, and automatic gamma correction does not suit them. Instead, we perform gamma correction manually to suit each picture. This often involves *stretching* (amplifying) an underexposed image so that it uses the full brightness range, as well as lightening its midtones to the desired extent. The combination of the two is called *nonlinear stretching*.

In practice, JPEG files are always gamma-corrected, and raw files always are not, but TIFF, PNG, FITS, and XISF files may be either kind. You can open a raw file from your DSLR and store it as TIFF, PNG, FITS, or XISF whether or not you have gamma-corrected it. Those are the formats we use to store astronomical images during processing.

2.1.3 Color Encoding

A color image normally has three numbers (called *channels*) for each pixel, giving the brightness in red, green, and blue (*RGB color*). The main alternative is *CMYK color,* which describes color in terms of cyan, magenta, yellow, and black printing inks; the two are interconvertible.

Further alternatives are the *Lab* and *L*a*b** systems, which use three coordinates based on the two-dimensional CIE chromaticity chart plus luminosity. Roughly, the dimensions are brightness, cyan versus magenta, and cyan–magenta versus yellow. *Lab* and *L*a*b** can represent colors that fall outside the gamut (Section 2.3.5) of the computer screen.

Astronomical images are sometimes constructed by combining the luminosity (L) from one image with the color (RGB) from another; this technique is called *LRGB*. It is not normally done with DSLRs, whose output is already in full color.

2.1.4 The Alpha Channel

In TIFF and some other file formats, the alpha channel is a fourth channel that is not a color but rather a measure of opacity; it allows parts of images to be partly or completely transparent. Thus, for example, a Windows icon can be round even though stored in a square digital image; the parts outside its round shape are transparent, and for greater smoothness, pixels on the edge of the transparent region can be partly transparent.

Astronomical images, in finished form, are always completely opaque, but file formats allow for an alpha channel. If included, it takes up extra space without adding anything to the image.

2.1.5 Frames

Individual exposures taken with a digital camera are called *frames,* a term borrowed from the movie industry. A motion picture consists of thousands of frames to be shown one after another. An astronomical picture often consists of many frames aligned and combined. Besides image frames, there are dark frames and other kinds of calibration frames to record and correct deficiencies in the sensor and optical system.

2.2 File Formats

2.2.1 File Size

Compared to word processing or spreadsheets, the files produced by DSLR imaging are enormous. Without compression, a file needs one byte for every 8 bits of data. For a 16-bit, 3-color, 24-megapixel image, that works out to $2 \times 3 \times 24$ megabytes = 144 megabytes. Lossless compression can easily get this down to a quarter of that, but each image is still big.

Note that you can shrink a DSLR image to an eighth of its linear size (1/64 of its area, 1/64 of the file size) and still put a decent-sized picture on a web page. That's an easy way to hide hot pixels and other defects without doing any other processing.

2.2.2 Compression

A compressed file does not store all the bytes of the image one after another, but instead encodes them in a more concise way. For example, many images contain repeated pixels; that is, adjacent pixels are alike. It is obviously more concise to store some code that means "the following pixel repeats 100 times," followed by the pixel value, than to store 100 pixels that are just alike. There are elaborate schemes for finding repetitions (across areas, not just along lines) and encoding them concisely.

What I've just described is *lossless* compression; after decoding, it gives back the pixel values exactly as they were put in. A familiar example of lossless compression (of all kinds of files, not just images) is the ZIP format commonly used with PC software.

Sometimes more compression is needed, even at the expense of image quality. That's where *lossy* compression comes in. By discarding some low-level fluctuations and fine detail, a lossy compression algorithm can make the pixels more repetitious so they can be compressed further; it may even make the image look less grainy. But it also makes the image less suitable for further processing, since the low-contrast details you want to bring out may have been discarded and replaced by errors of approximation ("artifacts," often looking like haloes and swirls).

2.2.3 Raw Files

As already noted, a raw file represents the number of photons that struck each pixel on the sensor during the exposure. It gives you what the camera captured, with no information lost. Some in-camera corrections may, however, be performed. Many raw files also contain a small JPEG version of the image for rapid previewing.

Cameras usually add a constant bias to the raw pixel values so that minor electrical fluctuations will not produce negative numbers, which cannot be represented. All standard image processing algorithms allow for the presence of this bias.

The term *raw* is not an abbreviation and need not be written in all capital letters; it simply means "uncooked" (unprocessed). Filename extensions for raw images include .CRW and .CR2 (Canon Raw) and .NEF (Nikon Electronic Format). The format depends on the kind of camera, even down to the exact model; for example, some software can read Nikon D810 .NEF files and can't read Nikon D850 .NEF files, which provide for larger images.

Raw images are compressed – their size varies with the complexity of the image – but the compression is lossless or nearly so; the exact value of every pixel in the original image is recovered when the file is decoded in the computer. Canon and Nikon raw images occupy between 1 and 2 megabytes per megapixel, varying with the complexity of the image. Uncompressed, a 14-bit-deep color digital image would occupy 5.25 megabytes per megapixel.

2.2.4 *dcraw* and Adobe DNG

What do you do if your software doesn't support the raw format of your new camera? There are two important workarounds.

First, there is the computer program *dcraw*, distributed free by Dave Coffin (https://cybercom.net/~dcoffin/dcraw), and frequently updated. It converts raw files to TIFF and other formats while preserving the pixel values and optionally decoding the Bayer matrix in any of several ways. Written in ANSI C, it is released as source code and can be compiled for many different operating systems.

In fact, *dcraw* is built into a number of astronomical image processing programs, including *PixInsight*. Why reinvent or reverse-engineer raw file decoding, and keep updating it for new cameras, when *dcraw* has done it for you?

The other approach to standardizing raw files is Adobe's DNG ("digital negative") file format. This is simply a standard format into which all raw files can be converted by Adobe's free *DNG Converter* (go to www.adobe.com and search for it by name).

The main purpose of *DNG Converter* is to keep Adobe users from having to get a new version of *Photoshop* every time a new camera is introduced with a new raw file format; instead, they can update the converter, which is free, and process the DNG files.

But you can also use DNG files with astronomical image processing software. Among others, *PixInsight*, *MaxIm DL*, *Nebulosity*, and *dcraw* all accept them.

2.2.5 JPEG

JPEG is as far as you can get from raw. It is the usual file format for displaying pictures on the World Wide Web, and many daytime photographers work with nothing else. It is a good format for displaying correctly exposed pictures of ordinary daytime subjects, but it is limited to 8-bit depth and involves lossy compression. The amount of compression can be chosen when saving the file, and too much compression can produce ripples around sharp features in the image.

JPEG is not a good format for images that will undergo further processing, for several reasons. JPEG images are already gamma-corrected, so operations such as dark-frame subtraction cannot be performed on them. The low bit depth and the lossy compression mean that the faint details you would like to bring out by processing have probably already been discarded. And every time you save the file, it suffers further loss, so repeated editing of the same JPEG file gradually degrades it.

I convert astronomical images to JPEG at the last step, to display them on the Web and possibly to make prints (though I prefer to make prints from 16-bit TIFFs).

2.2.6 TIFF

TIFF (Tagged Image File Format) is often the most convenient file format. You can store images that are raw, partly processed, or completely processed. Compression is optional and, if used, is completely lossless; you always get exactly the pixels that you saved.

There are several varieties of TIFF files. The pixels may be 8, 16, 32, or 64 bits per color (or even 32- or 64-bit floating-point); the color can be grayscale, RGB, or CMYK; the compression may be none, LZW, RLE (Packbits), or ZIP; and the image can consist of multiple layers (such as labels and arrows distinct from the underlying picture).

Because LZW (Lempel–Ziv–Welch) compression was protected by a patent until late 2006, some low-cost software cannot create compressed TIFFs. Also, not all software accepts 32- or 64-bit pixels.

Use ZIP compression unless your software doesn't support it. Uncompressed 16-bit TIFF files occupy about 6 megabytes per megapixel.

2.2.7 PNG

PNG (Portable Network Graphics) is another widely used format that features lossless compression, though it is less versatile than TIFF. The images have 8 or 16 bits per color per pixel.

2.2.8 FITS

The Flexible Image Transport System (FITS) is the standard file format for professional astronomical images. Unfortunately, non-astronomers rarely use it, and non-astronomical software seldom supports it. *PixInsight* and *MaxIm DL* support it well.

FITS is the most adaptable graphics file format. It allows you to use 8-bit, 16-bit, 32-bit, 64-bit, or floating-point pixels, monochrome or color, with or without lossless compression. In addition, the FITS file header has room for copious *metadata* (information about the file's contents), such as the date, time, telescope, observatory, exposure time, camera settings, and filters. The header consists of ASCII text and can be read with any editor.

2.2.9 XISF

Promoted by the makers of *PixInsight*, XISF (eXtensible Image Serialization Format) aims to offer all the functionality of FITS while solving practical problems. XISF requires the file to include all the information needed to interpret it, such as color space (not a requirement in FITS). At the beginning of the file, after some introductory bytes, there is an XML header that can be read with a text editor. If the file was processed with *PixInsight*, this header summarizes how it was processed and can be very handy.

2.3 Color Imaging in Detail

2.3.1 The Bayer Matrix (CFA)

Almost all digital cameras sense color by using a Bayer matrix of filters in front of individual pixels (Figure 2.2), also known as a color filter array (CFA). This system was invented by Dr. Bryce Bayer, of Kodak, in 1975.[1] Green pixels outnumber red and blue because the eye is more sensitive to fine detail in the middle part of the spectrum. Brightness and color are calculated by combining readings from red, green, and blue pixels.

At first sight, this would seem to be a terrible loss of resolution. It looks as if the pixels on the sensor are being combined, three or four to one, to make the finished, full-color image. How is it, then, that a 24-megapixel sensor yields a 24-megapixel image rather than a 6-megapixel one? Is it really as sharp as the number of pixels would indicate?

The answer is that for each pixel, one color value is exact, and the other two are computed by interpolating. For example, a B pixel's B value is as given; its

R	G	R	G	R	G
G	B	G	B	G	B
R	G	R	G	R	G
G	B	G	B	G	B

Figure 2.2. Bayer matrix of red, green, and blue filters in front of individual sensor pixels. Full color of each pixel can be computed by looking at its neighbors.

[1] Pronounced *BY-er*, as in German, not *BAY-er*.

R value is the average of the four R pixels around it; and its G value is the average of the four G pixels around it. This is a simplification; actual decoding algorithms also track gradients in the image to obtain more accurate results.

When you open a raw color image in image processing software, you have a choice. You can *demosaic* (*deBayer*) the image to compute the colors of virtual pixels; you can group every four pixels (R, G, G, and B) into one *superpixel* to produce an image only half as tall and half as wide; or you can treat each set of pixels (R, G, the other G, and B) as a separate layer as if you had four images. You can even ignore the Bayer matrix and see a grayscale image that seems to have a fine plaid pattern all over it. There are uses for all of these approaches.

2.3.2 Low-pass Filtering

The Bayer matrix assumes that light falling on each pixel will also spread, to some extent, to the pixels of other colors adjacent to it. A *low-pass filter* (so called because it passes low spatial frequencies) ensures that this is so. The low-pass filter is an optical diffusing screen mounted right in front of the sensor. It ensures that every ray of light entering the camera, no matter how sharply focused, will spread across more than one pixel on the sensor.

That sounds even more scandalous – building a filter into the camera to blur the image. But we *do* get good pictures in spite of the low-pass filter, or even because of it. The blur can be overcome by digital image sharpening. The low-pass filter enables Bayer color synthesis and also reduces moiré effects that would otherwise result when you photograph a striped object, such as a distant zebra, and its stripes interact with the pixel grid.

Some of the newest DSLRs, such as the Nikon D5300, do leave out the low-pass filter. Nikon's reasoning is that no lens is sharp enough to hit just one of their sensor's tiny pixels, so no low-pass filter is needed.

Remember that light spreads and diffuses in film, too. Film is translucent, and light can pass through it sideways as well as straight-on. We rely on diffusion to make bright stars look bigger than faint stars. My experience with DSLRs is that, even with their low-pass filters, they have less of this kind of diffusion than even the finest-grained films.

2.3.3 Nebulae are Blue or Pink, not Red

For a generation of astrophotographers, emission nebulae have always been red. At least, that's how they show up on Ektachrome film, which is very sensitive to the wavelength of hydrogen-alpha (656 nm), at which nebulae shine brightly.

But DSLRs see nebulae as blue or pinkish. There are two reasons for this. First, DSLRs include an infrared-blocking filter that cuts sensitivity to hydrogen-alpha. Second, and equally important, DSLRs respond to hydrogen-beta and oxygen-III emissions, both near 500 nm, much better than color

Figure 2.3. The Rosette Nebula, a faint emission nebula, photographed with an unmodified Canon 40D. Stack of eight 6-minute exposures with a Canon 300-mm $f/4$ lens. In color, the nebula is pinkish.

film does.[2] Some nebulae are actually brighter at these wavelengths than at hydrogen-alpha. So the lack of brilliant coloration doesn't mean that the DSLR can't see nebulae.

DSLRs can be modified to make them supersensitive to hydrogen-alpha, like an astronomical CCD, better than any film. The modification consists of replacing the infrared filter with one that transmits longer wavelengths, or even removing it altogether. For more about this, see Section 17.2. The Canon 20Da and 60Da and the Nikon D810A have extended red sensitivity as they come from the factory.

In the meantime, suffice it to say that unmodified DSLRs record hydrogen nebulae better than many astrophotographers realize. Faint hydrogen nebulae can be and have been photographed with unmodified DSLRs (Figure 2.3).

[2] The Palomar Observatory Sky Survey also didn't respond to hydrogen-beta or oxygen-III; those spectral lines fell in a gap between the "red" and "blue" plates. This fact reinforced everyone's impression that emission nebulae are red.

2.3.4 Color Balance (White Balance)

In daytime photography, the color balance of pictures is adjusted to make them look realistic in spite of changes in the coloration of the light source, such as sunlight versus incandescent light or bluish light from the sky. In each image file, the DSLR records a white balance setting, which is selected from a menu (daylight, tungsten incandescent, flash, etc.), automatically estimated for each picture (auto), or established by photographing a gray card (custom white balance).

When creating a JPEG file, the camera actually uses the white balance setting to adjust the color. In raw files, though, the setting is recorded but not used; the color adjustment is left to subsequent software. That's why it usually does not much matter what white balance you choose for astrophotography. When we do adjust color, the adjustment is usually stronger than anything the white balance setting can achieve. As a default setting, though, choose "daylight" for astronomical images, to avoid any unexpected color shift when previewing them.

2.3.5 Gamut

Computer monitors and printers reproduce colors, not by regenerating the spectrum of the original light source, but simply by mixing three primary colors. This works only because the human eye has three types of color receptors. Creatures could exist – and indeed some human beings *do* exist – for whom the three-primary-color system does not work.[3]

By mixing primaries, your computer screen can stimulate the eye's color receptors in any combination, but not at full strength. That is, it has a limited *color gamut.* Colors outside the gamut can only be reproduced at lower saturation, as if they were mixed with white or gray. Nothing on your computer screen will ever look quite as red as a ruby or as green as an emerald.

The gamut of an inkjet printer is also limited, more so than the screen (especially in the deep blue), and the whole art of print color management revolves around trying to get them to match each other and the camera.

To account for their limited gamut, digital image files are often tagged with a particular *color space* in which they are meant to be reproduced. The most common color spaces are *sRGB,* or standard red-green-blue, which describes the normal color gamut of a CRT or LCD display, and *Adobe RGB,* a broader gamut for high-quality printers. *Photoshop* can shift an image from one color space to another, with obvious changes in its appearance.

---

[3] Severely color-blind people have only two primary colors. There are also humans with normal color vision whose primary red is not at the usual wavelength, and it is speculated that a person who inherits that system from one parent and the normal system from the other parent could end up with a working four-color system. It is not clear how the optic nerve and brain would handle the fourth channel.

2.4 Image Size and Resizing

2.4.1 Dots per Inch

"Change the size of the image" can mean either of two things: set it to print with a different number of dots (pixels) per inch, or change the number of pixels. The first of these is innocuous; the second almost always removes information from the picture.

Consider dots per inch (dpi), or pixels per millimeter, first. A sharp color print, comparable to the best film photographs, requires 100 to 200 dots per inch (4 to 8 pixels/mm). That means an 8×10-inch print needs only $8 \times 200 \times 10 \times 200 = 3\,200\,000$ pixels (that is, 3.2 megapixels). If you can tolerate slightly less crispness, even one megapixel is enough for a full-page picture.

Clearly, then, the 24-megapixel output of a modern DSLR is far more than you need for a sharp print. The large number of pixels allows you to scale the picture down or use just a small part of it and still have a sharp result. Alternatively, you can leave it as-is, specify its resolution as 500 dpi, and get a print whose visible sharpness is definitely not limited by the number of pixels.

2.4.2 Resampling

To save file space, or to make the picture fit on a web page, you may wish to reduce the actual number of pixels in the image. This reduction should be done as the last step in processing because it throws away information. Once you shrink an image, you cannot enlarge it back to its original size with full sharpness.

Changing the number of pixels is called *resampling,* and Figures 2.4 and 2.5 show how it is done. Reduction and enlargement are called *downsampling* and *upsampling* respectively.

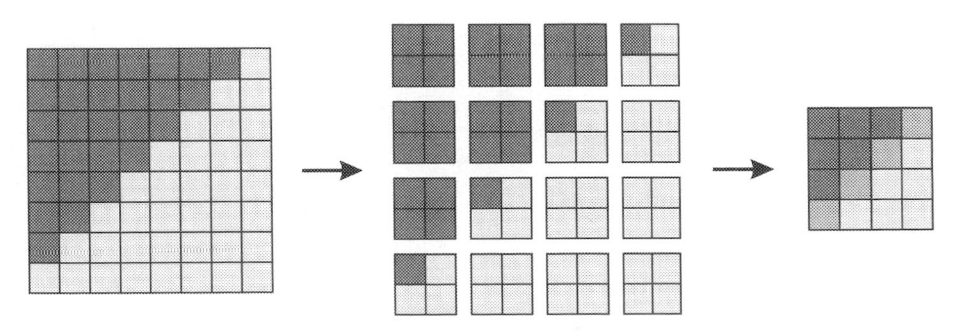

Figure 2.4. Resampling an image to shrink it. Pixels that are to be combined are averaged together. (From *Astrophotography for the Amateur.*)

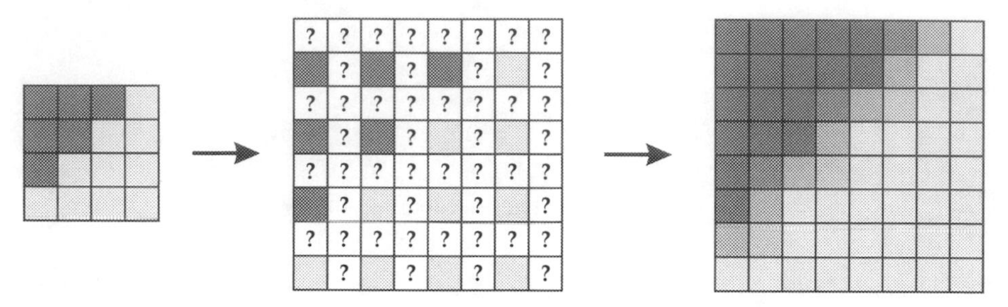

Figure 2.5. Resampling an image to enlarge it. Computer must fill in missing pixels by averaging their neighbors. (From *Astrophotography for the Amateur.*)

2.4.3 Binning

What Figure 2.4 actually shows is a particularly simple kind of resampling called *binning* – adjacent pixels are simply combined, in groups. The figure shows 2×2 binning. Reducing an image to an arbitrary fraction of its original size is not ordinarily that simple; instead, some calculation and interpolation has to be done.

The term *binning* most often denotes combining of pixels that is done at an early stage of the process, maybe even in the camera, rather than later, during editing.

2.4.4 The Drizzle Algorithm

The technique in Figure 2.5 assumes that every pixel covers a square area. In reality, pixels are spots; on the image sensor, pixels do not fill the squares allocated to them. When a set of images needs to be enlarged and stacked, as often happens in planetary imaging, it is unrealistic to treat each pixel as a square segment of the image. That is the key idea behind the Drizzle algorithm of Fruchter and Hook.[4]

To enlarge an image, the Drizzle algorithm treats each pixel as a small round spot and "drops" it (like a drop of rain) onto the output image. This reduces the effect of the square shape and large size of the input pixels. Missing pixels can be filled in by stacking additional images whose alignment is slightly different or by interpolating.

2.5 Histograms, Brightness, and Contrast

2.5.1 Histograms

A histogram is a chart showing how many of the pixels are at each brightness level. The histogram of a daytime picture is normally a hump filling the whole

---

[4] A. S. Fruchter and R. N. Hook (2002) Drizzle: a method for the linear reconstruction of undersampled images. *Publications of the Astronomical Society of the Pacific* 114:144–152.

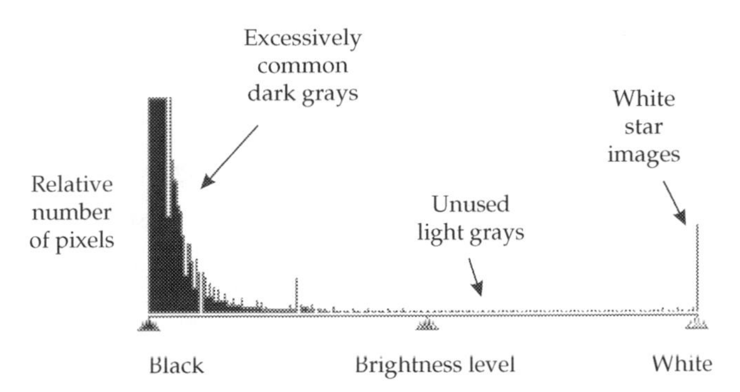

Figure 2.6. Histogram shows how many pixels are at each level of brightness.

brightness range; deep-sky images usually look more like Figure 2.6, with bright stars at the right, faint nebulosity at the left, and not much in the middle.

2.5.2 Histogram Equalization

A common task is to equalize the histogram, i.e., spread out the pixel values so that more of the range is used. This can be done with the Levels adjustment in *Photoshop* or the corresponding adjustment in other software. It works much the same way in all software packages and is covered in more detail in Section 4.3.

In *Photoshop* and other software packages, under the histogram are three sliders. The left slider sets the black level; the right slider sets the white level; and the middle slider lightens or darkens the midtones nonlinearly. If your picture has not yet been gamma corrected, you will need to move the middle slider strongly toward the left; that is how gamma correction is done manually.

As long as you are moving only the left and right sliders, you are peforming a *linear stretch*, changing the range but keeping the numbers proportional to each other. When you move the middle slider, you are introducing nonlinearity.

2.5.3 Curve Shape

The *characteristic curve* of any imaging system is the relationship between input brightness and output brightness. Perfectly faithful reproduction is a straight line (after gamma correction).

Almost all image processing software lets you adjust curves. The equivalent to moving the middle slider to the left on the histogram is to raise the middle of the curve, so that midtones are lighter. The dashed lines in Figure 2.7 show characteristic curves that perform gamma correction. For a whole gallery of curve adjustments with their effects, see *Astrophotography for the Amateur* (1999), pp. 226–228.

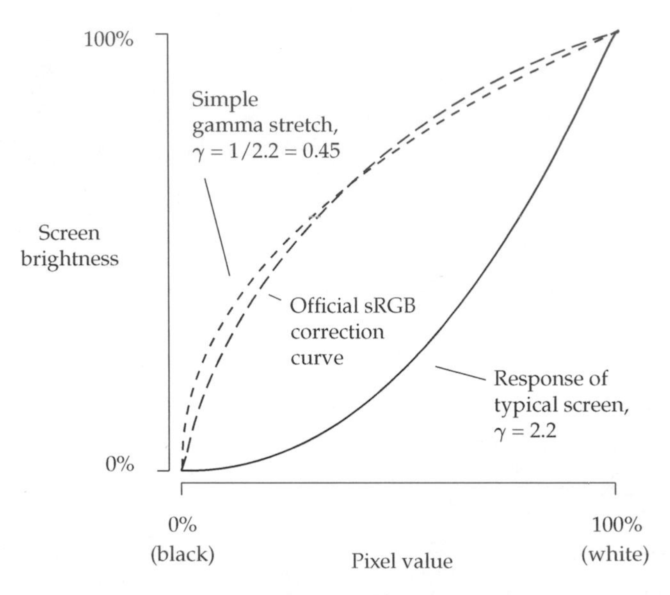

Figure 2.7. Gamma (γ) measures nonlinear relation between pixel values and brightness. Upper curves show correction applied to compensate for screen response.

2.5.4 Gamma Correction in Detail

When your camera saves a picture as JPEG, or when you decode a raw image file with ordinary (non-astronomical) software, the image undergoes gamma correction, which is needed because pixel values do not mean the same thing in a raw image as they do on a computer screen or printer. If you view an image without gamma correction, it looks very dark (Figure 11.4).

In the raw image, the pixel values are proportional to the number of photons that reached the sensor. But the brightness of a computer screen follows a power law that approximates the eye's logarithmic response to light. Specifically:

$$\text{Brightness (as fraction of maximum)} = \left(\frac{\text{Pixel value}}{\text{Maximum pixel value}}\right)^{\gamma},$$

where $\gamma \approx 2.2$.

Here γ (gamma) is a measure of the nonlinearity of the response. Printers, in turn, mimic the response of the screen. Some printers and displays are calibrated for $\gamma \approx 1.8$ instead of 2.2.

Figure 2.7 shows how this works. A pixel that displays on the screen at 50% of full brightness will have a pixel value, not 50%, but about 73% of the maximum value because $0.5^{1/2.2} = 0.73$. For example, if the pixel values are 0 to 255, a half-brightness pixel will have a value of 186. Monitor calibration test patterns test the gamma of your display by having you compare a patch of pixels at level 186 to a patch of alternating rows of 0 and 255 which blend together as you view the screen from far away.

That's why images taken straight from DSLR raw files generally have the midtones too dark. The upper curves in Figure 2.7 show how this is corrected. The simplest correction is a gamma stretch, defined as follows:

$$\begin{array}{c}\text{Output}\\ \text{pixel value}\end{array} = \text{Max. output pixel value} \times \left(\frac{\text{Input pixel value}}{\text{Max. input pixel value}}\right)^{1/\gamma}.$$

For example, if the input and output pixel values both range from 0 to 255, and $\gamma = 2.2$, then a pixel whose value was originally 127 (midway up the scale) will become

$$255 \times (127/255)^{1/2.2} = 255 \times (127/255)^{0.45} = 255 \times 0.73 = 186.$$

If $\gamma = 1$, this becomes the equation for a linear stretch.

The official correction curve for the sRGB color space is slightly different from a pure gamma stretch. As Figure 2.7 shows, it has slightly less contrast in the shadows (to keep from amplifying noise) and makes up for it in the midtones.

Since astronomical images do not strive for portrait-like realism in the first place, there is no need to follow a specific method of gamma correction; just lighten the midtones until the image looks right.

2.6 Sharpening

One of the wonders of digital image processing is the ability to sharpen a blurry image. This is done in several ways, and of course it cannot bring out detail that is not present in the original image at all. What it can do is reverse some of the effects of blurring, provided enough of the original information is still present.

2.6.1 Edge Enhancement

The simplest way to sharpen an image is to look for places where adjacent pixels are different, and increase the difference. For example, if the values of a row of pixels were originally

20 20 20 20 30 30 30 30 20 20 20 20

they might become:

$20\ 20\ 20\ \underline{15}\ \underline{35}\ 30\ 30\ \underline{35}\ \underline{15}\ 20\ 20\ 20$

(changes underlined). This gives the image a crisp, sparkling quality, but it is most useful with relatively small images. DSLR images have so many pixels that single-adjacent-pixel operations like this often do little but bring out grain.

2.6.2 Unsharp Masking

A more versatile sharpening operation is *unsharp masking* (Figure 2.8). This is derived from an old photographic technique: make a blurry negative (unsharp

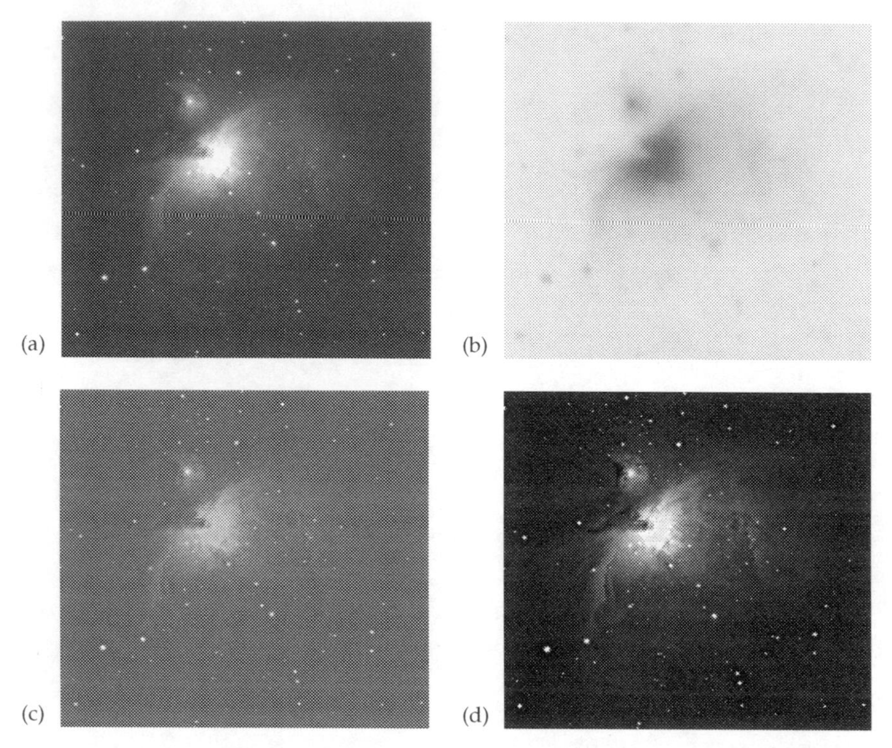

Figure 2.8. The concept of unsharp masking: (a) original image; (b) unsharp mask; (c) result of stacking (a) and (b); (d) contrast stretched to full range.

mask) from the original image, sandwich it with the original, and rephotograph the combination to raise the contrast. Originally, the unsharp mask was made by contact-printing a color slide onto a piece of black-and-white film with a spacer separating them so that the image would not be sharp.

The effect of unsharp masking is to reduce the contrast of large features while leaving small features unchanged. Then, when the contrast of the whole image is brought back up to normal, the small features are much more prominent than before. What is important about unsharp masking is that, by varying the amount of blur, you can choose the size of the fine detail that you want to bring out.

Today, unsharp masking is performed digitally, and there's no need to create the mask as a separate step; the entire process can be done in a single matrix convolution (*Astrophotography for the Amateur*, 1999, p. 237).

Note that *Photoshop* has a considerable advantage over some astronomical software packages when you want to perform unsharp masking – *Photoshop* can use a much larger blur radius, and a large radius (like 50 to 100 pixels) is often needed with large DSLR images.

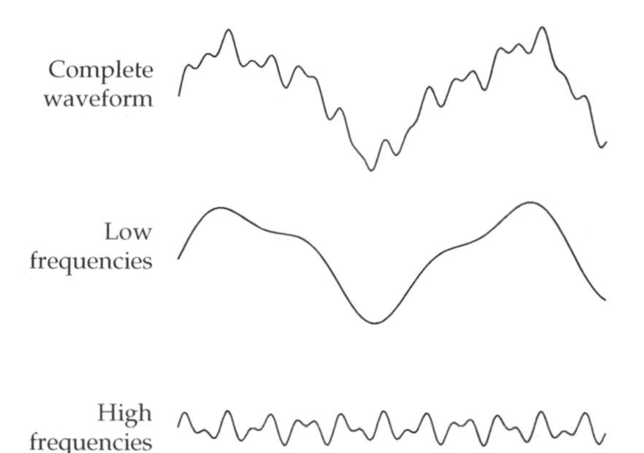

Figure 2.9. Sound waves can be separated into low- and high-frequency components. So can images. (From *Astrophotography for the Amateur.*)

2.6.3 Spatial Frequency and Wavelet Transforms

Another way to sharpen an image is to analyze it into frequency components and strengthen the high frequencies.

To understand how this works, consider Figure 2.9, which shows the analysis of a sound wave into low- and high-frequency components. An image is like a waveform except that it is two-dimensional; every position on it has a brightness value.

It follows that we can speak of *spatial frequency,* the frequency or size of features in the image. For example, details 10 pixels wide have a spatial frequency of 0.1. High frequencies represent fine details; low frequencies represent large features. If you run a *low-pass filter,* you cut out the high frequencies and blur the image. If you emphasize the high frequencies, you sharpen the image.

Any complex waveform can be subjected to *Fourier analysis* or *wavelet analysis* to express it as the sum of a number of sine waves or wavelets respectively. (A wavelet is a shape like that shown in Figure 2.10.) Fourier analysis into sine waves is more appropriate for long-lasting waveforms, such as sound waves; wavelets are more appropriate for nonrepetitive waveforms, such as images.

2.6.4 Multiscale Processing

Wavelets are the key to *multiscale processing,* which means processing an image in such a way that you can treat details of different sizes differently. This provides a way to bring out fine detail without strengthening the grain, which is

Figure 2.10. A wavelet (one of many types). A complex signal can be described as the sum of many wavelets of different heights and widths, shifted left or right as needed.

even finer. Multiscale processing, in turn, is the key to planetary video imaging, so the practical details are discussed in Chapter 14. Here I explain how it works.

Almost everybody who has ever done unsharp masking has almost immediately wanted to do it more than once to the same image, on different scales (with different sizes of blur). That is a primitive form of multiscale processing.

Recall that unsharp masking consists of subtracting a blurred image from the original to get a layer that consists only of fine detail, then combining this with the original image so that the detail is emphasized.

Multiscale enhancement means doing this repeatedly, blurring the image, subtracting it from the original, blurring the blurred image further, subtracting it from its predecessor, and so on, and keeping all of the difference layers, plus the last blurred image. Mathematically:

I_0 = original image

I_1 = result of blurring I_0 Layer 1 = $I_0 - I_1$

I_2 = result of blurring I_1 Layer 2 = $I_1 - I_2$

I_3 = result of blurring I_2 Layer 3 = $I_2 - I_3$

Residue = last blurred image = I_3.

Each of the layers corresponds to wavelets of a particular size, and the shape of the wavelet is the difference between two successive blurs.

Now, if the layers were combined with the residue, all in equal proportions, you'd get the original image back:

$$\text{Layer 1} + \text{Layer 2} + \text{Layer 3} + \text{Residue} = (I_0 - I_1) + (I_1 - I_2) + (I_2 - I_3) + I_3 = I_0.$$

But in fact we don't combine them in equal proportions. By increasing or decreasing the contribution of particular layers, we can emphasize detail of a certain size (such as belts and zones on Jupiter) while discarding detail of a different size (such as grain). It is even possible to do noise removal and other processing on individual layers.

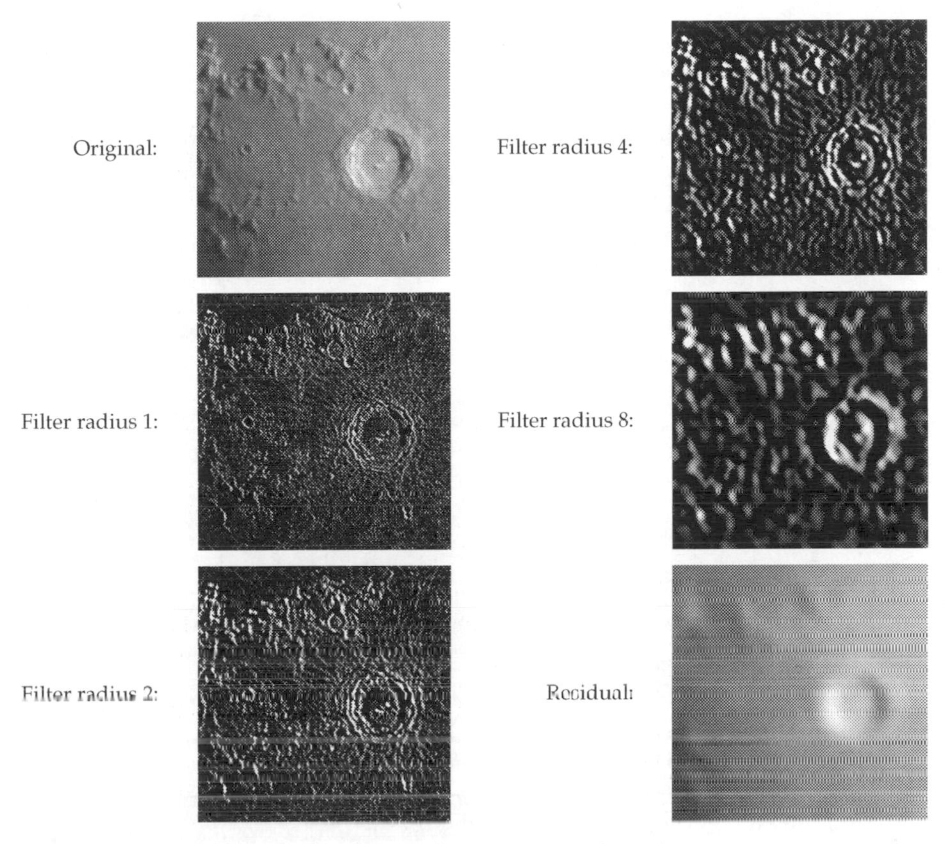

Figure 2.11. Image of lunar crater Copernicus decomposed into four layers of different sizes of fine detail, plus a residual layer.

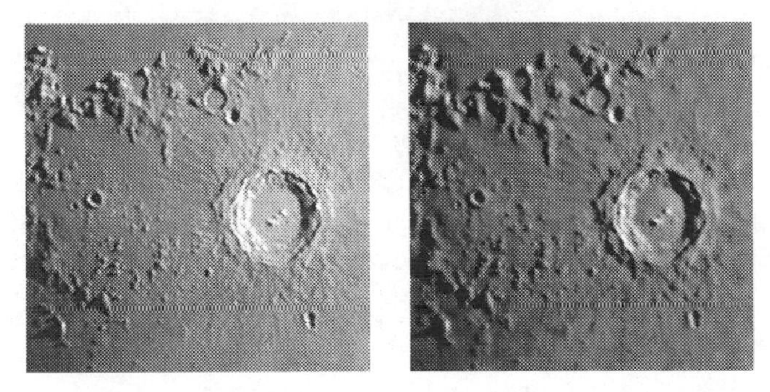

Figure 2.12. The image in Figure 2.11 reconstructed with enhancements. *Left:* Sharpened by increasing the contribution of the layers with filters of radius 1 and 2. *Right:* Contribution of residual layer reduced; note how the background gradient diminishes.

Figure 2.11 shows a picture of a moon crater decomposed this way, and Figure 2.12 shows two ways of putting it back together. Note that reducing the residual layer makes large gradients go away while preserving smaller details; that is how high-dynamic-range (HDR) images are made displayable.

The original form of this computation is called the *à trous wavelet transform* or the *starlet transform.* Each step applies a blur twice as large as the previous one. To speed computation, the blurring process at each step can actually skip half the pixels used by the previous step, so first you use every pixel, then one of every two, then one of every four, and so on. If this introduces inaccuracies, nothing is really lost because the subtractions guarantee that all the information in the image will end up in one layer or another. The skipped pixels are "holes" (in French, *trous*) in the data, hence the name of the algorithm.

It was quickly realized that, if you don't want to use that memory-saving trick, the blurs don't have to double in size each time. *Any* blurring algorithm will work, even a small blur applied over and over. The subtractions guarantee that the addition will give the correct result. For planetary work, a series of successive blurs, all about one pixel in radius, often works well.

The blurring function is normally some kind of average or weighted average of the pixels being blurred together. The result is, by definition, a wavelet transform, with the shape of the wavelet determined by the blurring function; it is also called a *multiscale linear transform.* If the blur is Gaussian (meaning an average weighted like a Gaussian bell curve), the wavelet is the classic one shown in Figure 2.10.

If you take the median of the pixels being blurred together, rather than a weighted average, an interesting thing happens. Details too small to appear in a layer are totally absent from it. (If an average were used, small areas with extreme values would still have some influence.) The arithmetic for reconstructing the image still works the same way as before – remember, it works for *any* blurring function – but there is no longer a risk of "ringing," dark rings around sharp bright features or bright rings around sharp dark ones.[5] What's more, all the numbers are integers, which speeds up computation. The multiscale median transform is available in *PixInsight*.

For much more about multiscale processing (and plenty of mathematics), see J.–L. Starck, F. Murtagh, and A. Bijaoui, *Image Processing and Data Analysis: The Multiscale Approach* (Cambridge, 1998) and, for even more detail, J.–L. Starck, F. Murtagh, and J. M. Fadili, *Sparse Image and Signal Processing* (second edition, Cambridge, 2015). The latter book introduces two new developments, ridgelets and curvelets – ways to make the computer look for features of particular shapes, not just spots of different sizes.

2.6.5 Deconvolution

Suppose you have a blurred image, and you know the exact nature of the blur, and the effects of the blur have been preserved in full. Then it ought to be possible to undo the blur by computer, right?

[5] Ringing is so called, not because it forms rings around stars, but because in one dimension, the waveform is like the ringing of a bell: sharp spikes turn into damped oscillations.

Indeed it is, but the process is tricky. It's called *deconvolution* and has two pitfalls. First, you never have a perfectly accurate copy of the blurred image to work with; there is always some noise, and if there were no noise there would still be the inherent imprecision caused by quantization.

Second, deconvolution is what mathematicians call an *ill-posed problem* – it does not have a unique solution. There is always the possibility that the image contained even more fine detail that was completely hidden from view.

For both of these reasons, deconvolution has to be guided by some criterion of what the finished image ought to look like. The most popular criterion is *maximum entropy*, which means maximum simplicity and smoothness. (Never reconstruct two stars if one will do; never reconstruct low-level fluctuation if a smooth surface will do; and so on.) Variations include the Richardson–Lucy and van Cittert algorithms.

Deconvolution is available in *MaxIm DL* and *PixInsight*. In Figure 2.13 you see the result of doing it. Fine detail pops into view. The star images shrink and become rounder; even irregular star images (from bad tracking or mis-collimation) can be restored to their proper round shape. Unfortunately, if the parameters are not set exactly right, stars are often surrounded by dark doughnut shapes.

To perform deconvolution, you must tell the computer what kind of blur to undo. This can be either a real blur, based on your best estimate of how the

Figure 2.13. Deconvolution shrinks larger star images, brightens smaller ones, and brings out fine detail. This example is slightly overdone. (Single 3-minute exposure of M17, the Omega Nebula, with a Canon 40D, 300-mm *f*/4 lens, ISO 800.)

image was degraded, or an imaginary blur which, if reversed, ought to bring out the detail you're interested in. In Figure 2.13 the latter was done.

Because deconvolution is tricky to set up and requires lots of CPU time, I seldom use it. My preferred methods of bringing out fine detail are unsharp masking and wavelet-based filtering. However, one striking use of deconvolution is to make elongated star images rounder, to make up for imperfect guiding. For more about this, see Section 9.8.5.

Chapter 3
DSLR Operation

This chapter covers camera settings and operation for astrophotography (Figure 3.1). In what follows, I'm going to assume that you have learned how to use your DSLR for daytime photography and that you have its instruction manual handy. No two cameras work exactly alike. Most DSLRs have enough in common that I can guide you through the key points of how to use them, but you should be on the lookout for exceptions.

3.1 Taking a Picture Manually

3.1.1 Shutter Speed and Aperture

To take a picture with full manual control, turn the camera's mode dial to M (Figure 3.2). Set the shutter speed with the thumbwheel.

You will find that available exposure times range from 1/8000 second all the way up to 30 seconds, which is long enough for serious deep-sky work. For longer exposures, use B (Bulb), which means "keep the shutter open as long as the button is held down," or, on Nikons, T (Time), which means "open the shutter when the button is pressed once and close it when it is pressed again." Some Canons feature a "Bulb timer" that lets you preset a long exposure time to be actuated with a single button press in B mode. Naturally, in all these situations, you will be pressing the button on a cable release or remote control rather than the camera itself.

To set the aperture on a camera lens, some cameras have a second thumbwheel, and others have you turn the thumbwheel while holding down a button. Either way, the aperture is controlled electronically by the camera body. Canon lenses normally have no aperture ring. Many Nikon lenses have an aperture ring, which you set to the smallest aperture (typically $f/22$) when you want electronic control.

Naturally, if there is no lens attached, or if the camera is attached to something whose aperture it cannot control (such as a telescope), the camera will not let you

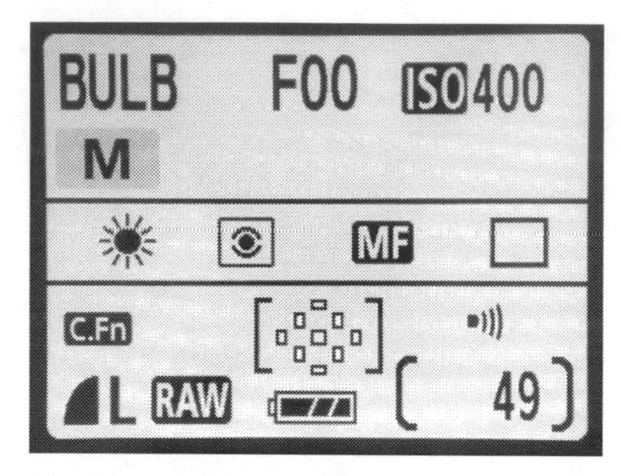

Figure 3.1. Display panel of a Canon XTi (400D) set up for deep-sky photography. *F00* means the lens is not electronic and the camera cannot determine the *f*-stop. *C.Fn* means custom functions are set.

Figure 3.2. For manual control of shutter and aperture, set mode dial to M.

set the aperture. It will generally be displayed as $f/0$ or F00. Nikons may also display a blinking question mark indicating a lack of electronic communication with the lens. You can still take pictures; the computer inside the camera just doesn't know what the aperture is.

3.1.2 Manual Focusing

In astrophotography, you must always focus manually; autofocus doesn't work on celestial objects. You must also *tell* the camera you want to focus manually,

Figure 3.3. Manual/autofocus switch is on camera body, lens, or both.

because if the camera is trying to autofocus and can't do so, it will refuse to open the shutter.

On Canon lenses, the manual/autofocus switch is on the lens, but on Nikons, it may be on the camera body (Figure 3.3), the lens, or both. If there are two switches, they should both be set the same way.

Astronomical photographs are hard to focus, but with a DSLR, you have a superpower – you can preview in Live View, and even if you don't have Live View, you can review each picture, magnified, right after taking it. For more about how this is done, see Section 3.6.

Do not trust the infinity (∞) mark on a lens. Despite the manufacturer's best intentions, it is probably not perfectly accurate, and anyhow, the infinity setting will shift slightly as the lens expands or contracts with changes in temperature.

3.1.3 ISO Speed

A point to which we will often return is that the ISO speed setting on a DSLR is actually the gain of an amplifier; you are using the same sensor regardless of the setting. The best ISO setting depends on the sensor (see Chapters 15 and 16), but if you want a quick starting point, set the ISO speed to 1600 for your first experiments. Lower settings, however, make it possible to cover a wider range of brightnesses (Figure 3.4).

Figure 3.4. High dynamic range in this picture of the globular cluster M13 was achieved by stacking four 3-minute ISO 400 exposures. Canon Digital Rebel (300D), 8-inch (20-cm) *f*/6.3 telescope, dark frames subtracted.

3.1.4 Do You Want an Automatic Dark Frame?

If you're planning an exposure longer than a few seconds, hot pixels and dark current will add to the noise, and you need to think about dark-frame subtraction. Done manually, this means that in addition to your picture, you take an identical exposure with the lenscap on, and then you subtract the dark frame from the picture. This corrects the pixels that are too bright because of electrical leakage.

Most DSLRs will do this for you if you let them. Simply enable long-exposure noise reduction (a menu setting on the camera). Then, after you take your picture, the camera will record a dark frame with the shutter closed, perform the subtraction, and store the corrected image on the memory card. *Voilà* – no hot pixels.

The drawback of this technique is that it takes as long to make the second exposure as the first one, so you can only take pictures half the time. The alternative is to take a set of dark frames manually and apply them during calibration.

3.2 Menu Settings

Like other digital cameras, DSLRs are highly customizable. It's worthwhile to work through the instruction manual, examine every menu setting, and make an intelligent choice, since what suits you will often be different from what suits an ordinary daytime photographer. What follows is a checklist of important menu settings from several different popular DSLRs. They may not go by exactly the same names on your camera, but they are an indication of what to check.

Note that the full menu is not likely to be visible unless the camera is set to one of its full-control modes, preferably M. If you turn on the camera in one of its "simple" modes, the more advanced parts of the menu will be inaccessible.

Note also two more things. Important settings may be hidden deep in "custom function" menus, or only visible if you make a setting to display full menus. Also, many cameras let you set up a "personal menu" of settings you use frequently, to pull them all together regardless of where they are located in the built-in menu system. I strongly recommend doing this.

3.2.1 Things to Set Once and Leave Alone

These are settings that you can probably leave unchanged as you go back and forth between astrophotography and daytime photography.

Auto Rotate (Auto Image Rotation): Off. Tell the camera not to rotate images if it thinks you are holding the camera sideways or upside down. Doing so can lead to very confusing astrophotos.

Highlight Tone Priority: Off. This Canon setting, if turned on, would cut the effective ISO speed in half, even when you are not using the exposure meter (see Section 16.7.3).

NEF Recording: 14 bits (or largest number). This Nikon setting lets you choose between 12- and 14-bit raw files on some cameras. Future cameras may support more than 14 bits.

Copyright or Image Comments: It is not a bad idea to enter your name and have it stored in the EXIF data of every image. This can help ensure you get credit for your images if someone distributes them without proper attribution.

Manual Movie Settings: Off. This Nikon D5300 setting shouldn't affect still photography, but it does. If turned on, it causes the shutter speed to go back to 1/60 second every time Live View is used. This behavior may change in a future version of the firmware.

3.2.2 Settings for an Astrophotography Session

Date and Time: I like to set the date and time, accurate to the second, at the beginning of every session, to ensure accurate times on the files that are created.

Picture Quality (Image Quality): Raw + smoothest, largest JPEG. If the memory card has room, I like to have the camera save each picture in both formats. Each file serves as a backup of the other, and the JPEG file contains exposure data that can be read by numerous software packages. This takes only 20% more space than storing raw images alone.

Image Size: Maximum. Many cameras give you the option of downsampling the image as it is taken, to save memory card space. That is rarely if ever a good idea in scientific work. You paid for all those megapixels; use them!

Aspect Ratio: 3:2 (default). Some cameras offer the option of trimming off part of the picture to change its shape, such as for pictures to be displayed on television. Again, you paid for all those megapixels...

Picture Style: Standard or Neutral. This Canon setting is important only if you make extensive use of JPEG output rather than raw files. It controls the processing performed to convert raw to JPEG in the camera.

White Balance: Daylight. This is the option that will give you the most consistent results, especially if you are going to use your JPEG image files. With raw image files, the white balance is recorded but not applied; it has an effect only if subsequent software chooses to use it.

Image Review: Off. At an observing session with other people, you'll want to minimize the amount of light emitted by your camera, as well as save battery power, so don't display every picture on the screen.

LCD Brightness (Monitor Brightness): Low. In the dark, even the lowest setting will seem very bright.

LCD Display When Power On: Off. On cameras that use their LCD screen to display camera settings, you probably don't want the display shining continuously at night. Instead, switch it on with the DISP or INFO (or Nikon *i*) button when you want to see it.

Auto Power Off: Never. You don't want the camera turning itself off during lulls in the action; at least, I don't.

High ISO Noise Reduction: Probably irrelevant. This setting controls a noise-reduction step that is applied when converting images to JPEG in the camera.

Like white balance, the setting is saved with raw files but not applied to them. If JPEGs are your finished product, you may find this setting useful.

Long Exposure Noise Reduction: Your decision. If turned on, this feature will eliminate hot pixels and dark current noise in your images by taking a dark frame immediately after each picture and automatically subtracting it. This takes time, and as you become more experienced, you'll prefer to take dark frames separately and subtract them during image processing.

3.3 How to See that Tiny Screen

To use a DSLR effectively without tethering it to a computer, you will need to see its LCD screen clearly, not just well enough to make out words, but well enough to judge focus precisely, without experiencing eyestrain. Even if you can read comfortably at normal distance, you may need optical help getting a critically sharp view of the camera's LCD screen.

The first challenge is seeing the screen at all while the camera is aimed upward. If the screen flips out and rotates, you have no problem. Otherwise, you can either crouch behind it or view it with the aid of a hand-held mirror (Figure 3.5).

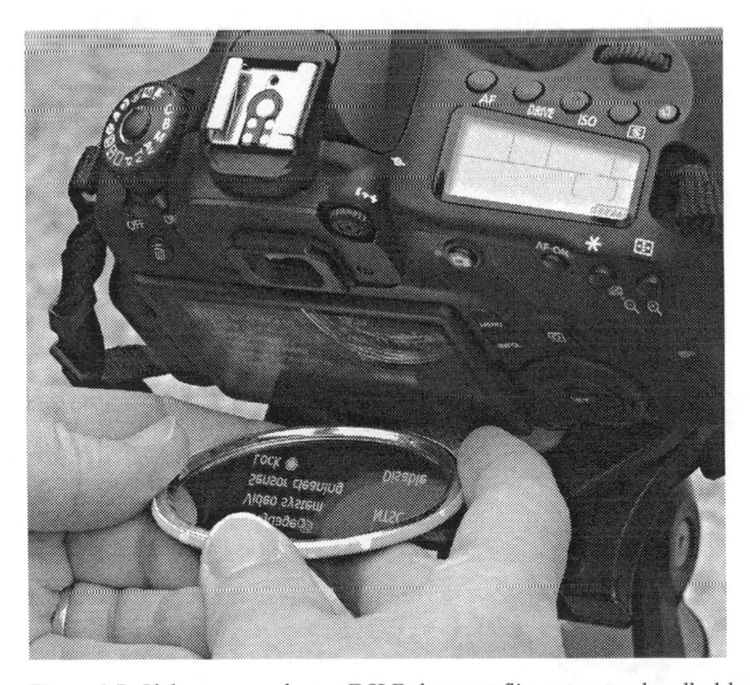

Figure 3.5. If the screen of your DSLR does not flip out, use a handheld mirror to view it while the camera is aimed upward. (Sharon Covington)

Next comes the challenge of focusing your eyes on it. I happen to have a superpower – I am severely nearsighted, so to look at the screen, I take my glasses *off* and move in close. To me it is then a giant screen.

If you're farsighted, one option is to carry a pair of +3.0-diopter reading glasses, maybe on a strap around your neck, and slip them on when you want to view the screen. Such reading glasses can be purchased without a prescription, and you can even slip them on over other glasses.

Another option is to carry a magnifying glass. It need not be very powerful, since its purpose is not to magnify, but to bring the screen into focus for farsighted eyes. A wide variety of magnifiers and supports can be found at a sewing-supply store, since many people who sew are presbyopic.

You can even attach an eyepiece to the viewing screen. A number of gadgets are on the market that consist of a housing and adjustable magnifying eyepiece through which to view the screen, but, unfortunately, they usually require the screen to be flat on the back of the camera, not swung out. That is challenging when the telescope is aimed upward.

If that's not enough, you have two more options. You can connect a video monitor and view the screen display greatly enlarged, or you can connect the camera to a computer and do all the viewing there (see Section 10.2.4).

3.4 More Features of the Camera Body

3.4.1 The Eyepiece Diopter

Although the camera eyepiece is not the primary way to focus modern DSLRs, it is still useful, and to get the most out of it, you'll need to make sure it is accurately focused for your eye. That is achieved with the eyepiece diopter adjustment (Figure 3.6). This adjustment is best made in the daytime, then refined in dim light because most people's eyes focus slightly differently in the dark.

Figure 3.6. Adjust the eyepiece diopter (*arrow*) so that you can see the focusing screen clearly.

Adjust the eyepiece so that you can see the center of the image with perfect sharpness. Don't focus on the digital display at the bottom or any indicator lights; they probably are not in the same plane as the focusing screen. Concentrate on the granularity of the focusing screen and the image that is on it.

3.4.2 The Strap and Eyepiece Cover

Should you use the neck strap that comes with the camera? I do; I drape it over the telescope to keep the camera from falling to the ground if its telescope adapter comes loose.

More importantly, the strap also holds a small cover for the camera eyepiece. This keeps light from entering the camera through the eyepiece and affecting the exposure meter or (much less likely) making its way to the sensor. It is for taking pictures through dark filters or the like, where the light coming into the camera from behind may be stronger than the light entering through the lens.

The eyepiece cover is not needed during routine astrophotography because, of course, your surroundings are dark. I use it when taking dark frames so that I can turn on lights and start taking down equipment without worrying about light getting into the camera. Having said that, I have never actually experienced any ill effects from light entering through the eyepiece. I suspect its effect is mainly on the exposure meter, which is not used in astrophotography.

3.4.3 Limiting Light Emission from the Camera

If you observe with other people, or if you want to preserve your night vision, a DSLR can be annoying. The bright LCD display lights up at the end of every picture, and on some models it glows all the time.

Fortunately, there are menu settings to change this. You can turn "review" off so that the picture isn't displayed automatically at the end of each exposure. (Press ▷ or "Play" when you want to see it.) And if the camera normally displays its settings on the LCD, you can set it not to do so.

That leaves the indicator LEDs. In general, red LEDs aren't bothersome; you can stick a piece of tape on those of other colors, such as the green power-on LED on top of the Digital Rebel XTi (400D) and the bright white LED on the front of many cameras that signals delayed shutter release (Figure 3.7).

To further protect your night vision, you can put deep red plastic over the LCD display. A convenient material for this purpose is Rubylith, a graphic arts masking material consisting of deep-red plastic on a transparent plastic base. What I do is cut out a piece of the right size, leaving it attached to its clear base, and secure it to the camera screen with pieces of transparent double-stick tape at the corners. When finished with it, I move it into a plastic envelope (actually a CD cover). The double-stick tape survives being transferred back and forth many times.

Figure 3.7. Dots punched out of red vinyl tape make good covers for the power-on LED and self-timer indicator.

Because it is designed to work with photographic materials, Rubylith is guaranteed to block short-wavelength light, so it's especially good for preserving dark adaptation. Many drafting-supply stores sell Rubylith by the square foot; one online vendor that sells pieces smaller than full rolls is www.dickblick.com. Other kinds of red plastic that can be used in similar ways are available from other sources.

A more rugged alternative that I have experimented with is a transparent red acrylic shield that fits over a flip-out screen; the front is flat and the back acts as a spring (Figure 3.8). This is made by carefully heating a piece of plastic with a hot-air gun and bending it between pieces of wood. As far as I know, nothing like it is commercially available, but I would like to see it come on the market.

3.5 Tripping the Shutter without Shaking the Telescope

Now it's time to take the picture. Obviously, you can't just press the button with your finger; the telescope would shake terribly. So what do you do? There are several options.

3.5.1 Self-timers and Remote Controls

If the exposure is going to be 30 seconds or less (so that you don't need to use B or T), you can use the self-timer (delayed shutter release). Press the button; the

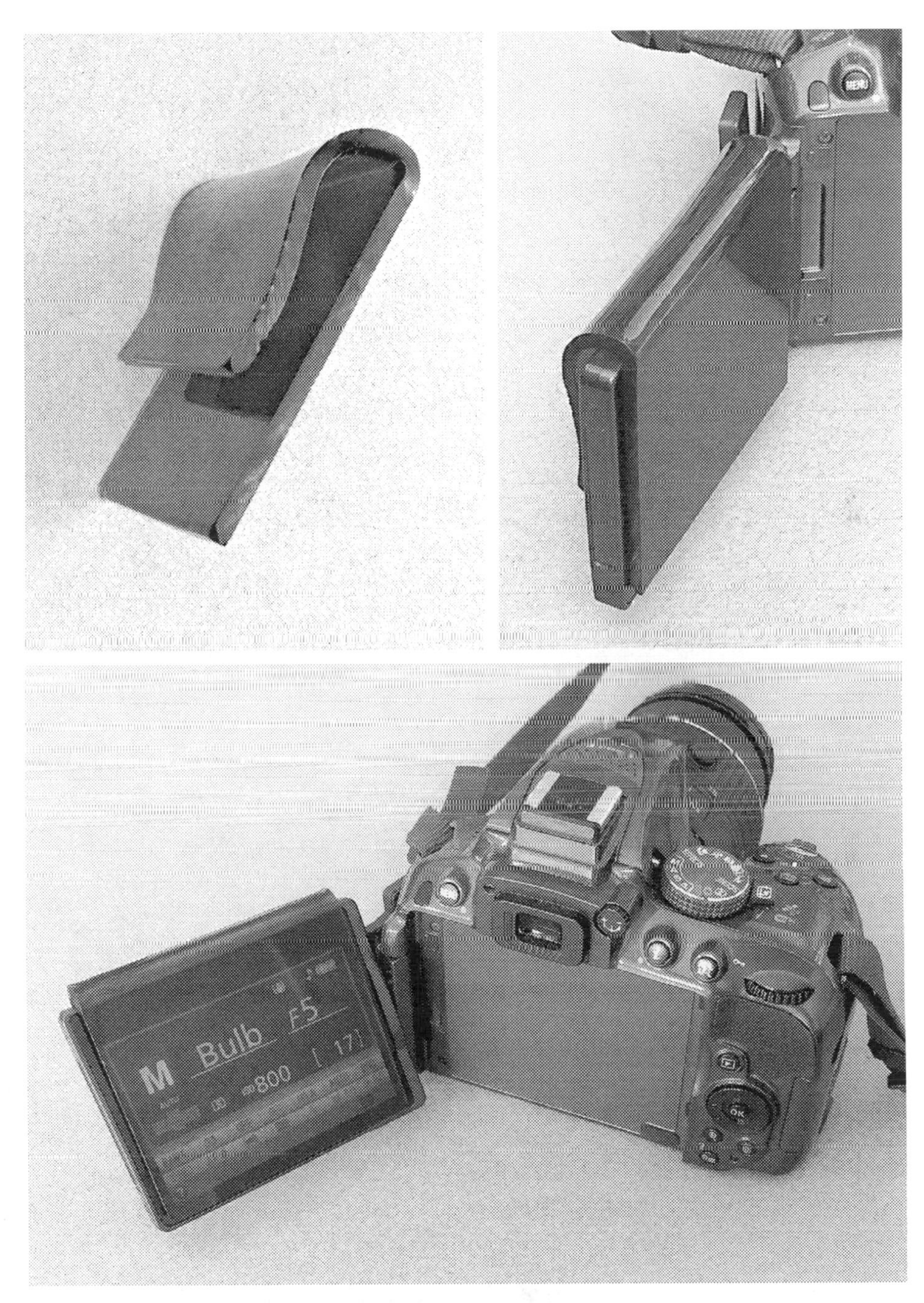

Figure 3.8. A night-vision-preserving red filter that clips on the flip-out screen can be made by heating and bending a piece of red transparent acrylic plastic.

camera counts down ten seconds while the telescope stops shaking; and then the shutter opens and the picture is taken.

That won't work if you're using B or T (the settings for time exposures). You need a remote shutter release of some kind.

Apart from connecting your camera to a computer, there are two alternatives, wired and infrared. That is, you can use a remote control connected by wire or an infrared remote control.

An infrared remote control is handy because there's no cable to keep up with. Its beam must shine on the front of the DSLR in order to reach the photocell that picks it up. If you've selected a shutter speed, the button on the remote control actuates the shutter. If the shutter is set to B (or T if available), then in general, one press of the button will start the exposure and a second press will end it, but check the instructions to make sure.

The advantage of a wired remote control is that it can contain an elaborate exposure timer (intervalometer). Figure 3.9 shows one common type, which is imported and sold under several names. It can be set to make any number of exposures, of any length, with any interval between them. For example, in deep-sky work I often take a series of 50 1-minute exposures with 30-second gaps between them. Using a remote control with a timer, I can start all this with the press of one button.

Figure 3.9. A wired remote control (intervalometer) can time sequences of exposures of any length at any interval.

3.5.2 Mirror Lock and Prefire

Much of the vibration of a DSLR actually comes from the mirror, not the shutter. If the camera can raise the mirror in advance of making the exposure, this vibration can die down before the shutter opens.

Mirror vibration is not a major issue in deep-sky work, though I do use mirror prefire to control it whenever possible. No great harm results if the telescope shakes for a few milliseconds at the beginning of an exposure lasting several minutes; that's too small a fraction of the total time. But in lunar and planetary work, it's a problem, and so is shutter vibration – so much so that eliminating mirror vibration alone is not enough. You need EFCS (see next section) or a dedicated astrocamera.

There are two main ways to control mirror vibration, *mirror prefire* and *mirror lock*. To further confuse you, "mirror lock" sometimes denotes a way of locking up the mirror to clean the sensor, not to take a picture. And on Schmidt–Cassegrain telescopes, "mirror lock" means the ability to lock the mirror in position so that it can't shift due to slack in the focusing mechanism. Neither of those is what I'm talking about here.

Mirror prefire is the usual Nikon approach. On the D5300, for instance, deep in the custom settings menu is an option called "Exposure Delay Mode." When enabled, this introduces a 2-second delay between raising the mirror and opening the shutter. This delay can be slightly disconcerting if you forget to turn it off when doing daytime photography. For astronomy, though, it can make a noticeable difference.

Canon offers mirror lock, which turns into mirror prefire when the self-timer is used. Turn mirror lock on, and you'll have to press the button on the cable release twice to take a picture. The first time, the mirror goes up; the second time, the shutter opens. In between, the camera is consuming battery power, so don't forget what you're doing and leave it that way. When you are using the self-timer, one button press is sufficient; the mirror goes up at the beginning of the time delay, and then, either 2 or 10 seconds later depending on your choice, the shutter opens.

In deep-sky work, I use 2 seconds of mirror prefire (via exposure delay mode on the Nikon and via the self-timer with mirror lock on the Canon); it seems to be sufficient. In lunar and planetary work, I use EFCS whenever possible.

3.5.3 Electronic First-curtain Shutter (EFCS)

The best way to eliminate shutter vibration is not to use the shutter at all by taking the picture in Live View with a camera that has electronic first-curtain shutter (see Section 1.3.2). I do lunar and planetary work this way. The benefit can be dramatic (Figure 3.10).

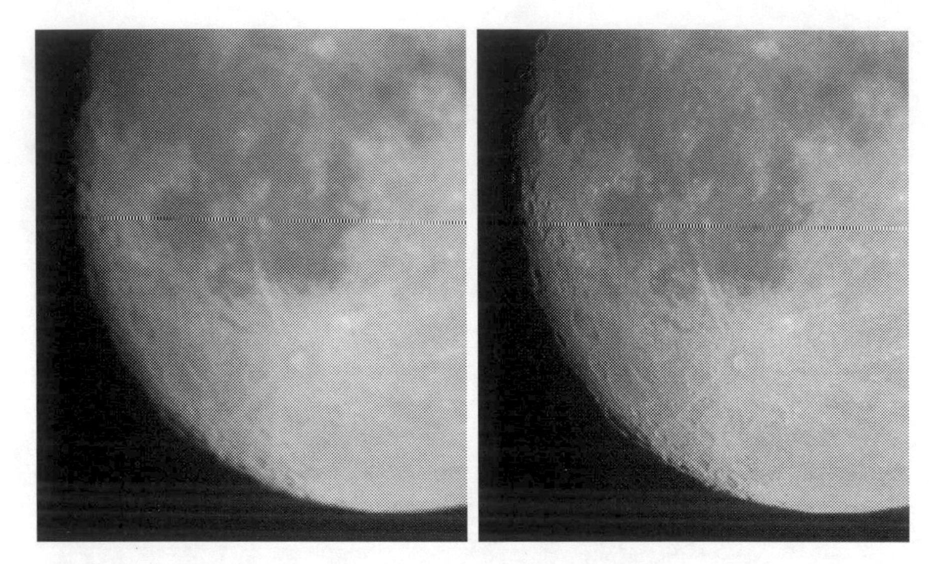

Figure 3.10. Electronic first-curtain shutter makes a big difference in lunar and planetary work. The moon through a Celestron 5 (12.5-cm aperture, *f*/10), 1/100 second at ISO 200, with a Canon 40D without (left) and with (right) Live View Silent Shooting. Unsharp masking, normal in lunar work, was not performed here.

Unless vibration is a serious problem, EFCS is not recommended for deep-sky work because the sensor heats up during Live View; it's better to keep it cool by only using it when the exposure is actually in progress.

3.5.4 Other Tricks

The *sure* way to take a vibration-free lunar or planetary image is called the "hat trick." Hold your hat (if you wear a black fedora), or any dark object, in front of the telescope. Open the shutter, wait about a second for vibrations to die down, and move the hat aside. At the end of the exposure, put the hat back and then close the shutter. Instead of a hat, I usually use a piece of black cardboard. I find I can make exposures as short as 1/4 second by swinging the cardboard aside and back in place quickly.

The "hat trick" was ideal during the film era. With a DSLR, it has the disadvantage that the sensor is on, consuming power and accumulating dark current, the whole time the shutter is open, not just when the hat has been moved aside. With newer sensors, that is not as much of a drawback as it used to be.

Another way to eliminate vibration almost completely is to use afocal coupling (that is, aim the camera into the telescope's eyepiece) with the camera and telescope standing on separate tripods. This technique is as clumsy as it sounds, but it's how I got my best planetary images during the film era. Today, a good option for afocal lunar and planetary work is not to use a DSLR or MILC at all, but rather a non-SLR digital camera with a nearly vibrationless leaf shutter.

3.5.5 Vibration-reducing Lenses

Unfortunately, vibration-reducing or image-stabilizing lenses (Nikon VR, Canon IS) are no help to the astrophotographer. They compensate for movements of a handheld camera; they do not help when the camera is mounted on a tripod or telescope. The same goes for vibration reduction that is built into some camera bodies. Just turn the vibration-reduction feature off.

In fact, some astrophotographers have found that vibration-reducing lenses are likely to be slightly out of collimation when used for deep-sky work. The reason is that the vibration reduction is implemented by shifting lens elements laterally, and there's a risk they won't be perfectly centered when the feature is turned off.

3.6 Focusing

3.6.1 Magnified Preview on the Screen

The first edition of this book devoted several pages to different focusing methods, practical and theoretical. Now, however, I want to go straight to the conclusion: The right way to focus a DSLR is to preview the image in Live View with as much magnification as possible.

Maximum magnification usually means $\times 10$, compared to showing the whole picture on the screen. Accordingly, Live View at maximum magnification is equivalent to looking at a small segment of a 20×30-inch (50×75-cm) enlargement. Although you won't be able to see individual pixels, you can see more detail than your lens or telescope can resolve.

The object you focus on should be a star or the moon, not an inherently blurry galaxy, nebula, or comet. If necessary, aim at a star of appropriate brightness and focus carefully before slewing to the object you actually want to photograph. With a computerized mount, you can use "precise go-to" mode, so that the telescope will slew to a star for you to focus and center, then move to the celestial object you've chosen. Some cameras will give you a better view of faint stars if you set the ISO higher while focusing.

If your camera doesn't have Live View, you can do the next best thing: take short test exposures (maybe 1 to 5 seconds) and immediately review each one on the screen, magnified. This process is effectively a slower version of Live View. I used it regularly with my old Canon 300D.

3.6.2 Stars and Spikes

Of course, the goal of focusing is to make bright stars as compact as possible, and to make faint stars visible. You will often see fainter stars pop into view just when you reach perfect focus. Around brighter stars you may see chromatic aberration in the form of a red or blue ring depending on the exact focusing error; it is then a matter of judgment which direction you should err in.

You're not limited to judging tiny star images. If you hold a piece of window screen wire mesh in front of the telescope or lens, the stars will have prominent diffraction spikes (see Section 7.6). The spikes become much longer and more prominent as the star comes into sharp focus. The brighter the star, the more it helps, and this method works well with stars that are too bright for focusing by just judging the compactness of the image. Because the window screen does not contain glass, it does not affect the focus of the system.

A *Scheiner disk* or *Hartmann mask* is an opaque piece of cardboard with two large holes in it, placed in front of the telescope or camera lens. Its purpose is to make out-of-focus images look double. Some people find that it helps them judge when the stars are in focus. I suggest making one out of a pizza box, which has a lip that enables you to hang it on the front of the telescope.

A *Bahtinov mask* is a combination of the two ideas, diffraction and multiple images.[1] Like a Hartmann mask, it has holes, but the holes are in the form of groups of parallel slits sloping in different directions. At the point of correct focus, not only are the diffraction spikes longest, but they are also symmetrical. Two of them form an X, and the third, longest, spike crosses the middle of the X only at perfect focus.

Of course, any of these aids (window screen wire mesh or mask) is removed from the telescope or lens before actually taking the pictures.

3.6.3 Computer Focusing

If you're using a computer to control your DSLR, the computer can also help you focus. Many software packages for DSLR control include the ability to take short exposures over and over, download them, and display each image immediately together with an analysis of its sharpness. We always focus on stars, which are perfect point sources, and the computer can analyze exactly how compact each image is.

Figure 3.11 shows a typical star image analysis from *MaxIm DL,* which includes focusing among its many other functions. A sharply focused star is a tall, narrow peak and has a small FWHM ("full-width-half-maximum," diameter of the portion of the image that is at least 50% as bright as the central peak).

If the telescope has an electric focuser, the computer can even adjust the focus for you. *MaxDSLR, ImagesPlus,* and other packages support automatic focusing with electric focusers.

The autofocus system inside the DSLR is not useful for astronomy. It usually does not respond to stars at all.

[1] The inventor, Pavel Bahtinov or Bakhtinov, pronounces his name Bakh-tin-off, with the first syllable like the name of the composer Bach. You can see his original proposal, in Russian, at www.astronomy.ru/forum/index.php/topic,10421.0.html. It replaced an earlier modified Hartmann mask that used triangular holes to create weak diffraction spikes, differently oriented on the two images that need to be superimposed.

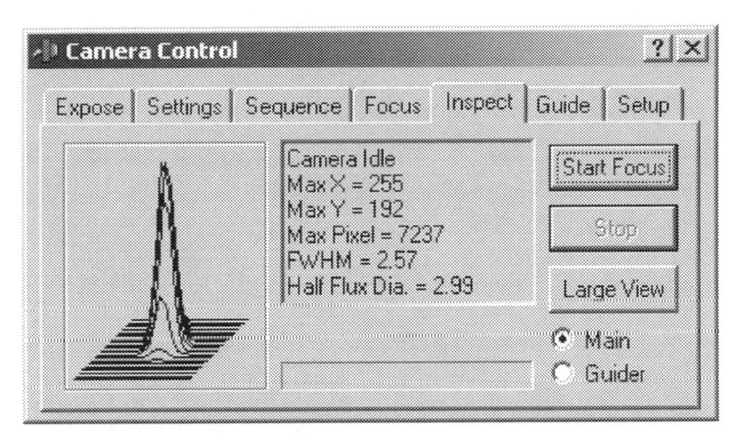

Figure 3.11. Analysis of a focused star image in *MaxIm DL*. Camera takes short exposures over and over and downloads them immediately for computer analysis.

3.6.4 Focusing Telescopes with Moving Mirrors

Because Schmidt–Cassegrains and Maksutov–Cassegrains usually focus by moving the main mirror, you may observe a couple of annoying effects. One is *lateral image shift* – the image moves sideways a short distance as you focus. This problem generally diminishes if you run the focuser through its range a few times to redistribute the lubricants.

The other problem is that if you are moving the mirror backward, it may continue to subside for a few seconds after you let go of the focuser knob. More generally, you cannot "zero in" on perfect focus by turning the knob first one way and then the other. There's a "dead zone," and when you try to undo a movement, results are somewhat hard to control.

For best results, always turn the knob clockwise first, to overshoot the desired position, and then do your final focusing counterclockwise. That way, you are pushing the mirror away from you, working against gravity and taking up any slack in the system.

Fastidious astrophotographers sometimes add a second focuser, either manual or electric, at the back end of the telescope so that the mirror does not have to be moved for fine focusing adjustments.

3.7 Other Image Quality Issues

3.7.1 Grain

Like film, DSLR images have grain, though the origin of the grain is different. In film, it's due to irregular clumping of silver halide crystals; in the DSLR, it's due to random noise of all kinds, including random variation in the number of photons picked up, noise in the amplifier, and small constant differences between pixels. Just as with film, grain is most prominent in an underexposed image whose contrast has been increased to make it look normal. Unfortunately, in astronomy that is often the only way to process an image.

Combining multiple images reduces grain. It reduces random noise by a factor of $\sqrt{N}$ where N is the number of images stacked. Grain that is due to constant differences between pixels is reduced by using dark frames and flat fields.

Finally, grain can be reduced using noise-reduction algorithms in software. Used judiciously, noise reduction helps a great deal. Overdone, it turns images into surrealistic swirls and clouds.

3.7.2 Star Eaters

A few DSLRs over the years have had a problem with "star eaters." The origin of the problem is that cameras try to get rid of hot or deviant pixels in the image by checking for pixels that differ drastically from their neighbors. The rationale is that genuine details in the image will always spill over from one pixel to another, so if a pixel is bright all by itself, it must be "hot" and should be eliminated.

All cameras probably do this to some extent, but if a camera does it very avidly, and the sensor has rather large pixels, and you're photographing stars with a very sharp lens, stars may get mistaken for hot pixels. In that situation, small, faint star images vanish.

Star eating was a problem with some early Nikon DSLRs and has reportedly happened with some Sony MILCs. For the early Nikons, there was a workaround: turn on long exposure noise reduction, set the image mode to raw (not JPEG), take your picture, and then, while the camera is taking the second exposure with the shutter closed, turn the camera off. This seems like a foolish thing to do, but actually, a "truly raw" image has already been stored on the memory card. If you let the second exposure finish, that image will be replaced by one processed by the star eater. By powering the camera off, you keep this from happening.

"Star eaters" are now rare. The overwhelming majority of DSLRs and MILCs have pixels small enough, and/or low-pass filters strong enough, that stars are never mistaken for hot pixels.

3.7.3 Dust on the Sensor

A film camera pulls a fresh section of film out of the cartridge before taking every picture, but a DSLR's sensor remains stationary. This means that if a dust speck lands on the low-pass filter in front of the sensor, it will stay in place, making a dark blotch on every picture until you clean it off (Figure 3.12).

To keep dust off the sensor, avoid unnecessary lens changes; hold the camera body facing downward when you have the lens off; never leave the body sitting around with no lens or cap on it; and never change lenses in dusty surroundings. But even if you never remove the lens, there will eventually be some dust generated by mechanical wear of the camera's internal components.

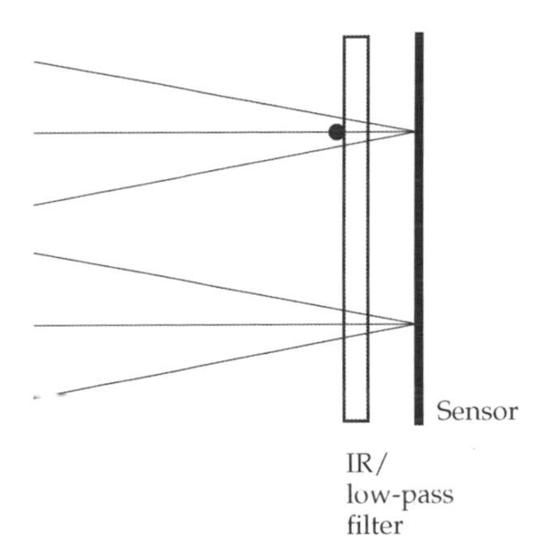

Figure 3.12. Effect of dust (enlarged). Diagram shows how light gets around the dust speck so that stars are visible through the blotch.

When dust gets on the sensor (and eventually it will), follow the instructions in your camera's instruction manual. Most DSLRs can vibrate the sensor to shake dust loose. By default, this happens when the camera is switched on and off.

Beyond that, the gentlest way to remove dust is to open the shutter (on "bulb," the time-exposure setting) and apply compressed air from a rubber bulb (*not* a can of compressed gas, which might emit liquid or high-speed particles). If you must wipe the sensor, use a Sensor Swab, made by Photographic Solutions, Inc. (www.photosol.com). And above all, follow instructions to make *sure* the shutter stays open while you're doing the cleaning. Many cameras have a menu option to power down with the shutter open. Better yet, see if a local camera repair shop will clean your sensor while you wait. You're paying them to take the risk; if they damage it, they have to fix it.

To a considerable extent, the effect of dust can be removed by image processing, either flat-fielding (Figure 5.2) or (in daytime photography) by "dust mapping" performed by the software that comes with the camera.

You can check for dust by aiming the camera at the plain blue daytime sky and taking a picture at $f/22$, then using software to view the picture with greatly increased contrast. You are looking for distinct dust blobs that show up with only a modest contrast increase.

Be forewarned that if you increase the contrast of an image too much, you will see trouble where there is none. If you take a picture of the blue sky with a perfectly clean sensor and view it with "auto screen stretch" in astronomical software, or with *Photoshop*'s Auto Tone, you will see specks and streaks that look awful but in fact are tiny, inconsequential quirks of the sensor, tremendously amplified. The software has been commanded to stretch the contrast until it finds something that looks like an image, and it does so, no matter how much stretching is needed.

Note that dust that you see in the viewfinder is *not* on the sensor and will not show up in the picture; it's on the focusing screen. Removing such dust is a good idea so that it does not eventually make its way to the sensor.

3.8 The Camera as Your Logbook

Digital cameras record a remarkable amount of information about each exposure. This is often called EXIF data (EXIF is actually one of several formats for recording it), and Windows itself, as well as many software packages, can read the EXIF data in JPEG files. *Photoshop* and other utilities can read the same kind of data in raw files. From this book's web site (www.dslrbook.com) you can get a software utility, EXIFLOG, that reads a set of digital image files and generates a logbook page that you can edit.

This fact gives you an incentive to do two things. First, keep the internal clock-calendar of your camera set correctly. Second, take each exposure as both raw and JPEG if your camera offers that option. That way, each exposure is recorded as two files that serve as backups of each other, and the JPEG file contains EXIF data that is easy to view (with any graphics program or even as file properties in the operating system).

Naturally, the camera doesn't record what it doesn't know. It can't record the focal length or *f*-ratio of a lens that doesn't interface with the camera's electronics. Nor does it know what object you're photographing or other particulars of the setup and conditions. Nonetheless, using EXIF data saves you a lot of effort. You don't have to write in the logbook every time you take a picture.

Chapter 4
Five Simple Projects

After all this, you're probably itching to take a picture with your DSLR. This chapter outlines five simple ways to take an astronomical photograph. Each of them will result in an image that requires only the simplest subsequent processing by computer.

All the projects in this chapter can be done with your camera set to output JPEG images (not raw), as in daytime photography. The images can be viewed and further processed with any picture processing program, such as *Photoshop* or *GIMP* (which is freeware). Special astronomical techniques are not needed.

We start the chapter by taking pictures of the moon, which are much like daytime terrestrial pictures. But in the middle of the chapter I introduce one very important processing technique, which you may already be using in daytime photography, though we use it more heavily in astronomy. It is stretching, the adjustment of contrast and brightness. You will need it to take successful pictures showing stars.

4.1 Telephoto Moon

The moon is a very good first target. Put your camera on a sturdy tripod and attach a telephoto lens with a focal length of at least 200 and preferably 300 mm (it can be a zoom lens). Take aim at the moon. Initial exposure settings are ISO 400, *f*/5.6, 1/125 second (crescent), 1/500 second (quarter moon), 1/1000 (gibbous), or 1/2000 (full); or simply take a spot meter reading of the illuminated face of the moon. An averaging meter will overexpose the picture because of the dark background.

If the camera has mirror lock (Canon) or exposure delay (Nikon), turn that feature on. If you have Silent Shooting or the ability to take pictures during Live View, so much the better; use that too. You can probably use autofocus successfully, but focusing manually in magnified Live View is better. After your

Figure 4.1. The moon. Canon 60Da and 300-mm *f*/4 telephoto lens set to *f*/8, with ×1.4 teleconverter, giving effective 420 mm at *f*/11. 1/80 second at ISO 640. This is just the central area of the picture, enlarged. Some unsharp masking was done in Photoshop to bring out detail.

first try, adjust the exposure for best results. Also try stopping down to *f*/8 and using a slower shutter speed.

Figures 4.1, 4.2, and 4.3 show what you can achieve this way. Use stretching to adjust the brightness and contrast. Images of the moon also benefit greatly from unsharp masking in *Photoshop* or multiscale sharpening in *Registax*; you'll be surprised how much more detail you can bring out.

To make the face of the moon fill the sensor, you'll need a focal length of at least 1000 mm. In the next example, we'll achieve that in a very simple way.

Figure 4.2. The moon passing in front of the Pleiades star cluster. Half-second exposure at ISO 200 with Canon Digital Rebel (300D) and old Soligor 400-mm telephoto lens at $f/8$. Processed with *Photoshop* to brighten the starry background relative to the moon.

4.2 Afocal Moon

Another easy way to take an astronomical photograph with a digital camera (DSLR or not) is to aim the telescope at the moon, hold the camera up to the eyepiece, and snap away.

This sounds silly, but as you can see from Figure 4.4, it works. The telescope should be set up for low power (50× or less, down to 10× or even 5×; you can use a spotting scope or one side of a pair of binoculars). Use a 28-, 35-, or 50-mm fixed-focal-length camera lens if you have one; the bulky design of a zoom lens may make it impossible to get close enough to the eyepiece.

Set the camera to aperture-priority autoexposure (*A* on Nikons, *Av* on Canons) and set the lens wide open (lowest-numbered *f*-stop). Focus the

Figure 4.3. The earthlit moon and Comet C/2011 L4 (PANSTARRS) just after sunset. The comet was visible in binoculars but not with the naked eye. Canon 60Da on fixed tripod, 1/3 second at ISO 400 with 105-mm telephoto lens at *f*/2.8. Cropped and stretched in *Photoshop*.

telescope by eye and let the camera autofocus on the image. What could be simpler?

This technique also works well with fixed-lens digital cameras, even smartphones. Many kinds of brackets exist for coupling the camera to the telescope, but (don't laugh) I get better results by handholding the camera or putting it on a separate tripod so that it cannot transmit any vibration.

This setup is called afocal coupling. The effective focal length is that of the camera lens multiplied by the magnification of the telescope – in Figure 4.4, $39 \times 50 = 1950$ mm.

4.3 Stretching – The Processing Technique to Learn Now

Before moving on to pictures with stars in them, we need to introduce an important processing technique.

Astronomical photographs usually don't use the full brightness range of the camera the way daytime photographs do. To put it bluntly, they're usually underexposed, as well as having other quirks. Accordingly, in order to get good images we have to control how the brightness range is actually used. Many images are hardly viewable until this is done (Figure 4.5). In the film era, this is why we had to have our own darkrooms and make our own prints; today we can control brightness and contrast digitally.

Figure 4.4. Lunar seas and craters. Nikon D70s with 50-mm $f/1.8$ lens wide open, handheld at the eyepiece of a 5-inch (12.5-cm) telescope at ×39, autofocused and autoexposed. Image was unsharp-masked with *Photoshop*.

Digitally, we control contrast and brightness with a technique known variously as stretching, level adjustment, or histogram adjustment. We've already met it, in a theoretical way, in Section 2.5, but now it's time to actually do it.

Figure 4.6 shows how the adjustment is done. Almost all software does it the same way. In *Photoshop,* it's under Image, Adjustments, Levels; in *GIMP,* under Colors, Levels; in *Nebulosity*, under Image, Levels; in *PixInsight*, under Process, Intensity Transformations, Histogram Transformation.

MaxIm DL is a bit different: there, you first adjust *screen stretch,* which affects only the screen display and not the actual image file, and then you apply it to the actual image with Process, Stretch. *Nebulosity* and *PixInsight* also have screen stretch options. We return to this in Chapter 12.

Figure 4.5. Stretching is often necessary to make an astronomical image viewable. (Telescope image of M31, before and after stretching.)

A few minutes of playing with this adjustment will teach you more than I could by writing ten more pages. For best results, do it in several steps rather than trying to get everything exactly right at once. Don't be afraid to move the middle slider a *long* way to the left to bring out deep-sky detail. To color-balance your image, you can adjust red, green, or blue separately from the other colors. Once you've mastered stretching, you'll find yourself using it (more subtly, of course) in your daytime photography too.

If you're going to do substantial stretching, it's best to start with a raw image from the camera, rather than a JPEG file, because of its greater bit depth, but if all you have is a JPEG, you can still use stretching effectively.

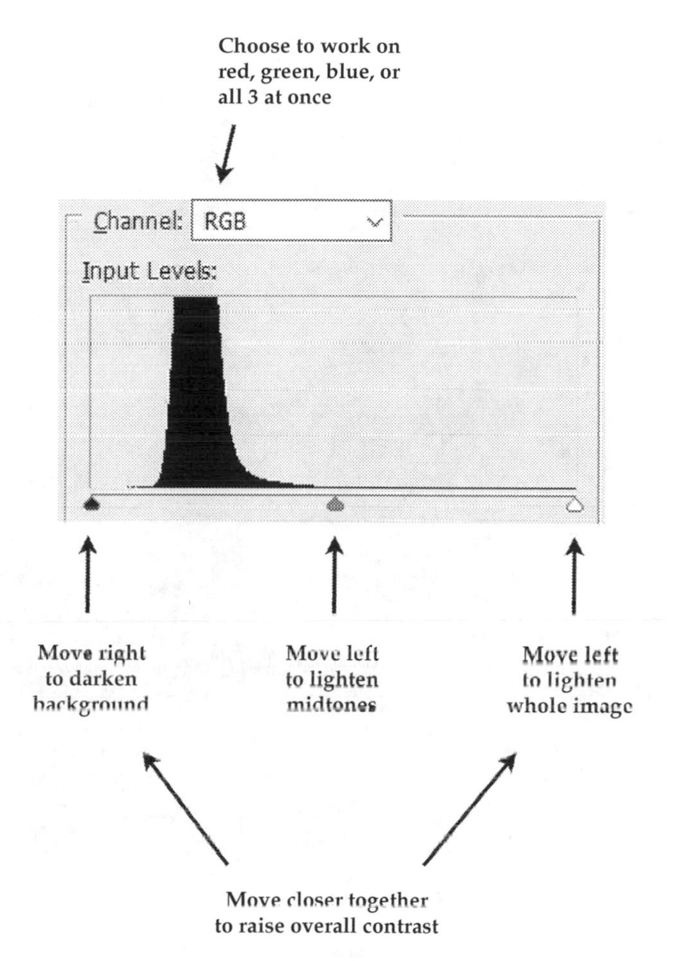

Figure 4.6. Stretching is done with the Levels adjustment in most software packages (*Photoshop* shown).

4.4 Stars from a Fixed Tripod

Now for the stars. On a starry, moonless night, put the camera on a sturdy tripod, aim at a familiar constellation, set the aperture wide open, focus on infinity, and take a 5- to 20-second exposure at ISO 1600 or higher. You'll get something like Figure 4.7.

For this project, a 50-mm $f/1.8$ lens is ideal. Second choice is any $f/2.8$ or faster lens with a focal length between 20 and 50 mm. Your zoom lens may fill the bill. If you are using a "kit" zoom lens that is $f/3.5$ or $f/4$, you'll need to make a slightly longer exposure (as much as 30 seconds) and the stars will appear as short streaks, rather than points, because the earth is rotating.

A good way to estimate the exposure time is to use the "rule of 300" – that is, the exposure time in seconds is 300 divided by the focal length in millimeters.

Figure 4.7. Orion setting over the trees. Canon EOS 40D at ISO 1600, Canon 50-mm $f/1.8$ lens wide open, 5 seconds.

That is for APS-C sensors; for full-frame sensors, it is more like a "rule of 500," but if the star images are too elongated for you, shorten the exposure.

If your camera has long-exposure noise reduction, turn it on. Otherwise, there will be a few bright specks in the image that are not stars, from hot pixels. And if you happen to have a Pentax Astrotracer, by all means put it to good use.

Focusing can be problematic. Autofocus doesn't work on the stars, and when you're using a small lens, it can be hard to see a star in Live View well enough to focus on it. Focusing on a very distant streetlight is a good starting point. Then, if Live View is still no help, you can refine the focus by taking repeated 5-second exposures and viewing them magnified on the LCD screen as you make slight changes.

The picture will benefit considerably from stretching; in fact, it may not look like much at all until you stretch it. You'll be amazed at what you can photograph; ninth- or tenth-magnitude stars, dozens of star clusters, and numerous nebulae and galaxies are within reach.

This is of course the DSLR equivalent of the method described at length in Chapter 2 of *Astrophotography for the Amateur* (1999). It will record bright comets, meteors, and the aurora borealis. In fact, for photographing the aurora, a camera with a wide-angle lens on a fixed tripod is the ideal instrument.

4.5 Nightscapes

A nightscape is a night landscape with stars, and usually the Milky Way, prominently visible in it. Arguably, Figure 4.7 is a nightscape, but Figure 4.8 is a much better one. The technique is the same as in the previous section, except that it's common to use wider-angle lenses and longer exposures.

If the lens is a fisheye, the horizon will only appear straight if it runs through the center of the picture. Putting it off center makes it bend outward, which may or may not be desirable. All wide-angle lenses, fisheye or not, have considerable darkening toward the edges of the field, which is usually used as part of the artistic effect but can be counteracted by flat-fielding if you want to use advanced techniques.

To include the Milky Way, you will need to go to a dark, or at least reasonably dark, location (not the middle of town) at the right time of year. The brightest part of the Milky Way is highest in the evening from August to October. Software such as *Stellarium* will tell you exactly how it's oriented at any particular place and time. You'll need to be at a reasonably southern latitude; if you're in Britain, the bright summer Milky Way will be elusive, and I suggest going for bright constellations such as Orion or Ursa Major instead.

One word of caution: I've seen entirely too many nightscape photographers trying to take the same picture. I chose Figure 4.8 because it's *not* the usual "country road with the Milky Way vertical at the end of it" or "mountain range with Milky Way horizontal above it." There's nothing wrong with replicating others' work as a technical challenge, but that's only a starting point. Be creative.

4.6 Piggybacking

If you have a telescope that tracks the stars (whether on an equatorial wedge or not), mount the camera "piggy-back" on it and you can take longer exposures of the sky. For now, don't worry about autoguiding; just set the telescope up carefully and let it track as best it can. Expose no more than five minutes or so (30 seconds may be plenty). As in the previous project, long-exposure noise reduction should be turned on if available. To keep things non-critical, use a lens shorter than 150 mm. Figure 4.9 shows what you can achieve.

With an equatorial mount (or a fork mount on an equatorial wedge), you can theoretically expose as long as you want, but both thermal noise and tracking errors start catching up with you if the exposure exceeds 5 minutes. If the telescope is on an altazimuth mount (one that goes up-down and left-right, with

Figure 4.8. The Milky Way over the distant lights of Jackson Hole, Wyoming. Nikon D3200 at ISO 1600 with 10.5-mm *f*/2.8 Fisheye-Nikkor wide open, 30 seconds, from fixed tripod. The fisheye lens makes the off-center horizon bend into a U shape. (Rex T. Smith)

Figure 4.9. Comet Machholz and the Pleiades, 2005 January 6. Canon Digital Rebel (300D) with old Pentax 135-mm $f/2.5$ lens using lens mount adapter, piggybacked on an equatorially mounted telescope; no guiding corrections. Single 3-minute exposure at ISO 400, processed with *Photoshop*.

no wedge), then you have to deal with field rotation (Figure 8.2). As a rule of thumb, this won't be a problem if you keep the exposure time under 30 seconds in most parts of the sky. You can expose up to 2 minutes if you're tracking objects fairly low in the east or west.

As you learn more advanced techniques later on, don't forget that you can piggyback while doing something else. That is, a second camera with a telephoto lens can ride piggyback on a telescope through which a guided long exposure is being taken. I have heard of professional astronomers doing this, clamping their personal cameras to great observatory telescopes. Why not?

Figure 4.10. Extreme fixed-tripod astrophotography of star clouds in the constellation Cygnus, including North America Nebula at the upper left. Sixty-nine separate 2-second exposures were taken with the camera on a fixed tripod, then calibrated and stacked as in Chapters 5, 11, and 12. Canon 60Da, Sigma 50-mm *f*/2.8 lens, ISO 6400.

4.7 Going Further

If your fixed-tripod or piggybacking experiments are successful, you can get better pictures by stacking multiple images. These can be the same JPEG images you're already capturing, but jump into the procedures in Chapters 11 and 12 at the appropriate point and combine them. Rotation during stacking will take care of the field rotation between (but not within) exposures taken on an altazimuth mount.

For even better results, switch your camera to raw mode and do the full procedure in Chapters 5, 11, and 12. Be sure to take some dark frames right after your series of exposures; leave the camera settings the same but put the lens cap on and expose for the same length of time. There is nothing wrong with stacking 100 or 200 two- or five-second fixed-tripod exposures; it is the best way to get the most out of a setup that can't track the stars. Figure 4.10 shows what can be achieved.

Part II
Equipment and Techniques

Chapter 5
Deep-sky Image Acquisition

Before going into the details of astrophotographic equipment, I need to outline how it is used. This chapter covers how to take pictures of deep-sky objects, including calibration frames. The actual processing and calibration are covered in Chapters 11–13, and lunar and planetary work in Chapter 14.

5.1 How to Avoid Most of This Work

You don't actually have to do all this work. A much simpler procedure is to let the camera do most of it for you, then finish up the picture as if it were a daytime photograph, as you did in Chapter 4. Here's how:

- Turn on long-exposure noise reduction in your camera. That way, whenever you take a celestial photograph, the camera will automatically take a dark frame and subtract it.
- Tell the camera to save the images as JPEG (not raw) so you can edit them with an ordinary photo editor.
- Take a single exposure, exposed generously at a high ISO setting.
- Open the resulting image file in *Photoshop* or any photo editor and adjust the brightness, contrast, and color balance to suit you.
- Downsample your image (make it smaller) to conceal flaws in guiding and in the optics and to reduce grain.

Why don't we always take the easy way out? For several reasons.

First, we usually want to combine multiple images. With digital technology, ten 1-minute exposures really *are* as good as one 10-minute exposure – they may even be better. They're certainly a lot better than *one* 1-minute exposure. Combining images improves the signal-to-noise ratio because random noise partly cancels out. Keeping the individual exposures short prevents overexposure of bright objects and sky fog.

Second, having the camera take automatic dark frames is time-consuming; the dark frames will take up half of every observing session. It's more efficient to take one set of dark frames that can be applied to all the sets of deep-sky images that you take that evening.

Third, we often want to do processing that wouldn't be easy with a photo editor, such as extreme nonlinear stretching, background flattening, or deconvolution (deliberate correction of a known blur). In subsequent chapters I'll tell you more about these operations.

If you don't want to avoid *all* the work, you can avoid *some* of it. You don't have to start with camera raw images; JPEGs can be aligned, stacked, and enhanced (though not calibrated).

5.2 How Long to Expose

Nearly 20 pages of *Astrophotography for the Amateur* were devoted to exposure calculations and tables. Guess what? We don't need to calculate exposures any more. With a DSLR or any other digital imaging device, it's easy to determine exposures by trial and error.

Nonetheless, it's useful to know where to start. That's what Table 5.1 is for. It is for deep-sky objects (stars, clusters, nebulae, and galaxies). For more scientifically based exposure tables, with an explanation of how they were computed, see Appendix B.

The best exposure is whatever best separates the objects of interest – stars, clusters, nebulae, and galaxies – from the sky background. Accordingly, it depends on the brightness of your sky (city, country, or desert). It also depends on the brightness of the image formed by your lens or telescope, which is determined by its *f*-ratio. For example, according to the chart, if you are in town (so the sky is only dark enough for you to see 5th-magnitude stars), and you're using an *f*/4 telescope, a good starting point is to expose for 30 seconds. That's a *single* exposure; you can take many 30-second exposures and stack them.

Those exposures are just starting points. Expose *more* if:

- You want to bring out the faintest areas while overexposing the brightest; or
- You are in exceptionally dark desert or mountain conditions; or

Table 5.1 *Suggested exposures for star clusters, nebulae, and galaxies.* Wide variation is possible. *Use ISO 800 unless you have established that a lower setting works well with your camera.*

Sky conditions	*f*/2.8	*f*/4	*f*/5.6	*f*/8	*f*/11
Dark country sky (6.5-mag. stars visible)	1 m	2 m	4 m	8 m	15 m
Country sky with some glow from town	30 s	1 m	2 m	4 m	8 m
Town sky (5th-mag. stars visible)	15 s	30 s	1 m	2 m	4 m

- You want to process the pictures like daytime photos, without calibration and stacking. (In that situation, you may also want to use a higher ISO setting.)

Expose *less* if:

- You are photographing star clusters and a dark background is acceptable; or
- Your tracking equipment does not permit longer exposures.

To judge an exposure, look at the LCD display just the way you do in the daytime. Better yet, get the camera to display a histogram (Figure 5.1) and look at the hump that represents the sky background. It should be in the left half of the graph, but not all the way to the left edge. Increasing the exposure moves it to the right; decreasing the exposure moves it to the left.

When viewing the histogram, you may also see a small preview image in which overexposed areas blink. It is normal for bright stars to be overexposed. For more about histograms, see Section 2.5.

If you are using a low-noise camera at a low ISO setting in order to maximize dynamic range, your picture may look considerably darker than this, and the histogram may be farther over to the left. Don't panic; in astronomy it is common to get good results with what would be, by daytime standards, severe underexposure.

Conversely, if you're exposing generously to pick up faint nebulosity, and you are not far out in the country, your histogram may be well to the right of the center. As long as it doesn't bump into the right-hand edge, you're OK; the picture may look very light on the camera screen, but it will clean up nicely when adjusted.

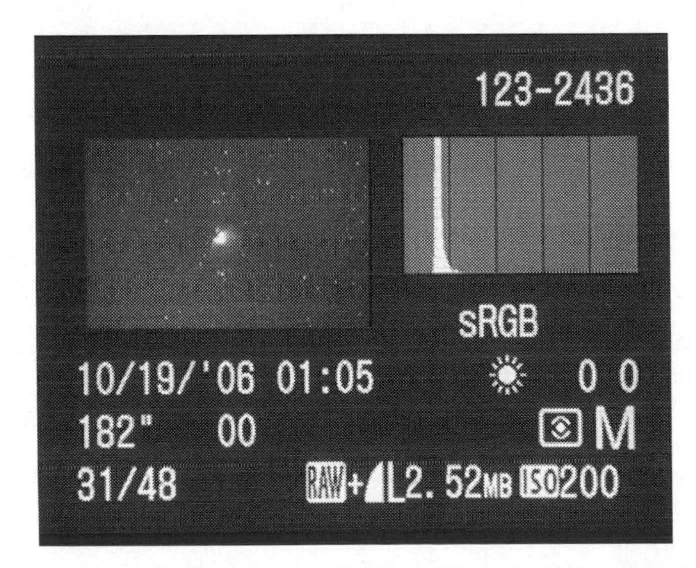

Figure 5.1. Histogram of a well-exposed deep-sky image as displayed on the camera's screen.

5.3 Dithering

Dithering means aiming the telescope or lens slightly differently for each exposure so that the same points in the image don't always fall on the same pixels. Then, when the images are aligned and stacked, bad pixels are ignored because they disagree with the same point in the image in other frames.

Dithering is optional, and with good sensors, it makes little difference, but if your sensor has conspicuous hot pixels, dithering can really help. You can dither manually by setting your slewing buttons to their slowest speed and then pressing one slewing button for about 2 seconds before each exposure, a different slewing button each time, but not in a cycle that would bring you back to the starting point. If you are autoguiding, you must of course tell the autoguider to stop before you move and re-acquire the guide star afterward. Automatic dithering is provided by camera control software (Section 10.2.4).

5.4 Taking Calibration Frames

Besides images of the object of interest, the thorough astrophotographer also takes three or four kinds of *calibration frames* (Table 5.2). These are recordings of the camera's and optical system's flaws, so that they can be corrected digitally.

Chapter 11 will go into how to *use* the calibration frames; here we only consider how to *make* them. We make more than one of each kind, then combine them by stacking to remove random noise. The dark frames should be as numerous as the lights (astronomical images). Flats and flat darks have an inherently lower noise level, and a dozen of each is sufficient, but it is easy to take more.

5.4.1 Dark Frames

The importance of dark frames is obvious: they record hot pixels, dark current (electrical leakage), and amp glow (if any) for subtraction from the astronomical images. The way to take them is obvious, too: under the same camera conditions (ISO setting and temperature), take exposures the same length as the lights, but with the lens cap on (and eyepiece cap on if you have one). I often do this at the end of an observing session, while shutting down the mount and starting to pack away accessories.

You can often reuse dark frames that were taken with the same camera at the same ISO setting and approximately the same temperature; that is, you don't have to take a new set every night. However, because sensors age, I am reluctant to use dark frames that are more than a few weeks old.

Dark frames may not be strictly necessary, especially with newer low-noise sensors under chilly conditions.

Table 5.2 *Types of astrophotographic exposures.*

Lights (Subs) Images of the celestial object itself. There are usually many, just alike, for stacking. "Subs" stands for "sub-exposures" or "subframes."

Dark frames (Darks) Exposures taken with no light reaching the sensor (lens cap on and camera eyepiece covered), at the same ISO setting, exposure time, and camera temperature as the images to be calibrated. This is to correct for hot pixels, dark current, and amp glow.

Flat fields (Flats) Images of a blank white surface taken through the same telescope as the lights, with the same camera at the same ISO setting, with the same lens or telescope at the same *f*-stop and focus setting, and preferably on the same occasion so that dust particles anywhere in the system will be in the same positions. These need not match the ISO setting of the lights, although some software works more smoothly if they do. Obviously, the exposure time will be different.

Flat darks (Dark flats) Dark frames that match the exposure time and ISO setting of the flats rather than the lights. They are used to remove offset from the flats for more accurate flat-fielding.

Bias frames (Offset frames) Dark frames with the exposure time as short as possible, so there will be no appreciable dark current; actually, any exposure under 1/10 second will do. With a DSLR, flat darks are almost always indistinguishable from bias frames. If your software asks for bias frames, give it the flat darks. The purpose of bias frames is to measure the nonzero offset of the pixel values. They are not needed (in fact, confer no benefit) if you have taken darks, flats, and flat darks according to the instructions above, all at the same ISO setting as the lights, because dark subtraction eliminates the nonzero offset.

5.4.2 Flats

Their Purpose

Flat fields do a lot of good (Figure 5.2) and should be made and used whenever possible. They record not only uneven illumination but also inequalities of sensitivity between pixels, so they reduce the noise in the calibrated image.

Flats duplicate the optical conditions of the lights, even down to the position of dust specks on the sensor. For that reason, I am hesitant to reuse flats from an earlier session but have done so successfully, especially with telescopes, where my main goal is to correct edge-of-field darkening. In that case, it is important to make sure the camera is attached to the telescope in the same orientation and that the focus setting is, as far as possible, the same.

Like lights and darks, flats are normally taken in a set of a dozen or more so they can be stacked to remove random noise.

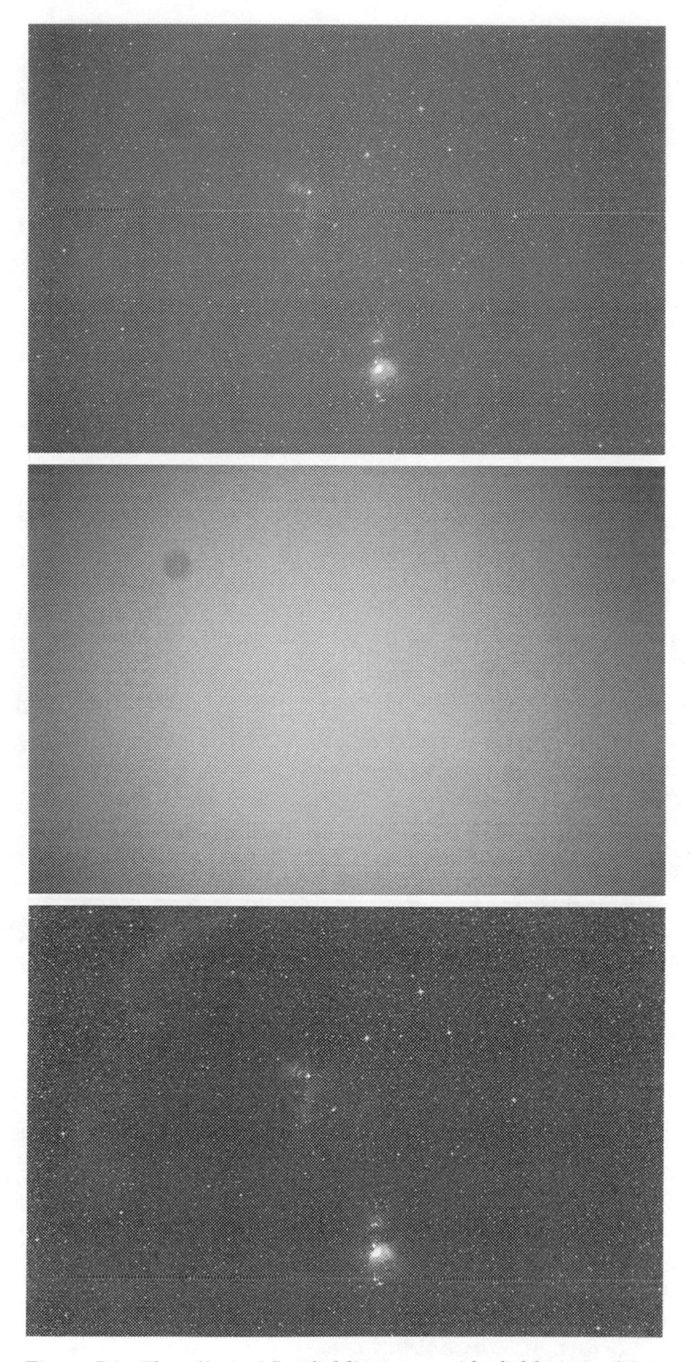

Figure 5.2. The effect of flat-fielding on a wide-field view of Orion with a Canon EOS 20Da and Hα filter. *Top:* Image calibrated with dark frames but not flat fields. *Middle:* A flat field taken through the same lens on the same evening. Note slight vignetting and prominent dust spot. *Bottom:* Image calibrated with dark frames and flat fields.

Acquiring Flat Fields

There are many ways to take flat fields. What you need is a uniform light source in front of the telescope or lens, and preferably as close as possible, so that any irregularities in it will not be in focus. Some possibilities include:

- Hand-holding a light box in front of the lens or telescope. That's what I often do, with a 10-cm-square battery-powered fluorescent panel originally made for viewing 35-mm slides. It works only with lenses and small telescopes smaller in diameter than itself, of course.

  Larger light boxes are used by artists to trace pictures and are available as big as 30 × 40 cm (12 × 16 inches), thin and lightweight. If the light box is very close to the telescope, the evenness of the illumination is not highly critical.

  If the light box is too bright, add layers of white cloth.
- Using a tablet computer, displaying a uniformly white screen, the same way. For that purpose I maintain a set of blank web pages, white and with various very pale tints, at www.dslrbook.com/blank.
- If using a camera with a lens, hand-holding it and aiming it at any uniformly illuminated surface, even a computer screen displaying uniform white (see previous item). The illuminated surface should be very close to the lens so it will not be anywhere near in focus.
- Photographing the sky in daylight or twilight. That is now my preferred technique, often with a layer or two of white cloth in front of the telescope. An embroidery hoop can hold the cloth conveniently.
- Illuminating the inside of the observatory dome (if you have one) and taking a picture of it through the telescope.

What is important is that the telescope or lens must be set up *exactly* the same as for the celestial images, down to the setting of the focus and the positions of any dust specks that may be on the optics or the sensor. Some software expects the flats to match the ISO setting of the celestial images, although this is theoretically not necessary. The exposure time obviously will not match.

The correct exposure may have to be determined by trial and error. The best flats are exposed to more than mid-scale, but not so much that any of the colors reaches maximum brightness. Look at the histograms in all three colors to make sure (Figure 5.3). One shortcut, if you can use your exposure meter, is to make a metered exposure and then triple it. Some software requires all the flats and flat darks in a set to have exactly the same exposure time.

Cautionary Notes

Two notes of caution. Exposures should not be too short, for two reasons. First, shutters are uneven at their fastest speeds; I recommend exposing 1/500 second or longer. Second, some light sources (even DC-powered ones) have a slight high-frequency flicker introduced by voltage conversion circuitry. If you have trouble getting good flats, you may have to expose as long as 1/15 second. To get an exposure that long, you may have to put layers of white cloth in front of

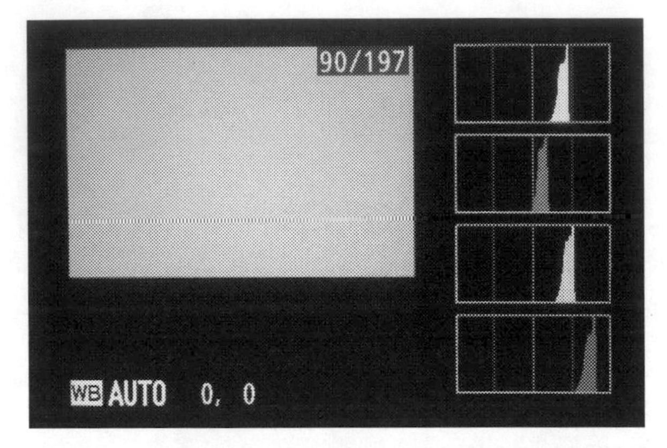

Figure 5.3. Correctly exposed flat field should be brighter than mid-range, but not so bright that any of the color histograms touches the right margin.

the lens. Remember not to change the *f*-stop or the focus of the lens; those are the conditions you are trying to duplicate.

Second, flat fields are powerful but not omnipotent. The areas of the picture that were blocked by dust motes or by vignetting are underexposed and will still be underexposed no matter what you do to them. Good flat-fielding brings them close to the brightness range of the original image, but they still have less dynamic range, and the difference is sometimes visible.

5.4.3 Flat Darks

Flat darks are dark frames that match the flats; just put the lens cap and eyepiece cover on, and repeat what you did when exposing the flats. This is very quick. You should take as many flat darks as you did flats.[1]

Why are flat darks necessary? Because the process of flat-field correction involves division. The computer has to find out what *fraction* of full illumination reached every pixel. Many DSLRs add a large offset, as much as 2000, to each pixel value, and if this is not subtracted out, the fractions come out wrong.

5.4.4 Bias Frames

If you have good lights, darks, flats, and flat darks, meeting all the conditions I've outlined, you don't need bias frames. Those are zero-length or minimum-length exposures, intended to measure the nonzero offset by itself. Dark-frame subtraction eliminates that nonzero offset.

[1] The terms *flat dark* and *dark flat* have been the subject of bitter debate. Arguably, both make sense; a flat dark is a dark that goes with a flat, just as a truck tire is a tire that goes with a truck, and a dark flat is a flat that happens to be dark rather than normally exposed. Also arguably, neither term quite makes sense.

Bias frames are needed if the software is going to scale the dark frames to different exposure times. The reason is, only the part of the dark frame above the offset is proportional to exposure time. However, this kind of scaling is no longer commonly done, especially with DSLRs.

More to the point, with a DSLR, if they are taken at the same ISO setting, bias frames and flat darks are the same thing. I have been unable to measure a difference between them. Unless your flats (and thus your flat darks) have unusually long exposure times (greater than 1/10 second), your flat darks *are* bias frames.

Some software packages use bias frames instead of flat darks; an example is *PixInsight*'s Batch Preprocessing script. In effect, flat darks are *called* bias frames in some software packages. If your software asks for bias frames, give it the flat darks.

Chapter 6
Coupling Cameras to Telescopes

How do you take a picture through a telescope? Any of numerous ways. This chapter will cover the basic optical configurations as well as some (not all) ways of assembling the adapters. The most important thing to remember is that it's not enough to make everything fit together mechanically; you must also consider the spacings between optical elements, and not all configurations work with all telescopes.

This chapter may seem out of place because it is usually easier and more satisfying to use a camera lens (covered in the next chapter) rather than a telescope. But the important optical concepts are easier to understand if we consider telescopes first.

6.1 Optical Configurations

6.1.1 Types of Telescopes

Figure 6.1 shows the optical systems of the most popular kinds of telescopes. Several new types have appeared on the market in recent years.

The diagram doesn't show whether the curved surfaces are spherical or aspheric, but that makes a lot of difference. Refractors normally use lenses with spherical surfaces; for higher quality, they sometimes use extra elements or extra-low-dispersion (ED) glass.

The Newtonian reflector, invented by Sir Isaac Newton, was the first aspheric optical device; its mirror is a paraboloid. (Newton could not actually make a parabolic mirror himself; like many amateur telescope makers, he had to settle for a sphere.)

The classical Cassegrain reflector has two aspheric mirrors, a paraboloidal primary and a hyperboloidal secondary. The classical Ritchey–Chrétien looks

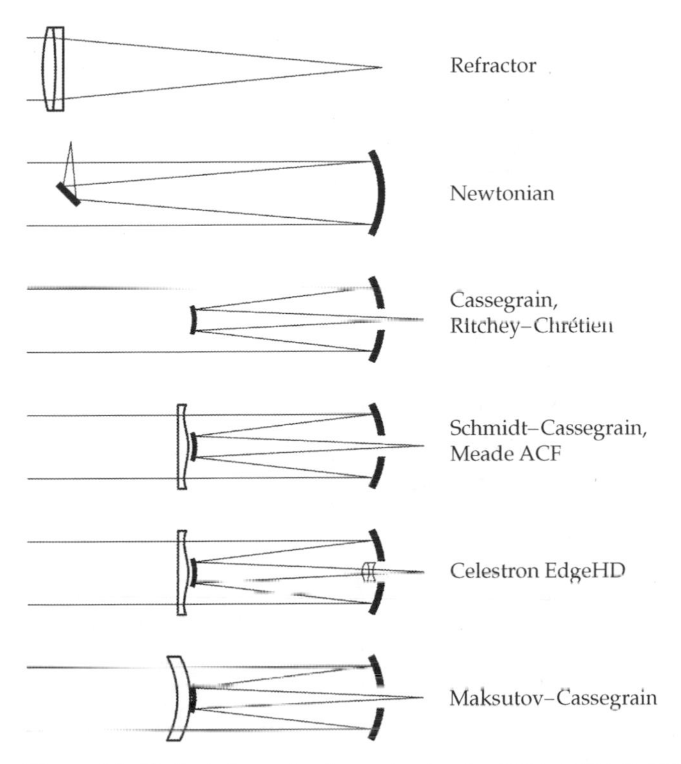

Figure 6.1. Optical elements of popular kinds of telescopes. Ritchey–Chrétien and ACF differ from related types only in use of aspherical surfaces.

just like it, but both mirrors are hyperboloidal, making it better at forming sharp images over a wide field on a flat sensor.[1]

Most amateurs use compact Schmidt–Cassegrain telescopes (SCTs). Here the two mirrors are both spherical, and the corrector plate makes up the difference between these and the desired aspheric surfaces. This type of telescope is sharp at the center of the field but suffers appreciable field curvature; that is, the periphery of the picture and the center are not in focus at the same time.

Recently, Meade Instruments introduced a design they call ACF (Advanced Coma-Free), which is a Schmidt–Cassegrain with an aspherized secondary. It reduces off-axis aberrations, especially coma. Initially they marketed it as a Ritchey–Chrétien derivative, but the name did not stick. Celestron responded

[1] The name *Cassegrain* is pronounced, roughly, *kahs-GRAN* in the original French but is nowadays normally *CASS-egg-rain* in English. (Curiously, it is not clear exactly who Cassegrain was, as he is known only from a third-hand reference by Jean-Baptiste Denys in 1672.) *Chrétien* is pronounced, approximately, *kray-TYAN*.

with the EdgeHD design, which is a Schmidt–Cassegrain with field-flattening lenses added, correcting aberrations even more effectively.

The Maksutov–Cassegrain is a classic high-resolution design for planetary work. It also works at least as well as the Schmidt–Cassegrain for deep-sky photography, except that the *f*-ratio is usually higher and the image is therefore not as bright.

6.1.2 Newer Telescopes

There's more. Figure 6.2 shows three innovative telescope designs. At the top is a four-element refractor, one of many new types of small refractor designed especially for astrophotography. Conventional two-element refractors have too much residual chromatic aberration (color fringing) to work well for astrophotography. This is corrected by adding more elements, field-flattening lenses, ED glass, and/or custom focal reducers. This enables them to cover a wider, flatter field than conventional refractors, with better correction of chromatic aberration. A four-element refractor with ED glass was used to take Figure 6.3.

At the middle and bottom of Figure 6.2 are two ways of getting an extremely low *f*-ratio (about *f*/2) with a modified Schmidt–Cassegrain. Back in the 1990s, Celestron introduced a system called Fastar that allows the user to remove the secondary mirror of a Schmidt–Cassegrain, insert an adapter into the hole, and mount the camera in front of the telescope. The telescope then works at the focal length and *f*-ratio of its primary mirror (usually *f*/2), and the

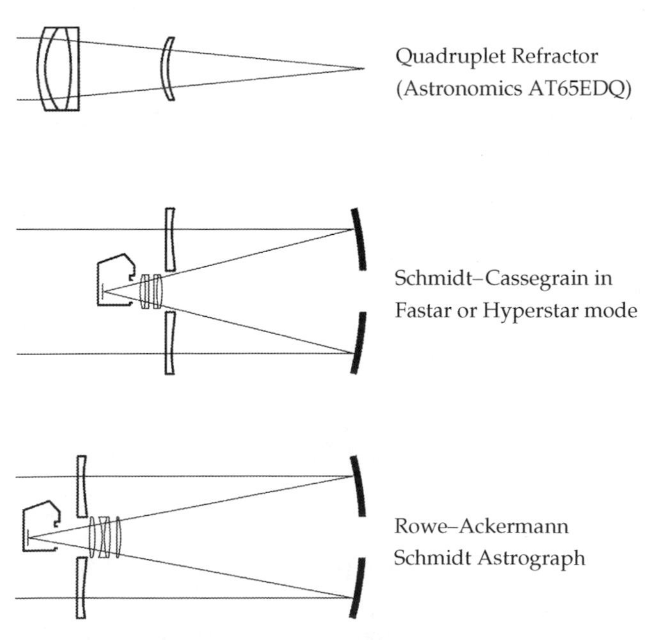

Figure 6.2. Innovation continues – here are three decidedly non-classical telescope designs. Two of them put the camera in front of the telescope.

Figure 6.3. Four-element AT65EDQ 6.5-cm $f/6.5$ refractor gives sharp stars over the entire frame of an APS-C sensor. Pleiades star cluster with nebulosity, stack of 50 1-minute exposures with a Nikon D5300 at ISO 400.

Figure 6.4. Orion Nebula (M42), Meade 14-inch (36-cm) ACF telescope with Starizona Hyperstar ($f/1.95$, focal length 700 mm), Hα-filter-modified Canon 40D with added Astronomik UHC filter, stack of 121 (yes, 121) 15-second exposures at ISO 400, on a portable tripod less than 10 km from downtown Los Angeles. Because of the short exposures, no guiding corrections were needed. (Blake Estes)

Figure 6.5. Globular cluster NGC 6723 and nebulae NGC 6726 and 6729 in Corona Australis, 11-inch (28-cm) Rowe–Ackermann Schmidt Astrograph (*f*/2.2, focal length 620 mm), stack of 32 1-minute exposures at ISO 1600 with a Canon 60Da. North is to the left. (Richard Jakiel)

lenses in the adapter correct the aberrations so that you get a good image. Since then, Celestron has discontinued the Fastar adapter, but their newest telescopes are still compatible with it, and similar adapters are made by Starizona (www.starizona.com) under the name HyperStar and fit some Meade telescopes, not just Celestrons.

The appeal of Fastar/HyperStar is that the *f*-ratio is low and the focal length is short; an 8-inch (20-cm) telescope has a focal length of just 400 mm, so that guiding is non-critical. Images are very bright, and 15-second exposures of deep-sky objects are often long enough. The field of view is wide, even with small sensors.

Fastar and HyperStar were designed for small astrocameras, and a DSLR body blocks an appreciable amount of the light coming in; the use of a DSLR is recommended only on larger Schmidt–Cassegrains. With sufficiently large telescopes, they enable observatory-quality deep-sky imaging with portable equipment (Figure 6.4).

A word to the wise: Using a Fastar or HyperStar configuration on a fork mount is perilous; with the adapter attached, the telescope cannot swing through the fork, but it is very front-heavy, and the camera will hit the fork if the declination brake becomes even slightly loose. Quite a few corrector plates have been broken this way. Use a German equatorial mount or use generous counterweights and great caution.

Fastar and HyperStar now have a distinguished descendant. Two Celestron consultants, David Rowe and Mark Ackermann, designed a Schmidt camera that resembles a permanent Fastar system (not convertible to a telescope) with more elaborate correcting lenses, designed to work well with DSLR bodies. The Rowe–Ackermann Schmidt Astrograph (RASA), shown at the bottom of Figure 6.2, is available in 11-inch (28-cm) and 14-inch (36-cm) apertures, both *f*/2.2. The focal lengths are respectively 620 and 790 mm, almost comparable to long telephoto lenses, and definitely within the range where guiding is not too difficult. Results can be spectacular (Figure 6.5).

6.1.3 Types of Coupling

Figure 6.6 shows, from an optical viewpoint, several ways to couple a camera to a telescope. Optical details and calculations are given in *Astrophotography for the Amateur* and in Table 6.1.

Not all of these modes work equally well, for several reasons. First, DSLRs excel at deep-sky work, not lunar and planetary imaging. Accordingly, we want a bright, wide-field image. That means we normally leave the focal length and *f*-ratio of the telescope unchanged (with direct coupling) or reduce them (with compression). The modes that magnify the image and make it dimmer – positive and negative projection and, usually, afocal coupling – are for lunar and planetary work.

Second, if you *do* want to increase the focal length, positive projection (eyepiece projection) is seldom the best way to do it. Positive projection increases

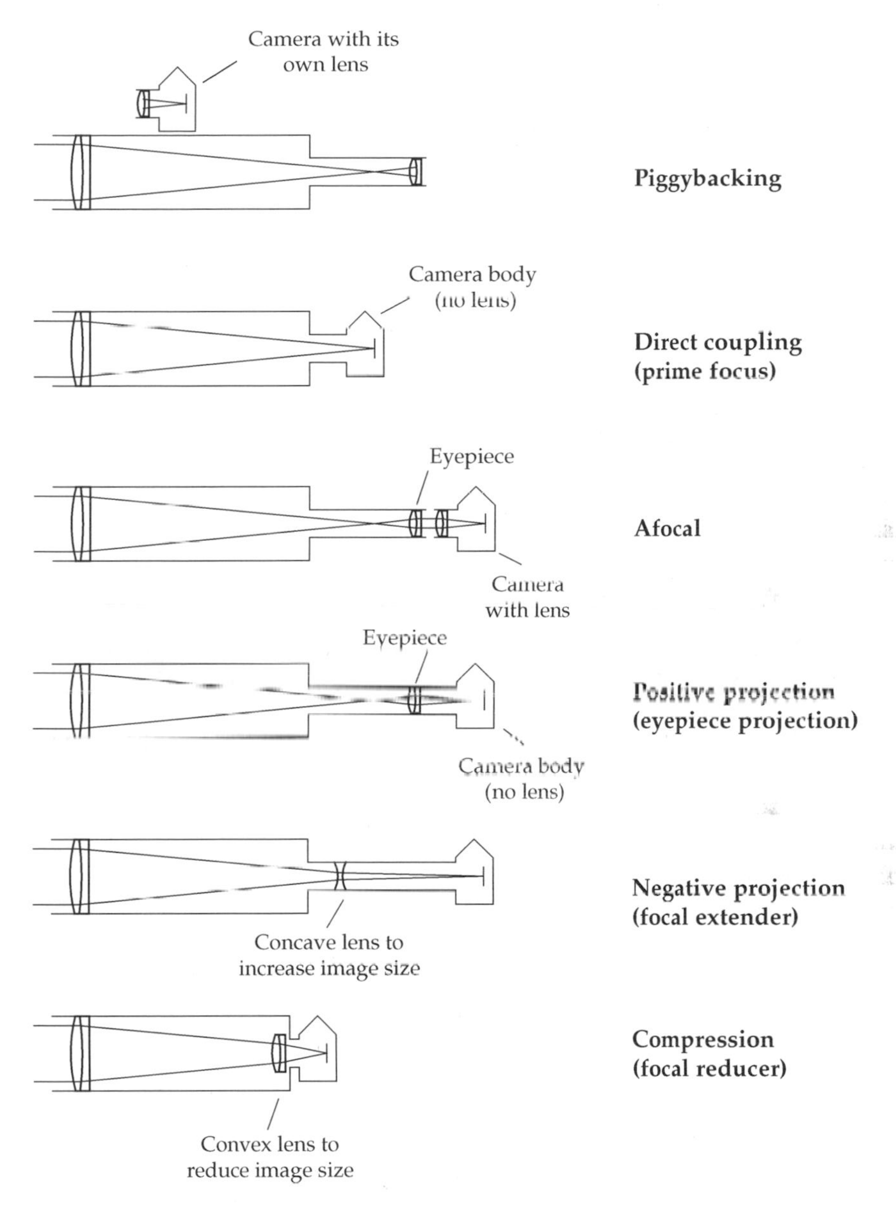

Figure 6.6. Ways of coupling cameras to telescopes. Piggybacking, direct coupling, and compression are main modes for deep-sky work.

the field curvature that is already our primary optical problem. Negative projection, with a Barlow lens or focal extender in the telescope or a teleconverter on the camera, works much better.

The appeal of positive projection is that, like afocal coupling, it works with any telescope that will take an eyepiece; you don't have to worry about the position of the focal plane (Figure 6.7). Indeed, positive projection with a 32- or

Table 6.1 *Basic calculations for camera-to-telescope coupling.*

Direct coupling

Focal length of system = focal length of telescope

f-ratio of system = *f*-ratio of telescope

Afocal coupling

$$\text{Projection magnification} = \frac{\text{focal length of camera lens}}{\text{focal length of eyepiece}}$$

Focal length of system = focal length of telescope × projection magnification

f-ratio of system = *f*-ratio of telescope × projection magnification

Positive projection, negative projection, and compression

If you get negative numbers, treat them as positive.

A = distance from projection lens to sensor or film

F = focal length of projection lens (as a positive number even with a negative lens)

$$\text{Projection magnification} = \frac{A - F}{F}$$ (or rated magnification of teleconverter or focal extender)

Focal length of system = focal length of telescope × projection magnification

f-ratio of system = *f*-ratio of telescope × projection magnification

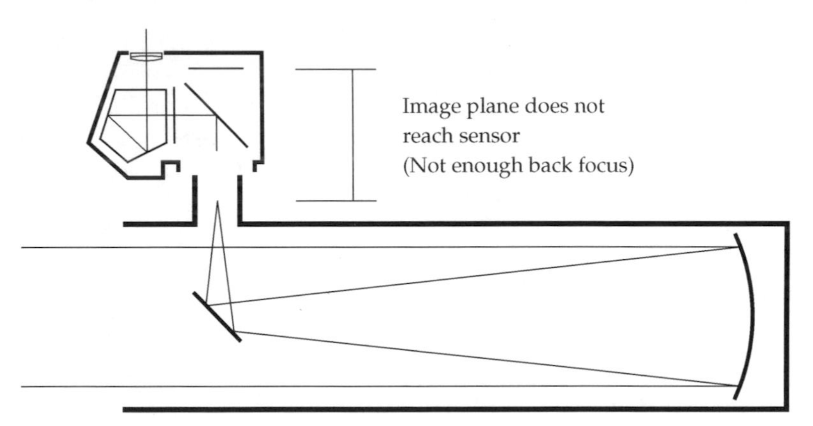

Figure 6.7. Why some Newtonians won't work direct-coupled to a camera. Solution is to use positive projection, or else modify telescope by moving the mirror forward in the tube.

40-mm eyepiece can give a projection magnification near or below 1.0, equivalent to direct coupling or compression. But it is better to move the mirror of a Newtonian forward so that it will reach focus, if it does not already do so, and to add a coma corrector to improve the off-axis sharpness.

Regarding negative projection, note two things. First, it is traditional to use a Barlow lens, but many Barlow lenses are nowadays actually telecentric focal extenders (Figure 6.8), and that's a good thing. The new-style focal extender

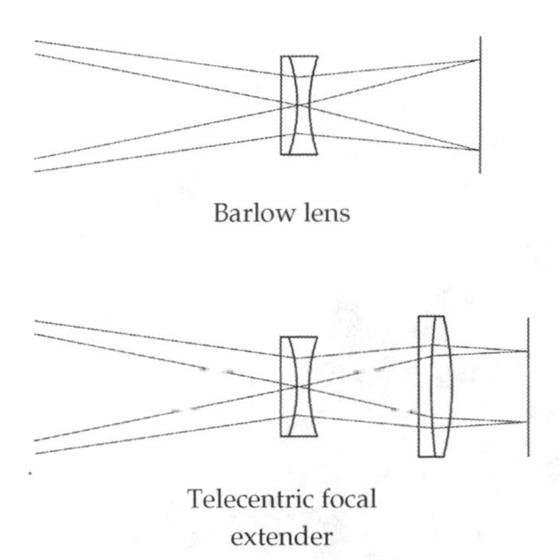

Figure 6.8. Barlow lenses are being replaced by telecentric extenders, whose magnification is not appreciably affected by the distance to the sensor.

produces a more highly corrected image and gives about the same magnification regardless of the distance to the sensor. A conventional Barlow may give as much as 50% to 100% more magnification with a camera than with an eyepiece, because of the increased spacing.

Second, another good way to do negative projection is with a teleconverter on the camera. The optical quality of teleconverters is often excellent, but the DSLR may refuse to open the shutter if the electrical system in the teleconverter isn't connected to a camera lens. The cure is to use thin tape to cover the contacts connecting the teleconverter to the camera body.

Compression is the opposite of negative projection; it makes the image smaller and brighter. Gone are the days when we had to improvise and make compressors for ourselves. Many telescope makers now supply focal reducers (compressors) matched to their telescopes, designed to correct aberrations and flatten the field of a particular optical system. Note especially the custom focal reducers that are available for many modern refractors and for the Celestron EdgeHD system (see Figure 6.9, and the advanced focal reducers made by Starizona, www.starizona.com). These devices work even better with DSLRs than with film SLRs, and they are very popular with DSLR astrophotographers. Note how far forward the compressor lens and camera have to sit; not all telescopes will accommodate compressors, and not all compressors work with all cameras.

6.2 Fitting it All Together

6.2.1 Types of Adapters

How do all these optical gadgets attach to the telescope? Figures 6.10–6.12 show a few of the most common kinds of adapters. The key to all of them is the *T-ring* or *T-adapter* that attaches to the camera body. Originally designed

Figure 6.9. Focal reducer specially made for Celestron 8 EdgeHD gives focal ratio of *f*/7 with a sharp image almost covering the field of an APS-C sensor. Galaxies M65 (right) and M66, stack of 15 1-minute exposures with a Canon 60Da at ISO 3200, slightly cropped.

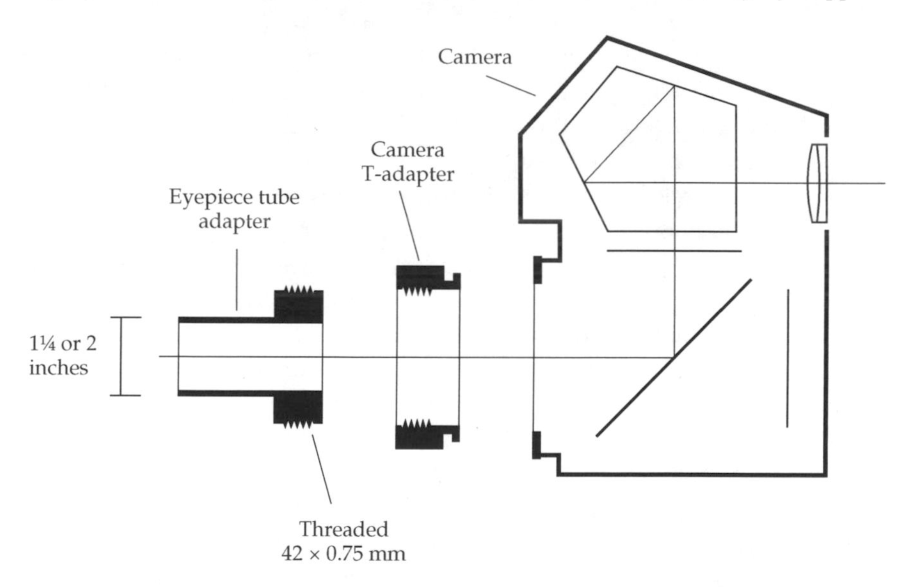

Figure 6.10. Simplest camera-to-telescope adapter fits into telescope in place of eyepiece.

for cheap "T-mount" telephoto lenses in the 1960s, the T-ring is threaded 42 × 0.75 mm so that other devices can screw into it, and its front flange is always 55 mm from the film or sensor.

T-rings differ in quality; some are noticeably loose on the camera body. All contain small screws that you can loosen to rotate the inner section relative to

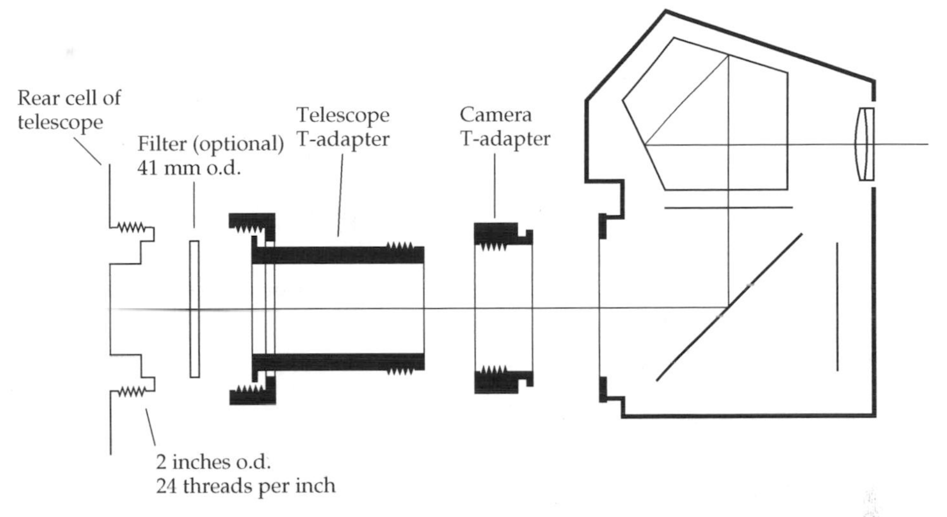

Figure 6.11. Schmidt–Cassegrains have a threaded rear cell and accept a matching T-adapter.

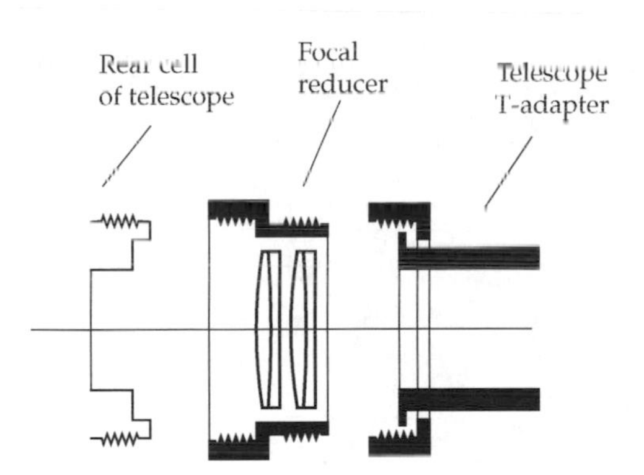

Figure 6.12. Meade or Celestron focal reducer screws onto rear cell of telescope.

the outer part; do this if your camera ends up upside down or tilted and there is nowhere else to make the correction.

You can get a simple eyepiece-tube adapter that screws into the T-ring (Figure 6.10) or a telescope T-adapter for other types of telescopes (Figure 6.11). Here you'll encounter the other screw coupling that is common with telescopes, the classic Celestron rear cell, which is 2 inches in diameter and has 24 threads per inch. Meade adopted the same system, and telescope accessories have been made with this type of threading for over 30 years. In particular, that's how the Meade and Celestron focal reducers attach to the telescope (Fig. 6.12).

Besides these common adapters, there are also adapters for many other configurations; check the catalogs or web sites of major telescope dealers. In particular, there are 48-mm T-rings for cameras with full-frame sensors, and other adapters that fit them.

6.2.2 Sensor Position Matters

One very important note: Optical systems designed to produce a flat field generally require the sensor to be a particular distance behind the telescope. This is true even though Schmidt–Cassegrains and Maksutov–Cassegrains are inherently able to move their focal plane through a wide range of positions.

For example, the Celestron 8 EdgeHD (without a focal reducer) produces best images when the sensor is 133 $\pm$10 mm from the back surface of the rear cell; other telescopes have similar requirements, and every focal reducer has a specific distance at which it works best. If these position requirements are violated, the image will be poor, and in some cases the telescope will not form an image at all.

6.3 Optical Parameters

6.3.1 Focal Length

The *focal length* of a telescope or camera lens is a parameter that determines the size of objects in the image (Figure 6.13). In conjunction with the film or sensor size, the focal length also determines the field of view.[2]

If your telescope is a refractor or Newtonian, the focal length is also the length of the tube (the distance from the lens or mirror to the sensor or film). Technically, the focal length is the distance at which a simple lens forms an image of an infinitely distant object, such as a star. The focal length of a more complex system is defined to be that of a simple lens that would form an image the same size.

Telescopes in the Cassegrain family have a focal length longer than the actual tube length because the secondary mirror enlarges the image. For example, the popular 8-inch (20-cm) $f/10$ Schmidt–Cassegrain packs a 2000-mm focal length into a telescope tube about 500 mm long.

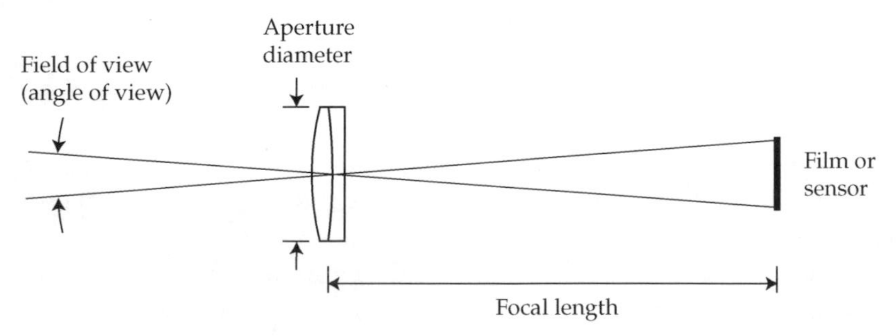

Figure 6.13. Every telescope or camera lens has an aperture, a focal length, and a field of view.

[2] If you're new to optical diagrams, you may wonder why Figure 6.13 shows rays of light spreading apart while Figure 6.1 shows them coming together. The answer is that Figure 6.1 shows two rays from the same point on a distant object, but Figure 6.13 shows one ray from each of two points some distance apart.

The way a camera is coupled to a telescope affects the focal length of the resulting system (Table 6.1). With direct coupling, you use the telescope at its inherent focal length, unmodified. Positive projection and afocal coupling usually increase the focal length; negative projection always does. Compression always reduces it.

6.3.2 Aperture

The aperture (diameter) of a telescope determines how much light it picks up. A 10-cm (4-inch) telescope picks up only a quarter as much light from the same celestial object as a 20-cm (8-inch), because a 10-cm circle has only a quarter as much surface area as a 20-cm circle.

Telescopes are rated for their aperture; camera lenses, for their focal length. Thus a 200-mm camera lens is much smaller than a 20-cm (200-mm) telescope.

In this book, apertures are always given in centimeters or inches, and focal lengths are always given in millimeters, partly because this is traditional, and partly because it helps keep one from being mistaken for the other.

In the context of camera lenses, "aperture" usually means *f*-ratio rather than diameter. That's why I use the awkward term "aperture diameter" in places where confusion must be avoided.

6.3.3 *f*-Ratio and Image Brightness

The *f*-ratio of a telescope or camera lens is the ratio of focal length to aperture:

$$f\text{-ratio} = \frac{\text{Focal length}}{\text{Aperture diameter}}.$$

One of the most basic principles of photography is that the brightness of the image, on the film or sensor, depends on the *f*-ratio. That's why, in daytime photography, we describe every exposure with an ISO setting, shutter speed, and *f*-ratio.

To understand why this is so, remember that the aperture tells you how much light is gathered, and the focal length tells you how much it is spread out on the sensor. If you gather a lot of light and don't spread it out much, you have a low *f*-ratio and a bright image.

The mathematically adept reader will note that the amount of light gathered, and the extent to which it is spread out, are both *areas* whereas the *f*-ratio is calculated from two *distances* (focal length and aperture). Thus the *f*-ratio is actually the square root of the brightness ratio. Specifically:

$$\text{Relative change in brightness} = \left(\frac{\text{Old } f\text{-ratio}}{\text{New } f\text{-ratio}}\right)^2.$$

So at $f/2$ you get twice as much light as at $f/2.8$ because $(2.8/2)^2 = 2$. This allows you to take the same picture with half the exposure time, or in general:

$$\text{Exposure time at new } f\text{-ratio} = \text{Exposure time at old } f\text{-ratio} \times \left(\frac{\text{New } f\text{-ratio}}{\text{Old } f\text{-ratio}}\right)^2.$$

For DSLRs, this formula is exact. For film, the exposure times would have to be corrected for reciprocity failure.

Low *f*-ratios give brighter images and shorter exposures. That's why we call a lens or telescope "fast" if it has a low *f*-ratio.

Comparing Apples to Oranges

A common source of confusion is that lenses change their *f*-ratio by changing their diameter, but telescopes change their *f*-ratio by changing their focal length.

For instance, if you compare 200-mm $f/4$ and $f/2.8$ telephoto lenses, you're looking at two lenses that produce the same size image but gather different amounts of light. But if you compare 8-inch (20-cm) $f/10$ and $f/6.3$ telescopes, you're comparing telescopes that gather the same amount of light and spread it out to different extents.

Lenses have adjustable diaphragms to change the aperture, but telescopes aren't adjustable. The only way to change a telescope from one *f*-ratio to another is to add a compressor (focal reducer) or projection lens of some sort. When you do, you change the image size.

That's why, when you add a focal reducer to a telescope to brighten the image, you also make the image smaller. Opening up the aperture on a camera lens does no such thing because you're not changing the focal length.

There are no focal reducers for camera lenses, but there are focal extenders (negative projection lenses). They're called *teleconverters*, and – just like a Barlow lens in a telescope – they increase the focal length and the *f*-ratio, making the image larger and dimmer.

What about Stars and Visual Observing?

The received wisdom among astronomers is that *f*-ratio doesn't affect star images, only images of extended objects (planets, nebulae, and the like). The reason is that star images are supposed to be points regardless of the focal length. This is only partly true; the size of star images depends on optical quality, focusing accuracy, and atmospheric conditions. The only safe statement is that the limiting star magnitude of an astronomical photograph is hard to predict.

The *f*-ratio of a telescope does not directly determine the brightness of a *visual* image. The eyepiece also plays a role. A 20-cm $f/10$ telescope and a 20-cm $f/6.3$ telescope, both operating at $\times 100$, give equally bright views, but with different eyepieces. With the same eyepiece, the $f/6.3$ telescope would give a brighter view at lower magnification.

Why Aren't All Lenses *f*/1?

If low *f*-ratios are better, why don't all telescopes and lenses have as low an *f*-ratio as possible? Obviously, physical bulk is one limiting factor. A 600-mm $f/1$ camera lens, if you could get it, would be about two feet in diameter, too heavy to carry; a 600-mm $f/8$ lens is transportable.

A more important limitation is lens aberrations. There is no way to make a lens or mirror system that forms perfectly sharp images over a wide field. This fact may come as a shock to the aspiring photographer, but it is true.

Consider for example a Newtonian telescope. A perfect paraboloidal mirror forms a perfect image at the very center of the field. Away from the center of the field, though, the light rays are no longer hitting the paraboloid straight-on. In effect, they are hitting a shape which, to them, is a distorted or tilted paraboloid, and the image suffers coma, which is one type of off-axis blur. A perfectly made Cassegrain or refractor has the same problem.

The lower the *f*-ratio, the more severe this problem becomes. An $f/10$ paraboloid is nearly flat and still looks nearly paraboloidal when approached from a degree or two off axis. An $f/4$ paraboloid is deeply curved and suffers appreciable coma in that situation. For that reason, "fast" telescopes, although designed for wide-field viewing, often aren't very sharp at the edges of the field.

Complex mirror and lens systems can reduce aberrations but never eliminate them completely. Any optical design is a compromise between tolerable errors. For more about aberrations, see *Astrophotography for the Amateur* (1999), pp. 71–73.

Incidentally, $f/1$ is not a physical limit. Canon once made a 50-mm $f/0.95$ lens. Radio astronomers use dish antennas that are typically $f/0.3$.

6.3.4 Field of View

Astronomers measure apparent distances in the sky in degrees (Figure 6.14); a degree is divided into 60 arc-minutes (60′), and an arc-minute is divided into 60 arc-seconds (60″).

The exact formula for field of view is:

$$\text{Field of view} = 2\tan^{-1}\frac{\text{Sensor width (or height, etc.)}}{2\times\text{Focal length}}.$$

For focal lengths much longer than the sensor size, such as telescopes, a much simpler formula gives almost exactly the same result:

$$\text{Field of view} = 57.3^\circ \times \frac{\text{Sensor width (or height, etc.)}}{\text{Focal length}}.$$

The sensor size and focal length are in the same units, usually millimeters.

Figure 6.15 shows the field of view of an APS-C sensor with various focal lengths, superimposed on an image the Pleiades star cluster. For the same concept applied to telephoto lenses, see Section 7.1.2, and for more about field of view, see *Astrophotography for the Amateur* (1999), pp. 73–75.

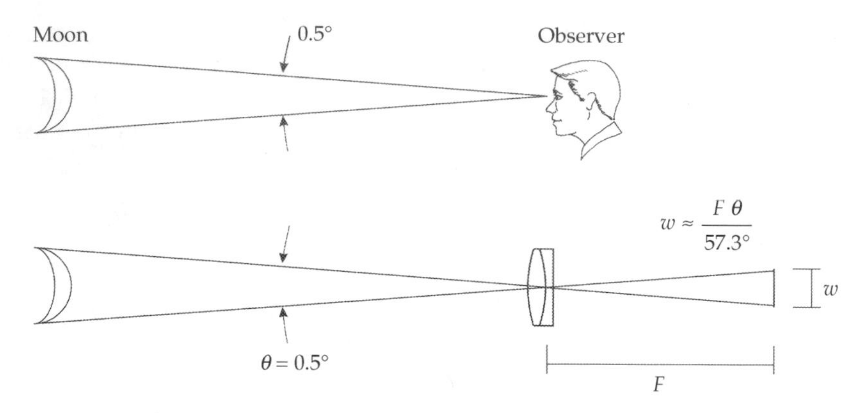

Figure 6.14. The apparent size of objects in the sky is measured as an angle. (From *Astrophotography for the Amateur.*)

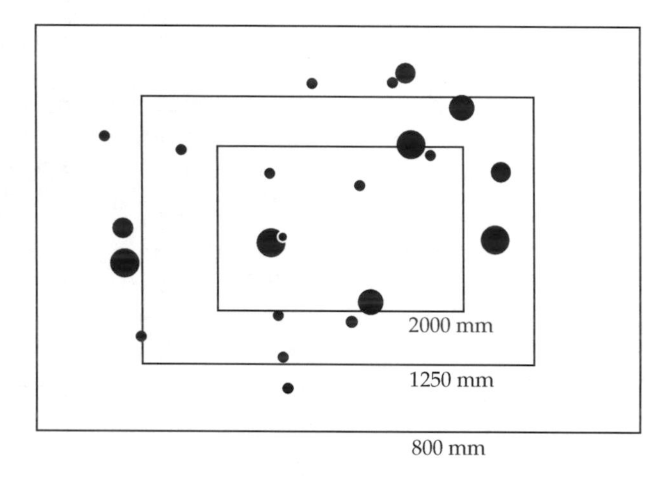

Figure 6.15. Field of view of an APS-C-size sensor with various focal lengths, relative to the Pleiades star cluster. Compare Figure 7.3.

Note however that the focal length may not be what you think it is. Schmidt–Cassegrains and similar telescopes change focal length appreciably when you focus them by changing the separation of the mirrors. Even apart from this, the focal length of a telescope or camera lens often differs by a few percent from the advertised value. To determine focal length precisely, use *Stellarium* or another computerized star atlas to plot the field of view of your camera and telescope, then compare the calculated field to the actual field.

6.3.5 Sensor Size

To determine the field of view, you must of course know the size of the sensor, and DSLR sensors are not all alike (Figure 6.16). I'm talking about the actual physical size, not the number of megapixels; we'll get to pixel count and pixel size shortly.

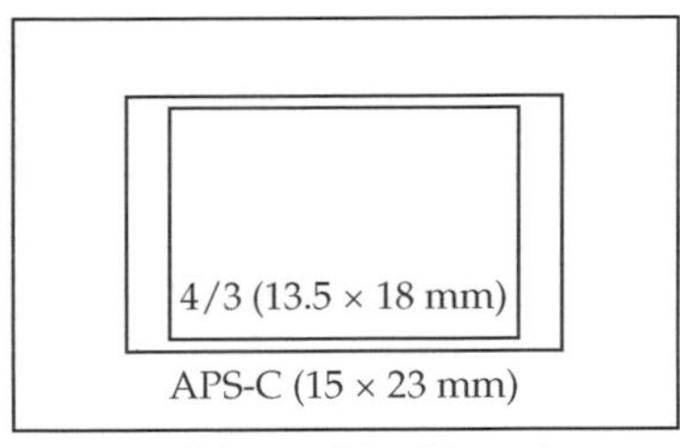

Figure 6.16. Relative sizes of commonly used sensors (shown larger than actual size).

Some high-end DSLRs have a sensor the size of a full 35-mm film frame (24 × 36 mm), but most DSLRs have sensors that are only two thirds that size, a format known as APS-C (about 15 × 23 mm; Nikon's is slightly larger than Canon's). In between is the APS-H format of a couple of high-end Canons (19 × 29 mm). APS-C originally denoted the "classic" picture format on Advanced Photo System (APS, 24-mm) film, and APS-H was its "high-quality" format.

The rival Four Thirds (4/3) system, developed by Olympus and Kodak, uses digital sensors that are smaller yet, 13.5 × 18 mm. Micro Four Thirds (MFT) is the mirrorless version of the same thing, with the same sensor size. "Four thirds" denotes an image size originally used on video camera tubes that were nominally 4/3 of an inch in diameter. Many video CCD sensors are still specified in terms of old video camera tube sizes, measured in fractional inches.

You shouldn't feel cheated if your sensor is smaller than "full frame." Remember that "full frame" was arbitrary in the first place. The smaller sensor of a DSLR is a better match to a telescope eyepiece tube (32 mm inside diameter) and also brings out the best in 35-mm camera lenses that suffer aberrations or vignetting at the corners of a full-frame image.

Sensor size is sometimes expressed as a "focal-length multiplier" or "zoom factor." This can be a source of confusion. Historically, at the end of the film era, when photographers switched from full-frame film to APS-C DSLRs, they felt they were getting bigger pictures because the field of view was narrower. Enlarging the picture to the same final size would therefore make the same object look bigger. It was as if 100-mm lenses had changed to 160 mm, and so forth. Thus, the APS-C sensor was billed as having a "zoom factor" or "focal-length multiplier" of 1.6. This has nothing to do with zooming (varying focal length) in the normal sense of the word. All you're doing is using the central part of the original field instead of the whole thing. It is more accurate to call it a "crop factor" because the smaller sensor does not enlarge the image but rather trims it.

6.3.6 Arc-seconds per Pixel

DSLR astrophotographers do not often deal with single pixels because there are too many of them. Even a pinpoint star image usually occupies six or eight pixels on the sensor. This is a good thing because it means that star images are much larger than hot pixels.

Still, it can be useful to know the image scale in arc-seconds per pixel. To find it, first determine the size of each pixel in micrometers (microns, μm, thousandths of a millimeter). For instance, the Canon 80D sensor measures 15.0×22.5 mm, according to Canon's specifications, and has 4000×6000 pixels. That means the pixel size is

$$\frac{15.0 \text{ mm}}{4000} = 0.00375 \text{ mm} = 3.75 \ \mu\text{m vertically}$$

and also

$$\frac{22.5 \text{ mm}}{6000} = 0.00375 \text{ mm} = 3.75 \ \mu\text{m horizontally.}$$

Now find the field of view of a single pixel; that is, pretend your sensor is the size of one pixel, and compute its field of view. We use the same field-of-view formula as before, except that the pixel size is in μm and the result is in arc-seconds. With appropriate adaptations, the formula is:

$$\text{Field of one pixel} = 206.3'' \times \frac{\text{Pixel size } (\mu\text{m})}{\text{Focal length (mm)}}.$$

For example, suppose the telescope is a common 20-cm (8-inch) $f/10$ Schmidt–Cassegrain with a focal length of 2000 mm, and the camera is the one just described. Then:

$$\text{Field of one pixel} = 206.3'' \times \frac{3.75 \ \mu\text{m}}{2000 \text{ mm}} = 0.39''$$

As often happens, the pixel size is smaller than the resolution limit of the telescope. If the steadiness of the air permits star images that are 3″ in diameter, they will be almost 8 pixels in diameter on this camera.

6.3.7 "What is the Magnification of This Picture?"

Non-astronomers seeing an astronomical photograph often ask what magnification or "power" it was taken at.

In astronomy, such a question almost has no answer. With microscopes, it does. If you photograph a 1-mm-long insect and make its image 100 mm long on the print, then clearly, the magnification of the picture is ×100.

But when you tell people that your picture of the moon is 1/35 000 000 the size of the real moon, somehow that's not what they wanted to hear. Usually, what they mean is, "How does the image in the picture compare to what I would see in the sky with my eyes, with no magnification?" The exact answer depends, of course, on how close the picture is to the viewer's face.

But a rough answer can be computed as follows:

$$\text{"Magnification" of picture} = \frac{45^\circ}{\text{Field of view of picture}}.$$

That is: If you looked through a telescope at this magnification, you'd see something like the picture. Here 45° is the size of the central part of the human visual field, the part we usually pay attention to, and is also the apparent field of a typical (not super-wide) eyepiece.

So a picture that spans the moon (half a degree) has a "magnification" of 90. Deep-sky piggyback images often have rather low "magnification" computed this way, anywhere from 10 down to 2 or less.

6.4 Edge-of-field Quality and Vignetting

When you consider quality at the edge of the field, there are two kinds of telescopes in the world: those designed to cover the full field of a camera, and those designed for visual use with eyepieces.

The vast majority of telescopes are of course the second kind. As *Astrophotography for the Amateur* explains at length, few telescopes produce a sharp, fully illuminated image over an entire 35-mm film frame or even an entire DSLR sensor. The reason is that in a visual telescope, you want maximum sharpness at the very center of the field, and performance away from center is much less important.

Telescopes designed for photography, especially small multi-element refractors, give superb image quality over the full field of a camera. Other telescopes do not; that was not their purpose.

A related problem is *vignetting*, or lack of full illumination away from the center. An APS-C-size DSLR sensor is slightly larger, corner to corner, than the inside diameter of a standard eyepiece tube. Thus, to reduce vignetting, you should avoid working through a $1\frac{1}{4}$-inch eyepiece tube if possible; switch to a 2-inch focuser or a direct T-adapter. But even then, a telescope with glare stops designed to work well with eyepieces will not fully illuminate the edges of the field.

Eyepiece tubes are not the only problem. In some cameras, part of the image can be cut off or darkened by the camera's own lens mount or mirror box. This happens with telescopes where no lens element is close to the camera, not with telephoto lenses, whose rear element is positioned to reach all of the sensor. Lens-mount vignetting is round, but mirror-box vignetting usually shows up as darkening of one edge or two opposite edges.

If not severe, vignetting is corrected by flat-fielding, but sometimes all you can do is crop the picture. In that situation, I take the optimist's point of view: it's not that the image is too small, but rather that the sensor is too big. Eyepieces are, after all, much smaller than DSLR sensors.

Chapter 7
Camera Lenses

The previous chapter was mostly about adapting telescopes to make them deliver images to cameras rather than eyepieces. There is an easier way. If you want to form a sharp, bright image on the sensor of a camera, why not use optics designed for the purpose – namely camera lenses?

Even professional astronomical research sometimes uses camera lenses. For example, Yale astronomer Pieter van Dokkum and his Dragonfly Project team regularly discover dwarf galaxies and intergalactic matter using an array of Canon 400-mm $f/2.8$ telephoto lenses (with dedicated astrocameras, not DSLRs). They found that commercially available telephoto lenses with the newest coatings scatter less light, and hence see fainter objects against a dark background, than even the best observatory telescopes.[1]

Most of my best deep-sky work has been done with medium to long telephoto lenses, see Figure 7.1. I joke that astrophotography is subsidized by high-school football – sports photographers in every town are why good 300-mm $f/4$ lenses are made in such quantities that we can afford them. These lenses compete head-on with small refracting telescopes and often outperform them.

7.1 Why You Need Another Lens

The "kit" lens that probably came with your DSLR is not very suitable for photographing star fields. It has at least three disadvantages:

- It is slow (about $f/4$ or $f/5.6$).
- It is a zoom lens, and optical quality has been sacrificed in order to make zooming possible.

[1] P. G. van Dokkum *et al.*, "Forty-seven Milky Way-sized, extremely diffuse galaxies in the Coma cluster," *Astrophysical Journal Letters* **798**:L45 (2015); R. G. Abraham and P. G. van Dokkum, "Ultra-low surface brightness imaging with the Dragonfly Telephoto Array," *Publications of the Astronomical Society of the Pacific* **126**:55–69 (2014).

Figure 7.1. Exploring the universe with a telephoto lens. The spiral galaxy M33 imaged with a Canon 300-mm $f/4$ telephoto lens; stack of four 5-minute exposures at $f/5$, Canon 60Da, ISO 1600, autoguided on a Celestron AVX mount.

- It is plastic-bodied and not very sturdy; the zoom mechanism and the autofocus mechanism are both likely to move during a long exposure.

Fortunately, you have many alternatives, some of which are quite inexpensive. One is to buy your camera maker's 50-mm $f/1.8$ "normal" lens; despite its low price, this is likely to be the sharpest lens they make, especially when stopped down to $f/4$. Another alternative is to use an inexpensive manual-focus telephoto lens from the old days. There are several ways of doing this; Nikon DSLRs take Nikon manual-focus lenses (though the autofocus and light meter don't work), and Canons accept several types of older lenses via adapters.

One note of caution: Since writing the first edition of this book, I have become less optimistic about using older "classic" lenses. Lens design and manufacturing have improved to the point that only the very best lenses from before 1990 compete with those in current production. But any lens can produce good pictures when used within its limits, and you may be quite happy to get 70% of the quality for 20% of the price. Besides, trying an older lens can help you evaluate the feasibility of using a particular size and focal length before you spring for its modern counterpart.

7.1.1 Big Lens or Small Telescope?

But wait a minute – instead of a lens, should you be looking for a small telescope, perhaps an $f/6$ or $f/8$ "short-tube" refractor?

If the refractor is designed for photography, and has the focal length you want, then the answer is very likely yes. You can get excellent results with such a refractor, one with three or four elements and maybe a separate field-flattening focal reducer (Figure 6.3).

That applies only to refractors designed for the job. Conventional two-element refractors do not perform well for astrophotography; they have too much chromatic aberration and do not cover a wide field. The lower the *f*-ratio, the worse the performance.

More to the point, camera lenses and small refractors don't quite compete head-on. Camera lenses offer you shorter focal lengths (from 300 mm down to 20 mm or even less) for wider fields, and lower *f*-ratios for brighter images. Camera lenses enable you to do wide-field imaging at $f/2.8$ (as in Figure 7.2) or even $f/2$. But if you want a focal length of 400 mm or more, and $f/6$ is fast enough (which it may well be, with modern sensors), the small refractor wins.

One advantage of the refractor is that, because it has a fixed aperture and fewer elements, it is less vulnerable to internal reflections. Its glare stops are positioned exactly as needed for its one and only aperture; it doesn't have to let light in for apertures you're not using.

7.1.2 Field of View

The first question to ask about a lens is of course its field of view – how much of the sky does it capture?

Figure 7.3 and Table 7.1 give the field of view of several common lenses with DSLRs that have an APS-C-size sensor. The numbers in Table 7.1 are exact for the Canon APS-C sensors; Nikon's are about 5% larger.

Figure 7.2. The Pipe Nebula, a dark nebula in front of the star clouds of Ophiuchus. Stack of five 5-minute exposures with a Canon 40D at ISO 800 with a Sigma 105-mm $f/2.8$ lens piggybacked on a Meade LX200 telescope, autoguided.

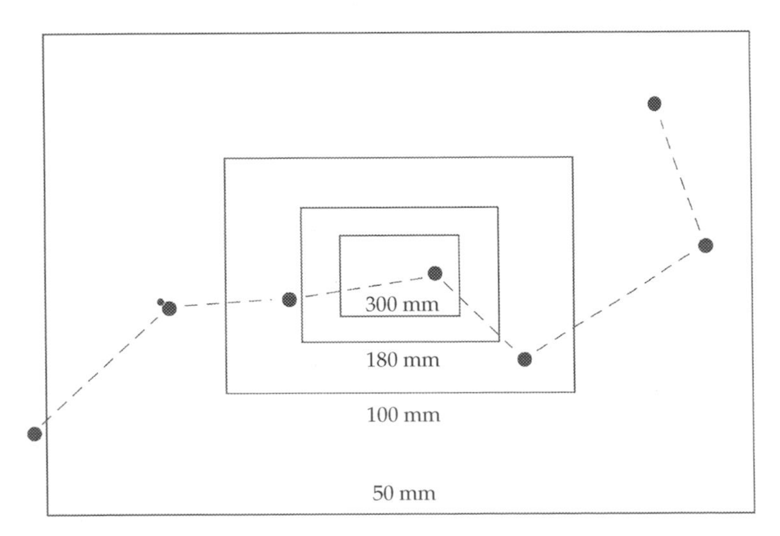

Figure 7.3. Field of view of various lenses with Canon or Nikon APS-C sensor compared to the Big Dipper (Plough).

Table 7.1 *Field of view of various lenses with APS-C sensor.*

Focal length	Field of view		
	Height	Width	Diagonal
28 mm	30°	43°	50°
35 mm	24°	35°	42°
50 mm	17°	25°	30°
100 mm	8.5°	12.7°	15°
135 mm	6.3°	9.4°	11°
180 mm	4.7°	7.1°	8.5°
200 mm	4.2°	6.3°	7.6°
300 mm	2.8°	4.2°	5.1°
400 mm	2.1°	3.2°	3.8°

Field of view is measured in degrees, and some lens catalogs give only the diagonal "angle of view" from corner to corner of the picture. The exact formula for field of view is given in Section 6.3.4.

7.1.3 *f*-Ratio

Faster is better, right? The *f*-ratio of a lens determines how bright the image will be, and hence how much you can photograph in a short exposure. Lower *f*-numbers give a brighter image.

But there's a catch: speed is expensive. Nowadays, fast lenses are quite sharp but cost a lot of money. In earlier years, the cost was in performance, not just price, and lenses faster than $f/4$ were not completely sharp wide open. Until about 1990, it was a rule of thumb that every lens performed best if you stopped it down a couple of stops from maximum aperture, but in this modern era of ED glass and aspheric designs, some lenses are actually sharpest at their widest aperture.

Because DSLRs do not suffer reciprocity failure, deep-sky photography is no longer a struggle to get enough light onto the film before it forgets all the photons. We no longer need fast lenses as much as we used to. Under dark skies, I find that $f/4$ is almost always fast enough.

7.1.4 Zoom or Non-zoom?

In my experience, some of the best zoom lenses are *just* acceptable for astrophotography. Most zoom lenses aren't. A zoom lens, even a high-technology zoom lens from a major manufacturer, is designed to be tolerable at many focal lengths rather than perfect at one.

Dealers have learned that a zoom lens gives a camera "showroom appeal" – it's more fun to pick up the camera, look through it, and play with the zoom. In the 1990s, people got used to extreme zoom lenses on camcorders, which were low-resolution devices not bothered by optical defects, and the makers of compact digital cameras have done their best to follow through. And some photographers just don't like to change lenses (especially if there's a risk of getting dust on the sensor). For these reasons, zoom lenses, including those with extreme ratios, have become ubiquitous.

One look at Nikon's or Canon's published MTF curves should make it clear that a mediocre fixed-length lens is usually better than a first-rate zoom. If you have nothing but zoom lenses, try a non-zoom ("prime") lens; you're in for a treat.

If you do attempt astrophotography with a zoom lens – as I have done – beware of "zoom creep." During the exposure, the zoom mechanism can shift. You may have to tape it in place to prevent this.

7.2 Lens Quality

7.2.1 Sharpness, Vignetting, Distortion, and *Bokeh*

Star fields are a very tough test of lens quality. The stars are point sources, and any blur is immediately noticeable. Every star in the picture has to be in focus all at once; there is no out-of-focus background. And we process the picture to increase contrast, which emphasizes any vignetting (darkening at the edges).

Having said that, I should add that the situation with DSLRs is not quite the same as with film. High-speed film is itself very blurry; light diffuses sideways through it, especially the light from bright stars. When you put a very sharp lens on a DSLR and this blurring is absent, the stars all look alike and you can no longer tell which ones are brighter.

For that reason, less-than-perfect lenses are not always unwelcome with DSLRs. A small amount of uncorrected spherical or chromatic aberration, to put a faint halo around the brighter stars, is not necessarily a bad thing. What is most important is uniformity across the field. The stars near the edges should look like the stars near the center.

All good lenses show a small amount of vignetting when used wide-open; the alternative is to make a lens with inadequate glare stops. Besides vignetting in the strict sense (caused by blockage of light rays), wide-angle lenses show some edge darkening for physical reasons – the effective focal length is longer at the corners, but the aperture is the same, so the light is spread thinner (*Astrophotography for the Amateur* (1999), pp. 124–126). Edge darkening can be corrected when the image is processed (Figure 5.2), so it is not a fatal flaw. Another way to reduce vignetting is to close the lens down one or two stops from maximum aperture.

Distortion (barrel or pincushion) is important only if you are making star maps, measuring positions, or combining images of the same object taken with different lenses. Zoom lenses almost always suffer noticeable distortion, as you can demonstrate by taking a picture of a brick wall; non-zoom lenses almost never do.

One lens attribute that does not matter for astronomy – except in a back-handed way – is *bokeh* (Japanese for "blur").[2] *Bokeh* refers to the way the lens renders out-of-focus portions of the picture, such as the distant background of a portrait.

The physical basis of "good *bokeh*" is spherical aberration. Years ago, it was discovered that uncorrected spherical aberration made a lens more tolerant of focusing errors and even increased apparent depth of field. But in astronomy, there is no out-of-focus background. The spherical aberration that contributes to "good *bokeh*" could help the bright stars stand out in a star field; apart from that, it is just a defect.

7.2.2 Reading MTF Curves

No longer can we say that a particular lens "resolves 60 lines per millimeter" or the like. This kind of measurement is affected by contrast; high-contrast film brings out blurred details. Instead, nowadays opticians measure *how much* blurring occurs at various distances from the center of the picture. That is, they measure the *modulation transfer function* (MTF).

Figure 7.4 shows how to read MTF graphs. Each graph has one or more pairs of curves for details of different sizes (such as 10, 20, and 40 lines per millimeter). Each pair consists of a solid line and a dashed line. Usually, the solid line indicates *sagittal* resolution (Figure 7.5) and the dashed line indicates *meridional* (*tangential*) resolution.

What should the MTF of a good lens look like? My rule of thumb is that everything above 75% is sharp. I'm more concerned that the sagittal and meridional curves should stay close together so that the star images are round, and that there shouldn't be a dramatic drop-off toward the edge of the picture. On that point, APS-C sensors have an advantage because they aren't as big as 35-mm film.

MTF curves do not measure vignetting or distortion. Also, most camera manufacturers publish *calculated* MTF curves (based on computer simulation of the lens design); the real MTF may not be as good because of manufacturing tolerances. An exception is Zeiss, which publishes *measured* MTF curves for all its lenses.

One last note. Lens MTF curves plot contrast against distance from center, with separate curves for different spatial frequencies (lines per mm). Film MTF

[2] Also transliterated *boke*. Another Japanese word with the same pronunciation but a different etymology means "idiot."

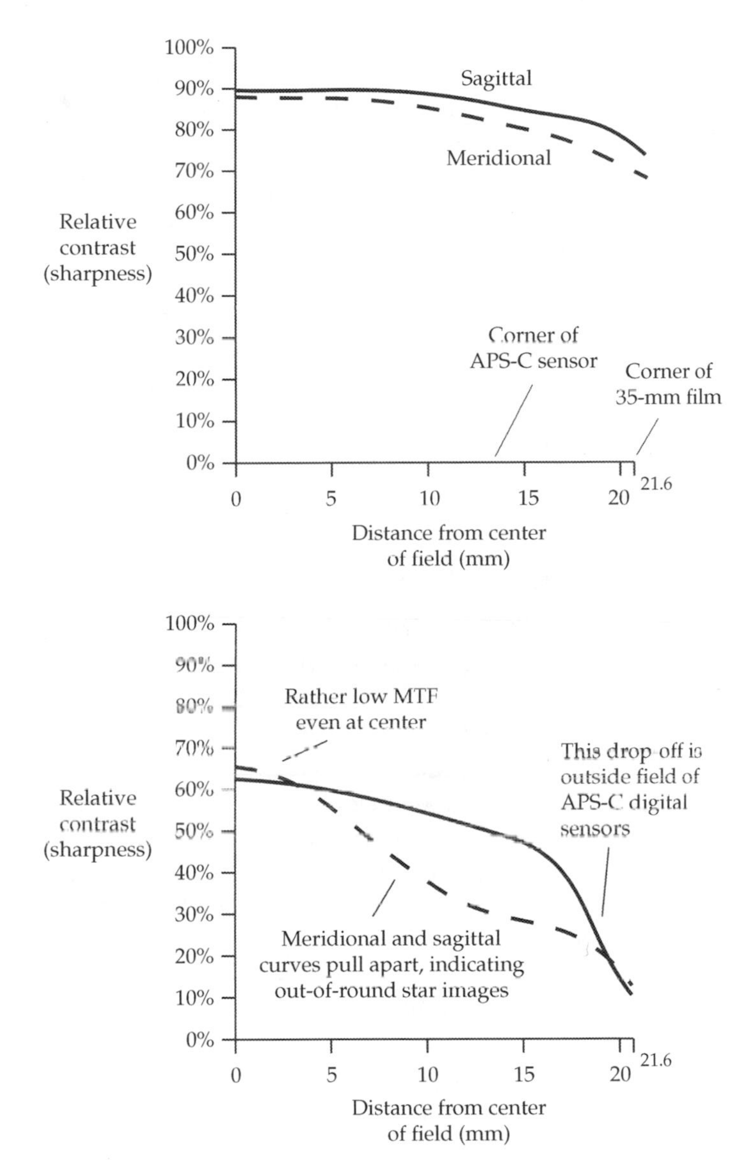

Figure 7.4. MTF curve from a good lens (top) and from one that warrants concern (bottom). Star fields are very demanding targets.

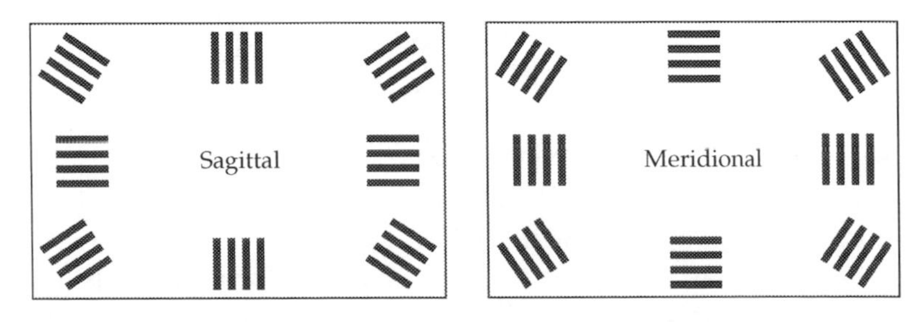

Figure 7.5. Orientation of targets for sagittal and meridional (tangential) MTF testing.

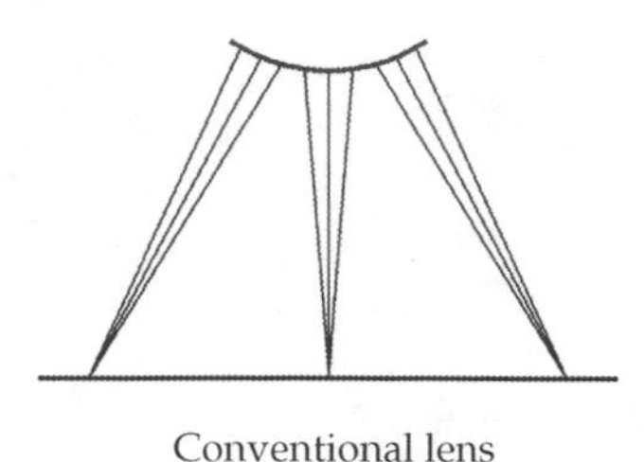

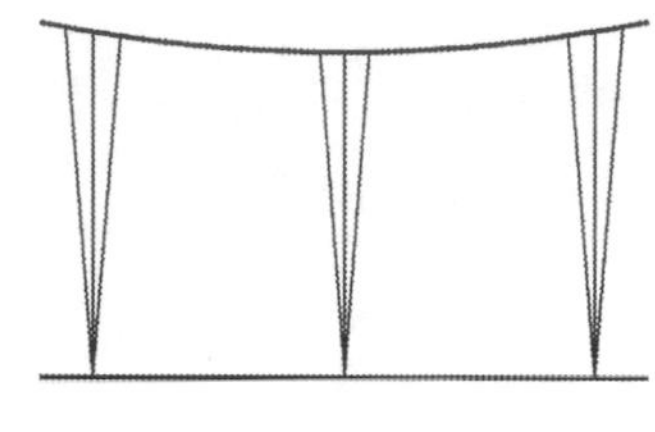

Figure 7.6. With a telecentric lens, light from all parts of the image arrives perpendicular to the sensor.

curves plot contrast versus spatial frequency. See *Astrophotography for the Amateur* (1999), p. 187. The two kinds of curves look alike but are not comparable.

7.2.3 Telecentricity

Digital sensors perform best if light reaches them perpendicularly (Figure 7.6). At the corners of a picture taken with a conventional wide-angle lens, there is likely to be color fringing from light striking the Bayer matrix at an angle, even if film would have produced a perfect image.

A *telecentric* lens is one that delivers bundles of light rays to the sensor in parallel from all parts of the image. The obvious drawback of this type of lens is that it requires a large-diameter rear element, slightly larger than the sensor.

Telecentricity is one of the design goals of the Four Thirds system, which uses a lens mount considerably larger than the sensor. Other DSLRs are adapted from 35-mm SLR body designs, and the lens mount is not always large enough to permit lenses to be perfectly telecentric.

MTF curves do not tell you whether a lens is telecentric, and some excellent lenses for film cameras work less than optimally with DSLRs, while some mediocre lenses work surprisingly well. It's a good sign if the rear element of the lens is relatively large in diameter and is convex (positive); see for example the Olympus 100-mm $f/2.8$ and the digitally optimized Sigma lens in the left-hand column of Figure 7.19. Telescopes and long telephoto lenses are always close to telecentricity because their exit pupil is so far from the sensor.

7.2.4 Construction Quality

It's remarkable how flimsy a lens can be and still produce good pictures when there's an autofocus mechanism holding it in perfect focus. When photographing the stars, that autofocus mechanism is turned off, and what's more, the camera and lens move and tilt as the telescope tracks the stars. Lenses that are very good for everyday photography can perform poorly in such a situation.

That's why older manual-focus lenses appeal to me, as well as professional-grade autofocus lenses that are built to survive rough handling. Almost any lens from the 1970s will seem to be built like a tank compared to today's products.

Another advantage of manual-focus lenses is that they are easier to focus manually. That sounds like a tautology, but it's important. Autofocus lenses *can* be focused manually – some more easily than others – but they are often very sensitive to slight movement of the focusing ring. Older manual-focus lenses are easier to focus precisely.

Of course, optical quality is also a concern, and it's where older lenses often fall down. Only the best lenses from yesteryear are likely to perform well by today's standards; too many of them had excessively simple designs or loose manufacturing tolerances.

7.3 Which Lenses Fit Which Cameras?

7.3.1 Canon

The Canon EOS DSLR lens mount has only two varieties, EF and EF-S; the latter is for lenses that only cover APS-C sensors, and it protrudes farther into the camera. Some light-pollution filters insert into the camera body and occupy the same space as this protrusion (Figure 17.7); you cannot use EF-S lenses with them. Of course, the EOS (EF) lens mount is completely different from the earlier Canon SLR lens mounts. EF-M, related to it, is the mount used on Canon MILCs.

If all you need is a T-ring, the situation is simple. Any T-ring marked "Canon EOS," "Canon AF," or "Canon EF" is the right kind for an EOS camera. The other kind, Canon F, FD, or FL, will not fit on the DSLR at all.

7.3.2 Nikon

With Nikon, there is only one kind of SLR lens mount – introduced on the Nikon F in 1959 – but there are many variations. Most Nikons today take only the "AI" version of the mount, introduced in 1977, and its descendants. If you try to mount a pre-AI lens on a modern Nikon, you can actually cause damage, so be careful. This is of interest because some pre-1977 Nikon lenses are good enough for serious use today.

The variations on the Nikon F lens mount have to do with how the lens communicates aperture information to the exposure meter in the camera. The "AI" (aperture indexing) system uses a notched ridge in the outermost ring of the lens mount (Figure 7.7) to engage a tiny lever or other mechanical sensor on the camera body in either of two places. The old "pre-AI" system uses a pin on the camera body above the lens mount, engaging a fork-like prong on top of the lens (also visible in Figure 7.7). The newest G lenses have no ring for setting the

Figure 7.7. On Nikon AI lenses, a notch or step in the outermost ridge of the lens mount engages a mechanical sensor in the camera. Pre-AI lenses have a raised rim without the step and can damage some newer cameras. G lenses lack both the step and the raised rim.

aperture manually; it is controlled from inside the camera body, as with Canon EOS lenses, and communication is electronic.

Some newer Nikons have provisions to accept pre-AI lenses. On the Nikon F3 film SLR, the AI sensing lever flips out of the way. On some DSLRs, such as the D5300, the AI sensor is a button designed to push down all the way without suffering damage when a lens is mounted that does not allow space for it. Other equally modern Nikon DSLRs would be damaged if you do this, so check your instruction manual carefully. Above all, never force a lens into place if it seems to be bumping into something.

T-rings on telescope adapters are not a problem for any Nikon body, nor are old lenses that have been converted to AI ("AI'd"). Also, Nikon adapters for Canon and other bodies take all F-mount lenses without regard to AI.

All Nikon autofocus lenses can be focused manually (and, in astrophotography, always are). Whether they autofocus with a particular camera is variable. Most newer DSLRs only autofocus with lenses that have built-in autofocus motors (AF-S, AF-I, AF-P; not plain AF or AF-D). That is good news for us because it makes excellent late-model AF-D lenses cheaper on the secondhand market. Newer Nikon AF lenses are designated FX if they cover the full frame or DX if they only cover APS-C sensors.

On most Nikon DSLRs, the light meter and auto exposure do not work if the lens does not contain any electronics. This is another reason manual-focus Nikon lenses are becoming cheaper secondhand. You might wonder why Nikon has this limitation and Canon doesn't. Basically, it's because Nikon lenses allow the aperture to be set manually, but DSLRs do not fully utilize the AI system to find out what this setting is. As with traditional film SLRs, the lens stays wide open for viewing and focusing, then stops down to the preselected aperture to take the picture, but the exposure meter does not know what this aperture is going to be. There is no such problem with Canon lenses because they do not have aperture settings; the aperture is always controlled by the camera body (like Nikon G lenses), or else the camera assumes the aperture isn't going to change (as with a telescope, or a manual lens on an adapter).

All of this just scratches the surface. Nikon is *complicated!* One good source of further information is www.kenrockwell.com.

7.3.3 Lens Mount Adapters

Besides the ubiquitous T-ring, originally designed for a line of cheap telephoto lenses, there are other adapters for putting one kind of lens onto another kind of camera body.

The lens going on the front of an adapter must, in general, be fully manual (in focusing and aperture) and not require electric power or signals from the camera body. (It can be an autofocus lens, such as a Nikon AF-D lens, provided you also have full manual control.) There are exceptions, but if you want to use a lens that is not fully manual, make sure the adapter contains circuitry to support its functions. Adapters to put Nikon G lenses on Canon EOS cameras and control the aperture electronically are available from Fotodiox and Novoflex.

Most adapters are simple, nothing more than a receptacle to fit the lens and a flange to fit the camera body. The trouble is, all of this takes up space, placing the lens farther from the film or sensor than it was designed to be. There are three ways to deal with this problem:

- Add a glass element (typically a 1.2× Barlow lens) so that the lens will still focus correctly. This is a common practice but degrades the image quality.
- Leave out the glass element and give up the ability to focus on infinity. The adapter is now a short extension tube and the lens can only take close-ups.
- If the camera body is shallower than the one the lens was designed for, then make the adapter exactly thick enough to take up the difference. In this case, the lens focuses perfectly, without a glass element, and the adapter is advertised as "maintains infinity focus."

The third kind of adapter is the only one that interests us, and it's only possible if the camera body is shallower, front to back, than the body the lens was designed for. In that case, the adapter makes up the difference. For example, a Nikon lens can fit onto a Canon EOS body with an adapter 2.5 mm thick.

The Canon EOS is one of the shallowest full-size SLR bodies, and adapters exist to put Nikon, Contax-Yashica, Olympus, M42, and even Exakta lenses on it. (Some late-model Contax-Yashica lenses are made by Zeiss and are excellent.) The one thing you can't do is put older Canon FD lenses on an EOS body, because the FD body was even shallower. Adapters to put other lenses on Canons often include a "focus confirmation chip" that allows the camera's autofocus system to be used to check the focus of a manually focused lens. This is not needed for astrophotography.

Nikon bodies generally can't take non-Nikon lenses because the Nikon body is one of the deepest in the business. Only the Leicaflex is deeper.

MILC and Four Thirds camera bodies are much shallower than film SLRs, and adapters exist to put a wide range of SLR and DSLR lenses on them, including sophisticated adapters that preserve electric power and signaling and even add manual aperture control.

One curious zero-thickness adapter does exist. Pentax's adapter to put screw-mount lenses on K-mount cameras simply wraps around the screw-mount threads, taking advantage of the K-mount's larger diameter.

7.3.4 What if there's no Aperture Ring?

If you are adapting a Canon EOS or Nikon AF-G lens to another camera (such as an MILC or a dedicated astrocamera), you may run into an obstacle: Those lenses have no aperture rings. How do you set the aperture?

One option is to use an adapter that includes aperture control. The other – worth trying, but not guaranteed – is to put the lens on a DSLR that it fits, set the aperture you want, hold down the depth-of-field preview button (which makes the lens stop down to the specified aperture), and remove the lens while doing so. Try this with any particular lens before relying on it.

7.3.5 Adapter Quality

Not all lens mount adapters are equally well made. The premier manufacturer of quality adapters is Fotodiox (www.fotodiox.com). On eBay you can buy adapters directly from the machinists in China who make them.

Figure 7.8 shows what to look for. Good adapters are usually made of chrome-plated brass or bronze and often include some stainless steel. Some are made of high-grade aluminum. For higher prices you get more accurate machining, more elaborate mechanisms to lock and unlock the lens, and more blackening of parts that could reflect light.

M42 screw mount to Canon adapters are the simplest, and my experience is that they always work well. I have also had good results with an inexpensive

Figure 7.8. Mid-grade and high-grade Nikon to EOS adapters. Avoid cheap imitations.

Olympus to Canon adapter. Each of these holds the lens tightly in place and is easy to attach and remove.

But Nikon to Canon adapters are tricky and always lack mechanical strength. When using one, *always let the lens support the camera* – don't let the adapter bear the weight of a heavy lens (Figure 7.11). And insist on a well-made adapter in the first place. I had a cheap one actually come apart in use as the screws stripped their threads.

The problem is that the Nikon lens mount contains springs. An adapter doesn't have room for good leaf springs like those in a Nikon camera body. Instead, the adapter relies on flexure of tabs that protrude from the lens mount itself. If these improvised springs are too weak, the lens and camera can pull apart slightly under the camera body's weight, and your pictures will be out of focus.

Another mark of a well-made Nikon adapter is that it makes it easy to remove the Nikon lens. Cheaper adapters require you to pry up a tab with your thumbnail; better ones give you a button to press.

7.3.6 The Classic M42 Lens Mount

Canon DSLRs with adapters have given a new lease on life to many classic lenses with the M42 (Pentax–Praktica) screw mount. Here "M42" means "metric, 42 mm" and has nothing to do with the Orion Nebula, Messier 42. It is the same diameter as a T-adapter, but with a thread pitch of 1.0 rather than 0.75 mm, and, obviously, a shorter flange-to-film distance. This mount was introduced by East German Zeiss on the Contax S in 1949 and promoted by Pentax as a "universal mount" in the 1960s. Several other camera makers used it, including Mamiya/Sekor and Vivitar. Pentax switched to a bayonet mount in 1976, but Zenit brand M42-mount cameras and lenses were made in Russia until 2005.

These lenses are usually mechanically sturdy and free of flexure in long exposures. Optical quality varies from brand to brand, but East German (Jena) Zeiss lenses and Pentax SMC Takumars are often quite good by the standards of their era, while off-brand imports vied for the bottom. Other M42-mount lenses range from mediocre to very poor by current standards, and today I use a couple of them as paperweights.

7.4 Supporting and Mounting a Lens

Any lens that is heavier than the camera body should support the camera, not vice versa. That implies the use of a tripod collar (Figure 7.9). If no collar is made for your lens, sometimes you can use the collar for a different lens, padded with extra cork or felt; at other times you'll have to improvise. Figure 7.10 shows Jerry Lodriguss' method of supporting the Nikon 180-mm *f*/2.8 manual-focus ED lens, and Figure 7.11 shows my hand-crafted solution for the AF version.

Generally, internal-focus (IF) lenses are easier to support because they focus by moving some of the internal elements while the outer frame remains fixed. More traditional lenses focus by moving all the elements together, and often, almost all of the barrel moves and turns as you focus. That's where rings may be the only solution.

It is common for the camera and lens to ride "piggyback" on a telescope, but there's an alternative. When a lens is mounted on a dovetail, the lens can go on a German equatorial mount with no telescope under it. (Under a slot in the dovetail, use a 1/4″-20 socket head screw; such screws are often included when you buy dovetail bars.)[3] You can still set up the mount, finding and centering stars, by using the lens with the camera in Live View mode (Section 8.5.5).

Figure 7.9. Tripod collar of heavy lens can mount directly to a dovetail so the lens can go on an equatorial mount as if it were a telescope.

[3] In fact, 1/4″-20 is perhaps the only English/American thread that has become internationally standard; it fits the tripod socket on almost all cameras. There is no metric equivalent, but M6×1.25 is close.

Figure 7.10. Some lenses can be mounted in rings like a small refractor. With this lens, the front ring screws cannot be tightened until after focusing. Take care not to de-center the front of the lens.

Figure 7.11. Homemade tripod collar for Nikon 180/2.8 AF lens consists of wooden block with large round hole, slot and bolt to allow tightening, and a 1/4″-20 threaded insert on the underside.

7.5 Testing a Lens

7.5.1 How to Test

Every lens, especially one bought secondhand, should be tested as soon as you get it. The stars make excellent test targets. What you need is a well-guided (or short) exposure of a rich star field. I often use the field of α Persei or the Pleiades.

Exposures need not be long, nor do you need a dark country sky. Take several 1/2- to 30-second exposures of the star field, both in focus and slightly out of focus. The shorter exposures serve to distinguish lens problems from tracking problems. To take tracking completely out of the picture, use Polaris or take very short exposures (1/2 second or less).

7.5.2 Limitations of the Lens Design

There are two things to test for, limitations of the lens design and malfunctions or defects of a particular lens. Let's tackle the first one first. On the test exposures, look for:

- Excessive degradation of star images away from the center;
- Color fringing of star images at the center of the field, varying with focus;
- Excessive vignetting (darkening at the edges and corners);
- Internal reflections (showing up as large haloes or arcs).

The quality of star images at the corners is what distinguishes great lenses from merely good ones (Figure 7.13); none are perfect, but some are better than others. Star images at the corners are often poor at the widest aperture, but improve greatly at $f/4$ and smaller.

Poor star images at the edges and corners can also be caused by telecentricity issues (Section 7.2.3), which are not affected by aperture. If there are "comet tails" on stars around the edge of the picture, and they persist at $f/5.6$ or $f/8$ as well as wide open, then what you are seeing is fringing within the sensor. With film or with a different sensor, the same lens may perform much better.

Color fringing from residual chromatic aberration is often noticeable even with relatively good lenses; in long lenses from before the ED-glass era, it can be severe. With modern lenses, the fringes are reddish on one side of focus and greenish on the other. My experience has been that the sharpest stars usually result from tolerating a bit of red fringing. Older lenses produced large amounts of blue or purple fringing that could be reduced with pale yellow filters.

Vignetting is inherent in the lens design, but diminishes greatly as you stop down, and can be corrected with flat-fielding. Some vignetting can be a good sign – that a lens is well protected against internal reflections.

Internal reflections can limit what you can photograph. All lenses have them to some extent. Two astronomical targets that particularly provoke internal reflections are the Pleiades star cluster (with very bright stars amid faint

Figure 7.12. The Orion Nebula (M42). Stack of twenty 30-second exposures with a Canon 300-mm $f/4$ lens and Canon 60Da camera at ISO 3200, on a Celestron AVX mount. Because of the short exposures and PEC, no guiding corrections were needed.

Figure 7.13. This is not field rotation from bad tracking – it's a lens aberration that distorts stars away from the center of the field. It goes away at smaller apertures. (Film image with Olympus 40-mm $f/2$ lens, from *Astrophotography for the Amateur.*)

nebulosity) and total solar eclipses (with the bright inner corona, surrounded by much fainter outer detail).

7.5.3 Defects of a Particular Lens

In the category of defects and malfunctions, first let's start with something common but easy. Whenever you use a lens that has not been used for several years – even if it's one of your own – check for a sticky diaphragm. Set it to its smallest f-stop and move the aperture lever (on the lens mount) back and forth gently to see if the diaphragm opens and closes correctly. If it's sticky, sometimes all you need to do is keep doing the same motion repeatedly, perhaps 500 times, to redistribute lubricants. More often, it needs to go to a repair shop for cleaning and lubrication.

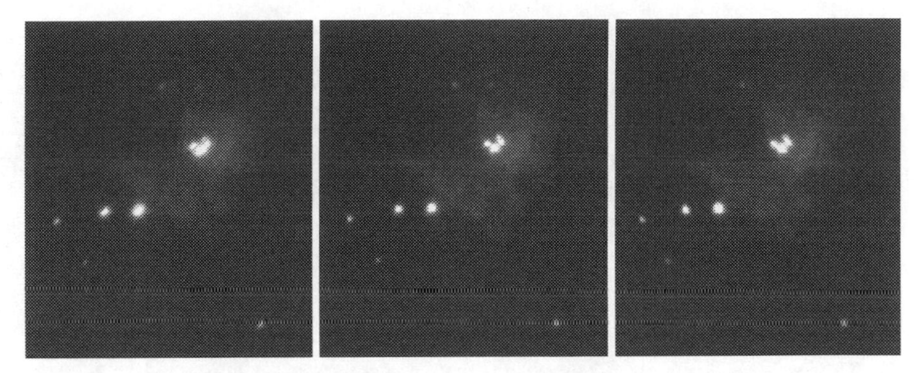

Figure 7.14. Astigmatism causes star images to be elongated in one direction on one side of focus and in the perpendicular direction on the other side of focus; it is easily mistaken for poor tracking. Many good lenses have slight but detectable astigmatism.

More worrying are indications that a lens element is off-center or tilted. Symptoms include out-of-round star images at the center of the field, or one side of the picture being in focus while the other side is not. To rule out trivial causes, if you see either of these problems, first take the lens off the camera, give it a bit of a shake, run it through its focusing range, and put it back on. Sometimes these problems disappear as mysteriously as they came, and sometimes they are a sign that a lens needs lubricating.

Even in good telephoto lenses, a 24-megapixel DSLR can often detect a trace of astigmatism (Figure 7.14). As long as the star images are round at the point of best focus, it doesn't matter that they're a bit elongated on either side of correct focus. What you don't want is misshapen stars like those in Figure 7.15 that cannot be focused into their proper shape.

7.6 Diffraction Spikes around the Stars

Bright stars stand out better in a picture if they are surrounded by diffraction spikes. Normally, a high-quality lens, wide open, will not produce diffraction spikes because its aperture is circular. If you close it down one stop, you may be rewarded with a dramatic pattern. Figure 7.16 shows an example, from a Canon lens whose aperture is, roughly speaking, an octagon with curved sides.

Another way to get diffraction spikes is to add crosshairs in front of the lens (Figure 7.17). The crosshairs can be made of wire or thread; they should be opaque, thin, and straight. Figure 7.18 shows the result. A piece of window screen wire mesh, serving as multiple crosshairs, would produce a similar but much stronger effect.

Whatever their origin, diffraction spikes are a stiff test of focusing accuracy. The sharper the focus, the more brightly they show up. For that reason, many people find crosshairs or window screen wire mesh useful as a focusing aid even if they don't use them for the actual picture.

Figure 7.15. Asymmetrical star images (enlarged) from lens with decentered element. At center of field, even out-of-focus stars should be round.

Figure 7.16. Dramatic diffraction patterns from the diaphragm of a Canon 300-mm $f/4$ EF L (non-IS) lens at $f/5.6$.

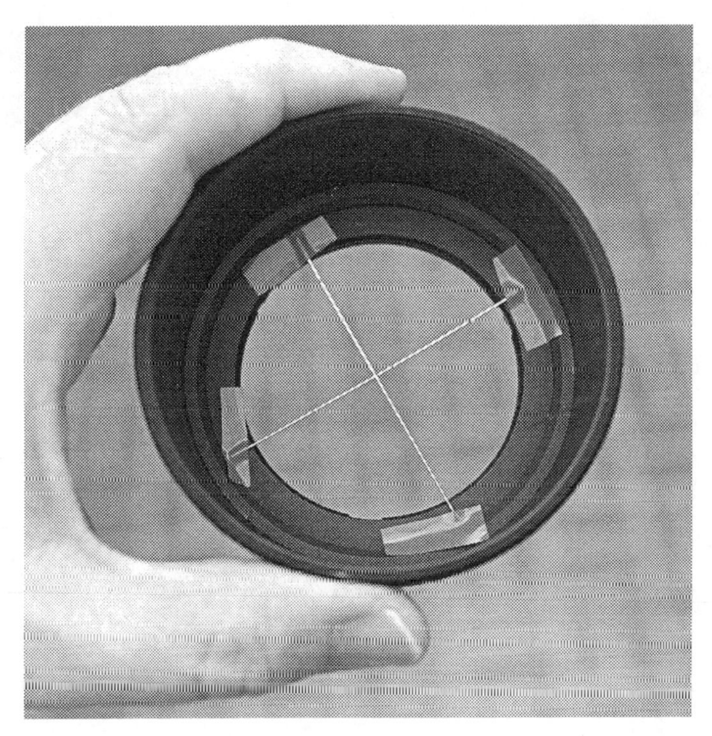

Figure 7.17. Homemade wire crosshairs mounted in the lens hood of a Sigma 105-mm $f/2.8$ lens.

If you want to get rid of diffraction spikes from the lens diaphragm, there is a way, provided the lens is relatively long (300 mm or longer). Stop it down, not with its own diaphragm, but with a smaller-diameter filter adapter placed in front of the lens. For example, if you want to use a 300-mm lens at $f/6$, install a filter adapter whose round opening is 50 mm in diameter. Unlike the diaphragm, this opening is circular and won't produce diffraction spikes. On a shorter lens, it might well produce vignetting.

7.7 Understanding Lens Design

7.7.1 How Lens Designs Evolve

Most of us have seen dozens of diagrams of lens designs without any explanation of how to understand them. Figure 7.19 is an attempt to make lens diagrams comprehensible. It shows how nearly all modern lenses belong to four groups.

Looking at the historical development, you can see that the usual way to improve a design is to split an element into two or more elements. Thus the 3-element Cooke Triplet gradually develops into the 11-element Sigma 105-mm f/2.8 DG EX Macro, one of the sharpest lenses I've ever owned.

Figure 7.18. Result of using crosshairs in Figure 7.17. North America Nebula region, with diffraction crosses on the brightest stars. Stack of ten 30-second exposures with a Canon 60Da at ISO 3200 and a Sigma 105-mm $f/2.8$ lens on an iOptron SkyTracker. No guiding corrections were necessary (or possible).

The reason for adding elements is that every lens design is, fundamentally, an approximate solution to a system of equations describing the paths of rays of light. By adding elements, the designer gains more degrees of freedom (independent variables) for solving the equations.

But there are also other ways to add degrees of freedom. One is to use more kinds of glass, and many new types have recently become available. Another is to use aspheric surfaces, which are much easier to manufacture than even 20 years ago.

Perhaps the most important recent advance is that anti-reflection coatings have improved. It is now quite practical to build a lens with 20 or 30 air-to-glass surfaces, which in the past would have led to intolerable flare and reflections.

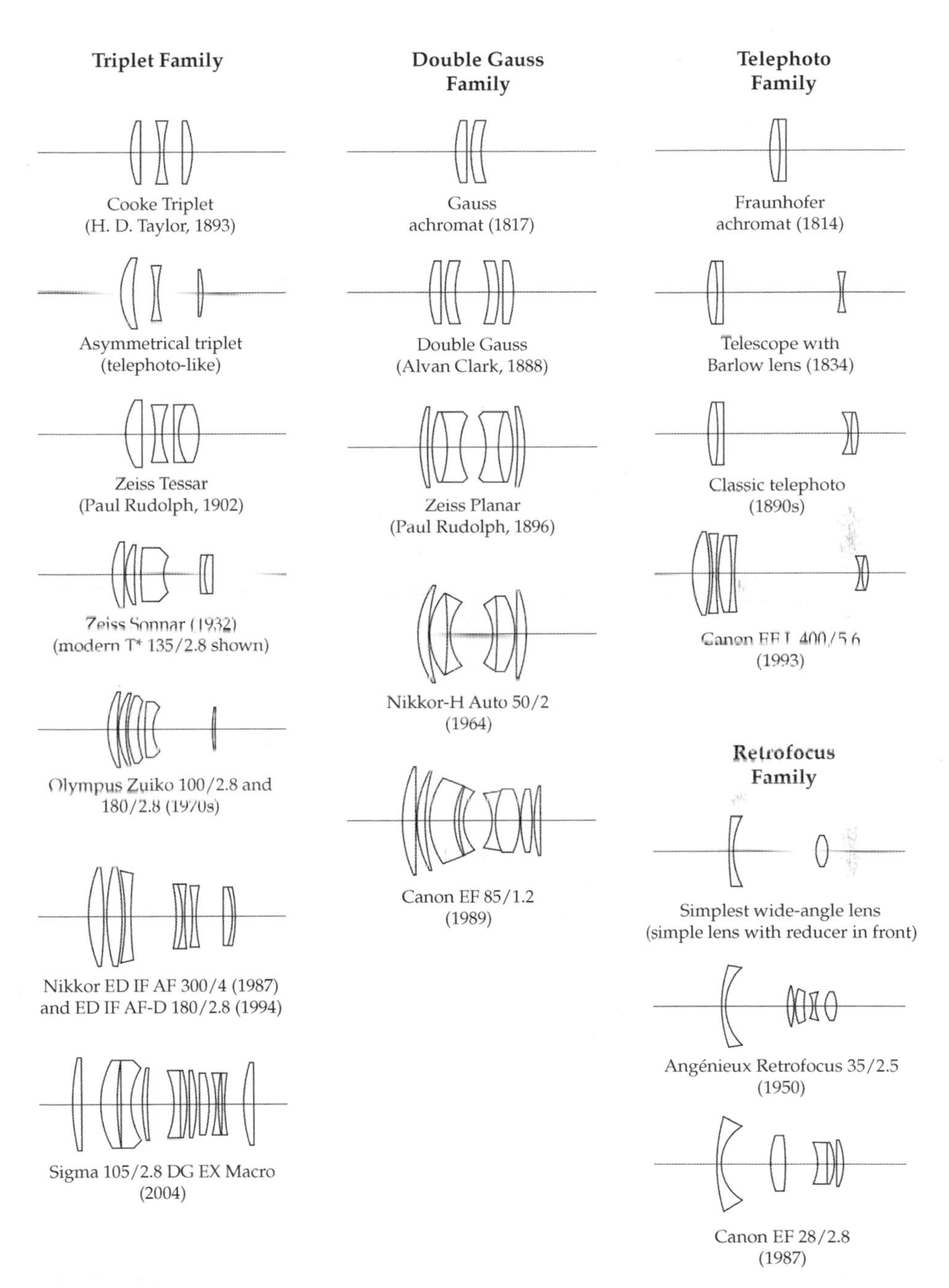

Figure 7.19. Lineage of many familiar lens designs. In all these diagrams, the object being photographed is to the left and sensor or film is to the right.

There have also been changes in the way designs are computed. Fifty years ago, designers used a whole arsenal of mathematical shortcuts to minimize aberrations one by one. Today, it's easy to compute the MTF curve (not just the individual aberrations) for any lens design, and by dealing with MTF, the designer can sometimes let one aberration work against another.

Manufacturers usually describe a lens as having "*X* elements in *Y* groups," where a group is a set of elements cemented together. How many elements are enough? It depends on the *f*-ratio and the physical size of the lens. Faster lenses need more elements because they have a wider range of light paths to keep under control. Larger lenses need more elements because aberrations grow bigger along with everything else.

Zoom lenses (Figure 7.20) need numerous elements to maintain a sharp image while zooming. In essence, the designer must design dozens of lenses, not just one, and furthermore, all of them must differ from each other only in the spacings between the groups! That is why zooms are more expensive and give poorer optical performance than competing fixed-focal-length lenses.

7.7.2 The Triplet and its Descendants

Multi-element lens design began in 1729 when Chester Moore Hall discovered that convex and concave lenses made of different kinds of glass could neutralize each other's chromatic aberration without neutralizing each other's refractive power. The same discovery was made independently by John Dollond in 1758. Later, the mathematicians Fraunhofer and Gauss perfected the idea, and telescope objectives of the types they invented are still widely used.

Early photography led to a demand for lenses that would cover a wider field of view, and in 1893 H. Dennis Taylor designed a high-performance triplet for Cooke and Sons Ltd. (www.cookeoptics.com).

Modern medium-telephoto lenses are often descendants of the Cooke Triplet. The key parts of the design are a convex lens or set of lenses in the front, a set of concave lenses in the middle, and a convex lens at the rear. A particularly important triplet derivative is the Zeiss Sonnar, whose second element is very thick and may consist of multiple elements cemented together to minimize the number of air-to-glass surfaces. Lenses of this type have been made ever since; you can recognize them by their weight. The famous Nikon 105-mm *f*/2.5 telephoto is a Sonnar derivative with a thick element.

Around 1972, Yoshihisa Maitani, of Olympus, designed a very compact, lightweight 100-mm *f*/2.8 lens which is a Sonnar derivative with the thick central element split into two thinner ones. (Its design alludes to the Sonnar's lighter-weight predecessor, the Ernemann Ernostar, which was the world's first *f*/2 camera lens.) Other camera manufacturers quickly brought similar designs to market, and even the Sonnar itself has shifted toward thinner, air-spaced elements. Over the years, Sonnar derivatives have become more and more complex.

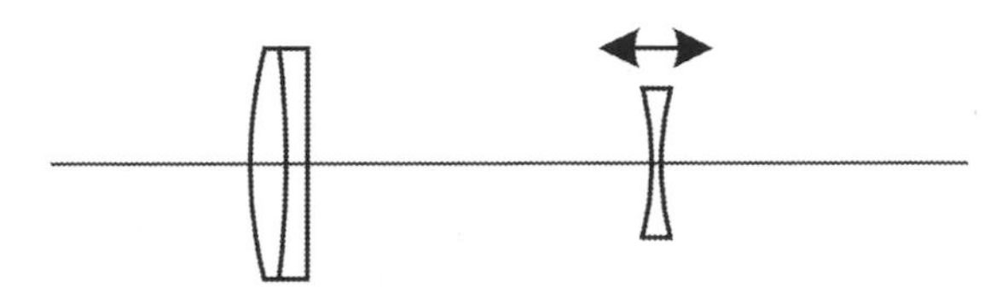

Zoom principle: moving Barlow lens
changes focal length of telescope

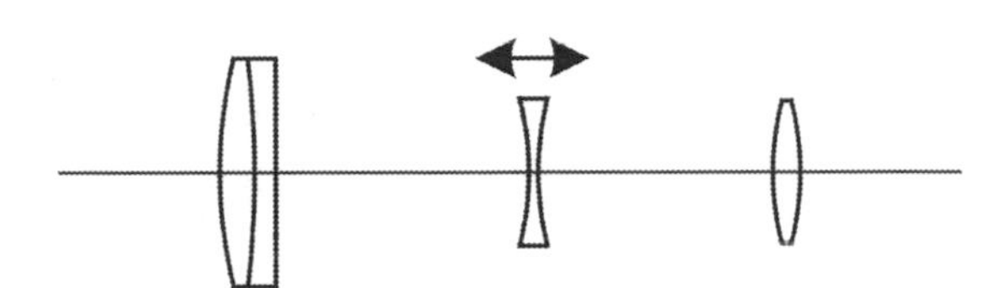

Adding positive element reduces
focus shift when zooming

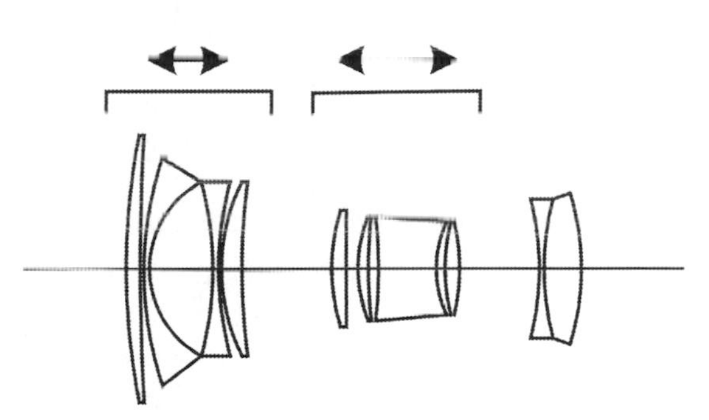

Canon EF 28-80/3.5-5.6 zoom lens
(2 sets of elements move separately)

Figure 7.20. A zoom lens is like a telescope with a variable Barlow lens. A simple, low-cost zoom is shown here; others have as many as 22 elements moving in 6 or 8 independent sets.

7.7.3 The Double Gauss

Present-day 50-mm "normal" lenses are almost always the double Gauss type (Figure 7.19, middle column), derived from the Zeiss Planar of 1892. Back in the 1800s, several experimenters discovered that if you put two Gauss achromats back-to-back, each cancels out the distortion introduced by the other. The same thing was tried with meniscus lenses and other kinds of achromats, resulting in various "rectilinear" lens designs.

Double Gauss designs were not practical until the advent of anti-reflection lens coatings in the 1950s; before then, their sharply curved air-to-glass surfaces led to internal reflections. Previously, the standard camera lens had usually been a variant of the Zeiss Tessar, which is classified as a triplet derivative but was actually invented as a simplified Planar.

7.7.4 Telephoto and Retrofocus Lenses

Very long telephoto lenses often work like a telescope with a Barlow lens (Figure 7.19, upper right). Technically speaking, *telephoto* means a lens whose focal length is much longer than its physical length, and the classic achromat-with-Barlow design is the standard way of achieving this, although asymmetrical triplets and asymmetrical double Gauss designs can do the same thing to a lesser degree.

The opposite of a telephoto is a *retrofocus* wide-angle lens, one whose lens-to-film distance is longer than its focal length. To leave room for the mirror, the lens of an SLR can be no closer than about 50 mm from the sensor. To get an effective focal length of, say, 28 mm, the wide-angle lens has to work like a backward telescope; it is a conventional lens with a large concave element in front of it.

7.8 Special Lenses

7.8.1 Macro Lenses

A macro lens is one whose aberrations are corrected for photographing small objects near the lens rather than distant objects. This does not imply a different lens design; it's a subtle matter of adjusting the curvatures and separations of elements.

Since the 1970s, many of the best macro lenses have had *floating elements*, elements whose separation changes as you focus. As a result, they are very sharp at infinity as well as close up. I have had excellent results photographing star fields with Sigma 90-mm and 105-mm macro lenses.

At the other end of the scale, cheap zoom lenses often claim to be "macro" if they will focus on anything less than an arm's length away, even if they don't perform especially well at any distance or focal length.

7.8.2 Mirror Lenses

Starting with the Reflex-Nikkor in 1960 and some Russian lenses that attracted attention around the same time, there have been lightweight telephoto lenses that are tiny Maksutov–Cassegrain telescopes (Figure 6.1) with additional glass elements. Their aperture is not adjustable, and they are surprisingly inexpensive.

My astrophotographic results with mirror lenses have been disappointing. Testing a low-end Spiratone 300-mm *f*/5.6 lens back in the 1980s, I was pleased by the lack of chromatic aberration, but there were severe reflections whenever a bright star was in the picture. Since then, mirror lenses have declined in popularity. Daytime photographers don't like them because out-of-focus images of bright spots look like doughnuts rather than disks or blobs.

7.8.3 Image Stabilization (Vibration Reduction)

Canon "IS" and Nikon "VR" lenses offer image stabilization (vibration reduction), which means that vibration of the lens while you are holding it is detected and corrected by moving a lens element automatically in real-time. Other cameras stabilize the image by moving the sensor.

This is a boon for daytime photography, but it is no help in astrophotography; we're already holding the camera very steady. Although most astrophotographers do not have problems, in theory the movable element can introduce trouble by shifting slightly during the exposure. This happened with the Dragonfly Project (p. 108), and the lenses were modified by Canon to immobilize the movable element. Others have attributed slight miscollimation to the fact that the movable element may not stay perfectly centered. I prefer to use non-IS lenses, but many of the newest and best lenses are made only with IS (VR).

If you use an image-stabilizing lens or camera for astrophotography, make sure the image stabilization is turned off. Otherwise it will fight the tracking motion, producing elongated star images because it doesn't know the camera is supposed to be moving.

7.8.4 Diffractive Optics

Both Canon and Nikon have introduced long telephoto lenses using diffractive optics (DO); Nikon calls them Phase Fresnel (PF) lenses. They work as well as conventional lenses but are smaller and lighter, and I have seen good astrophotography done with them, although there seems to be more scattered light around bright stars than with conventional optics – which actually helps make the stars stand out. The first version of Canon's 400-mm $f/4$ DO lens left people unimpressed, but the second ("II") version is superb; unfortunately, it costs as much as a large telescope.

How DO works is widely misunderstood. Despite diagrams that suggest so, a DO element is not a Fresnel lens, understood as an ordinary lens that has been flattened by dividing it up into ring-shaped zones and making each zone no thicker than it needs to be (Figure 7.21). That is not a diffractive element at all; it gives poor image quality and is not used in cameras but is a familiar sight in theatrical spotlights and lighthouses.

Diffractive optics are derived from a different invention of the same Augustin-Jean Fresnel (1788–1827, pronounced *freh-NELL*), the Fresnel zone plate, which is essentially a diffraction grating that forms an image.

Recall that a diffraction grating bends light according to its wavelength by passing it across narrow parallel stripes that affect its wave structure. A Fresnel zone plate (Figure 7.22) is a circular diffraction grating that brings light to a

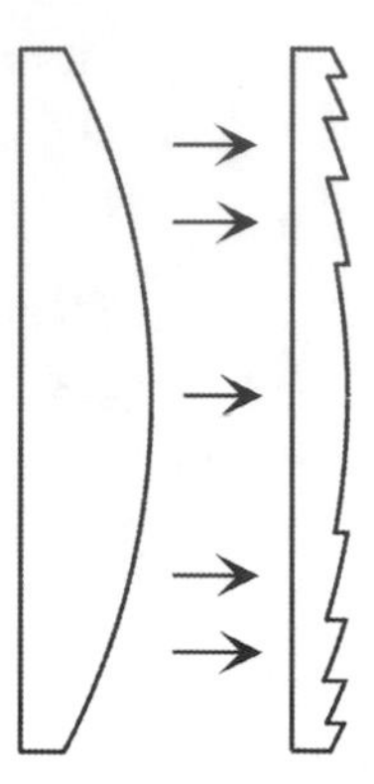

Figure 7.21. Not diffractive optics: This kind of Fresnel lens is just a lens collapsed into concentric zones to save glass, used in theater spotlights and lighthouses, not image-forming systems. Zones can be any diameter.

Figure 7.22. Diffractive optics: Fresnel zone plate is a diffraction grating that forms an image. Black zones can be replaced by clear zones with slight extra thickness to add phase shift. Diameter of zones follows a square-root formula and continues for thousands of zones in a real lens.

focus. Although it looks like a Fresnel lens, the stripes are spaced according to a precise formula and are very narrow, on the order of 0.001 mm.

A classic Fresnel zone plate is 50% black; the black zones block light that would diffract into the wrong places and destroy the image. That wastes too much light for use in a camera lens.

Another way to deal with those zones is not to make them opaque, but to make the glass slightly thicker in those zones so that the light undergoes a phase shift and then diffracts as desired. That was the insight of Kenro Miyamoto (*Journal of the Optical Society of America* **51**:17–20, 1961). The added thickness is achieved by depositing a thin film. In modern practice, the film is deposited on all the zones, not just the ones that would have been black, and slopes from the inner to the outer edge of each zone because the outer part of each zone needs slightly less phase shift. Greatly magnified, the construction ends up looking rather like Figure 7.21 but works entirely on diffractive principles.

The chromatic aberration of diffractive optics is opposite that of glass. Glass bends shorter wavelengths more, and longer wavelengths less; diffraction is the other way around. Accordingly, the main use of diffractive optics in camera lenses so far is to counteract chromatic aberration rather than to add diffractive power.

Chapter 8
Tracking the Stars

To take exposures longer than a few seconds, you must track the stars. That is, the telescope must compensate for the earth's rotation so that the image stays in the same place on the sensor while the earth turns.

Currently, tracking and guiding is an area of rapid technical progress, particularly as regards inexpensive equipment. This year's mounts are better than last year's, and next year's will be better yet. Accordingly, this book is not the place to give a complete guide to star tracking, and if I tried, it would not stay up-to-date long. In this chapter and the next, I'll hit the high points. For fundamentals, see also *Astrophotography for the Amateur* and *How to Use a Computerized Telescope.*

8.1 Two Ways to Track the Stars

Figure 8.1 shows the two major kinds of telescope mounts, altazimuth and equatorial. Until the 1980s, only an equatorial mount could track the stars; it does so with a single motor that rotates the telescope around the polar axis, which is parallel with the axis of the earth. In order to use an equatorial mount, you have to make the polar axis point in the right direction, a process known as *polar alignment* and long considered somewhat mysterious, although actually, correct polar alignment is not hard or time-consuming.

It is important to understand why an equatorial mount works. Computers can easily give you the impression that the sky is a tangle of whirling motions that no mere mortal can understand. In fact, however, the apparent motion of the sky is very simple. Except for lunar and planetary orbits, *the whole sky moves together,* due to the rotation of the earth, and the whole sky seems to rotate around a single axis. Everything twirls around the celestial pole. To track the sky *perfectly,* all you have to do is rotate your mount at the correct speed around an axis that points exactly at the celestial pole. We use computer-controlled drives today, but in the 1800s, astronomers tracked the stars with clockwork and

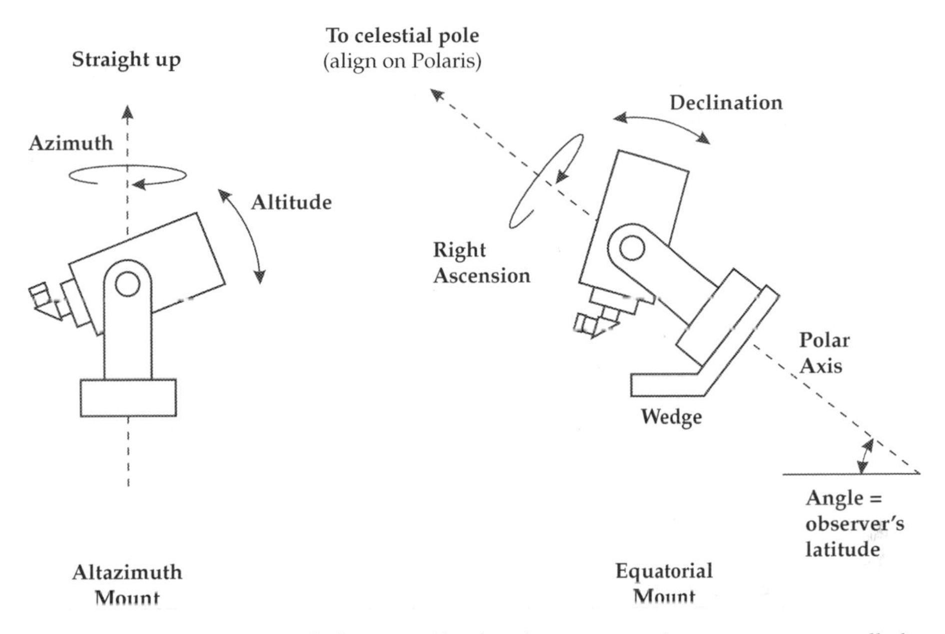

Figure 8.1. Two ways to track the stars. Altazimuth mount requires computer-controlled motors on both axes; equatorial needs only one motor, no computer. (From *How to Use a Computerized Telescope.*)

weights, and in the pre-telescopic era, Tycho Brahe did it by manually turning his great equatorial armillary, a position-measuring device.

Altazimuth mounts were not able to track the stars until computers were added. The computer computes how fast and in which direction to move each axis, depending on where in the sky the telescope is pointed. During setup, the computer has to be told the exact positions of at least two stars so that it can get oriented.

Altazimuth mounts can track the stars, but they can't stop the field from rotating (Figure 8.2) because they don't rotate the same way the sky is rotating. With an altazimuth mount, "up" and "down" are not the same celestial direction throughout a long exposure. As a result, the image twists around the guide star.

Some large altazimuth telescopes include field de-rotators, which are motors that rotate the camera to keep up with the image. Field de-rotation made its debut on the Soviet 6-meter telescope (the BTA-6) in 1975. It is cost-effective only with relatively large telescopes; otherwise the de-rotator costs more than an equatorial mount would.

More commonly, especially as a beginning experimenter, if you're tracking with an altazimuth mount, you simply take short exposures and combine them with software that can rotate images as it stacks them (as almost all stacking software does). Exposures as long as 2 minutes are possible low in the east and west.

Altazimuth mounts are at a distinct disadvantage because two motors are running, not just one. The smoothness of tracking suffers. Successful astrophotography has been done in altazimuth mode, but I consider it a

TRACKING WITH ALTAZIMUTH MOUNT
Image rotates; long exposure photographs are not possible

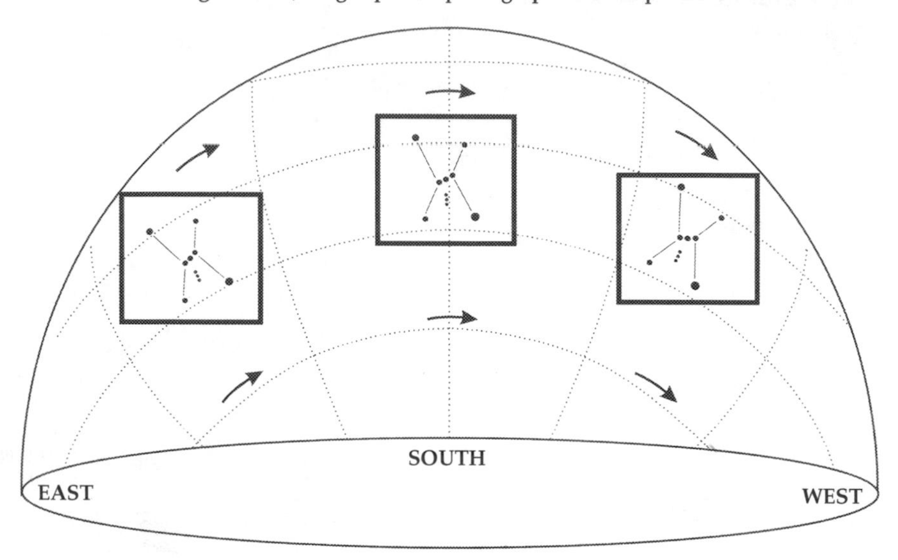

TRACKING WITH EQUATORIAL MOUNT
Telescope rotates with image; long exposures work as intended

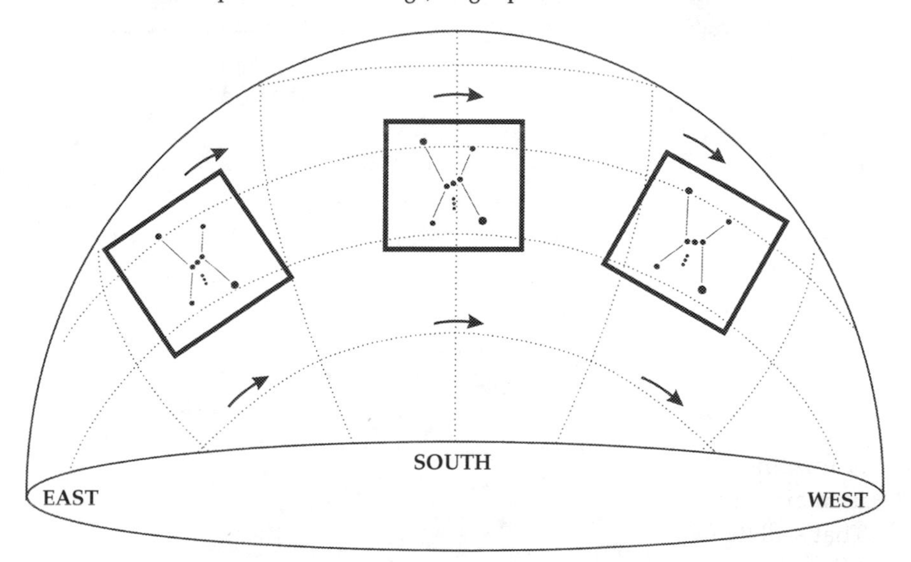

Figure 8.2. Tracking the stars with an altazimuth mount causes field rotation, which can be overcome by taking very short exposures and doing a rotate-and-stack. (From *How to Use a Computerized Telescope*.)

specialized, somewhat unreliable technique. Nonetheless, if you have an altazimuth-mounted telescope, see what you can do with it before moving on.

The rest of this chapter and the next will focus on equatorial mounts, which rely on only one motor turning at a constant speed. You can turn an altazimuth fork mount into an equatorial by putting it on a wedge and making appropriate settings in its computer.

8.2 The Rules Have Changed

The rules of the tracking and guiding game are not what they used to be. In the film era, the telescope had to track perfectly for 20 or 30 minutes at a time. Guiding corrections had to be made constantly, either by an autoguider or by a human being constantly watching a star and pressing buttons to keep it centered on the crosshairs. One slip and the whole exposure was ruined.

It was also important to guard against flexure and mirror shift. During a half-hour exposure, the telescope and its mount could bend appreciably. Also, notoriously, the movable mirror of a Schmidt–Cassegrain telescope would shift slightly. For both of these reasons, guiding was usually done by sampling an off-axis portion of the image through the main telescope.

Today, we commonly take 30-second to 2-minute exposures and combine them digitally. That makes a big difference. Tracking errors that would be intolerable over half an hour are likely to be negligible. If there is a sudden jump, we can simply throw away one of the short exposures and combine the rest. Many mounts can track adequately during a single exposure without guiding corrections.

At the same time, though, the rules have also changed in the opposite way. DSLR pixels are so small that we can be *very* picky about the quality of the tracking in a picture. Measuring the roundness of star images has become a common practice. Accordingly, it is easier than ever to get into a situation where no mount and guider seem good enough. Chapter 9 addresses these challenges.

8.3 Types of Equatorial Mounts

8.3.1 Fork Mounts on Wedges

A fork mount becomes equatorial when you place it on a wedge that points its main axis toward the pole (and tell its computer that you have done so). That's all there is to it. Figure 8.3 shows my astrophotography setup from about a decade ago: a fork-mounted telescope on a permanent pier with various accessories mounted on it. The main telescope cannot be removed from the fork, but sometimes it is not used; the guidescope and a camera "ride piggyback."

Well-made wedges are sturdy, easy to adjust, and *heavy*. To save weight and to take advantage of the vibration-damping properties of wood, I built the wooden wedge shown in Figure 8.4 for portable work. Its obvious drawback is that there

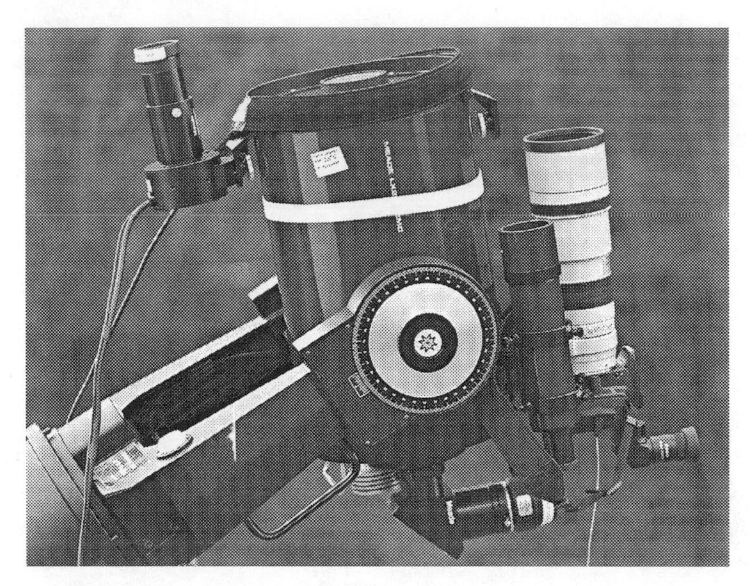

Figure 8.3. The author's DSLR astrophotography setup, circa 2006. Meade LX200 telescope with small homemade guidescope (leftmost), 8 × 50 finderscope, Canon 300-mm *f*/4 lens, and Canon XTi (400D) camera, mounted equatorially on a permanent pier in the back yard.

are no adjustments; the only way to align its axis with the pole is by moving the tripod legs or altering their length, and it only works at latitudes close to 34° north.

Would-be wedge builders should note that the mounting bolts of a Meade or Celestron telescope are not in an exact equilateral triangle. Measure their positions carefully before you drill the holes.

8.3.2 Sky Trackers

If you don't already have a telescope, or if you need maximum portability, consider using a tripod-mounted sky tracker such as the iOptron SkyTracker shown in Figure 8.5. This is a tiny equatorial mount (complete with polar alignment scope) that uses a worm drive with gears comparable in quality and size to those used in telescope mounts. It's good for loads up to about 2.5 kg (five or six pounds) and tracks accurately enough for 1-minute exposures with focal lengths up to 150 mm. I have done fine work with it (Figure 7.18).

For the experimenter on a budget, there is an even cheaper option: a barn-door tracker, which consists of two pieces of wood joined together by hinges whose axle is pointed at Polaris, powered by a human being turning a knob to keep it in sync with the second hand of a watch. For plans, see *Astrophotography for the Amateur* (1999), pp. 120–121.

Figure 8.4. The author's homemade wooden wedge, held together with large screws whose heads were countersunk and filled with putty before painting. Polar alignment is done by moving the tripod legs.

8.3.3 German Equatorial Mounts (GEMs)

The best-performing mounts these days are German equatorial mounts (GEMs), so called because the German instrument maker Joseph von Fraunhofer built the first one (installed in 1824 in Dorpat, now Tartu, Estonia). Unlike a fork, a single GEM can easily accommodate many different telescopes at different times. You can even mount a big camera lens on it without a telescope. But a GEM is considerably more complicated than a fork mount.

Anatomy of a GEM

Figure 8.6 shows a GEM and gives the names of the parts. Setting it up involves making the right ascension axis point toward the pole, and there are altitude and azimuth adjustments for the purpose. Here are some practical hints for setting up a GEM:

- Transport the mount head with the brakes unlocked, if possible, so that nothing will apply stress to it.
- If the mount head is heavy, consider using a cart to carry it.
- When lifting the mount head, grip it by the two ends of the declination axis housing, under the saddle and near where the counterweight bar emerges. These do not move relative to each other. If you grip the declination and R.A. housings, you may drop the mount as it swivels around the R.A. axis.

Figure 8.5. iOptron SkyTracker is a portable battery-powered equatorial mount that carries just a camera. Built-in polar scope enables good polar alignment.

- Always put the counterweights on the mount before the telescope, and put them farther down the bar than you think they need to be. You want the mount to be counterweight-heavy until you balance it, so that the telescope won't swing down suddenly if a brake is loosened.

There are two ways to put a telescope into a dovetail saddle; you can either slide it in from the end or lay it in from the side. I usually do the latter so that neither of the safety screws has to be removed. (They prevent the telescope from sliding out of a slightly loosened saddle.) In order to balance properly, one of my telescopes needed to sit farther forward in the saddle than its rear safety screw would allow, so rather than do without the safety screw, I built an extender to place it farther back (Figure 8.7).

A notorious peculiarity of the GEM is the need for "meridian flip" (Figure 8.8). Because the telescope is beside the right ascension axis rather than concentric

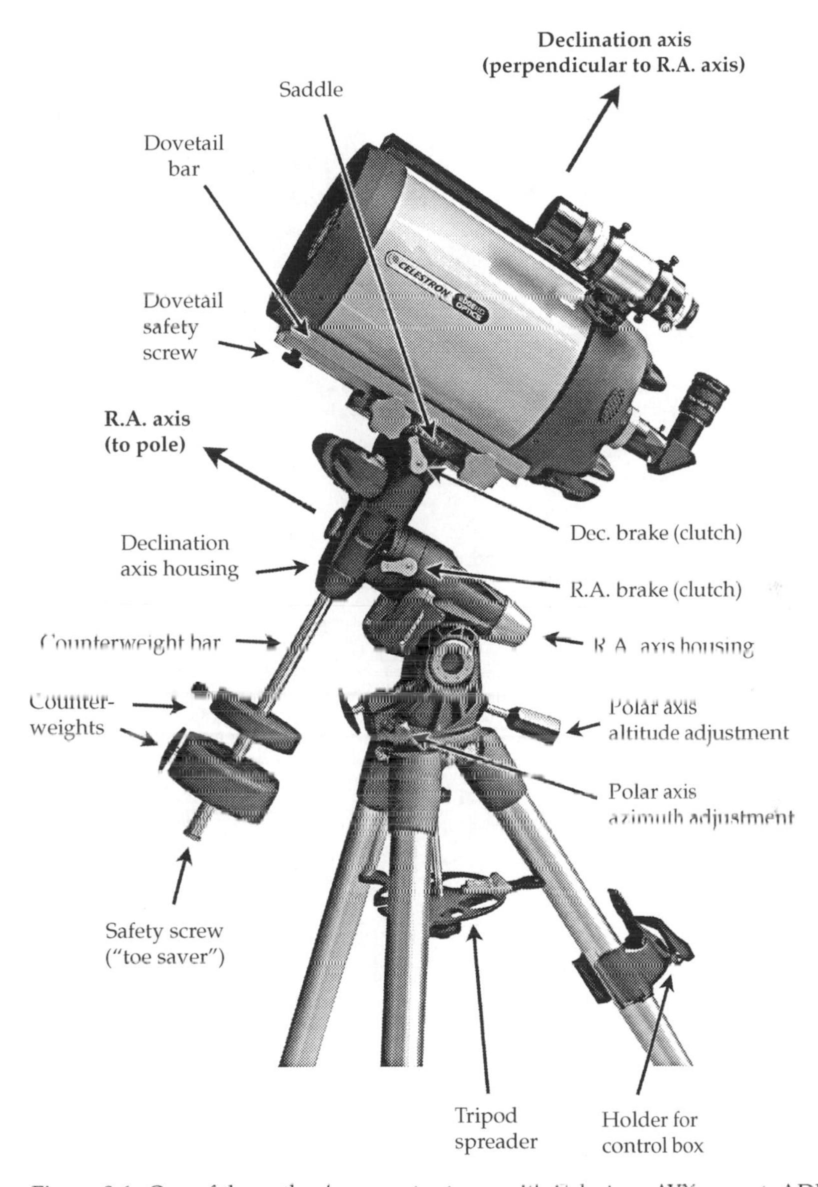

Figure 8.6. One of the author's present setups, with Celestron AVX mount, ADM saddle, and Celestron and iOptron counterweights. Cables and control box omitted for clarity. This is a visual configuration; for guided photography, this telescope requires a larger mount.

with it, it cannot swing around to cover the entire sky without bumping into the tripod or pier. In particular, it cannot track a celestial object across the meridian for more than a short distance (maybe as much as an hour's worth of terrestrial rotation, but not more). The solution is to "flip" the telescope by aiming it at the pole, rotating 180° eastward in right ascension, and then aiming it at the object again. When this is done, pictures will be upside down relative to the way they were before the flip. Computerized mounts usually do not perform meridian

Figure 8.7. Improvised extender, made from aluminum bar, repositions rear safety screw so the telescope can slide farther forward.

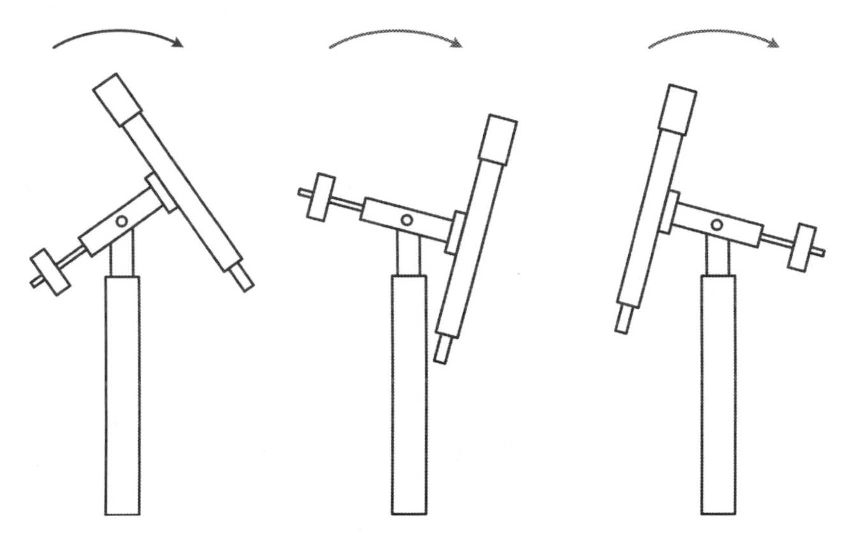

Figure 8.8. Meridian flip. Left: Starting to track a celestial object. Middle: Shortly after crossing the meridian, telescope cannot track further. Right: Tracking continues after telescope is flipped.

flip automatically while tracking, but they do perform it as needed to go to the objects you specify.

8.4 Hardware

8.4.1 Dovetails

Part of the appeal of GEMs is that you can attach almost any combination of equipment easily and sturdily to the mount using dovetail bars and related hardware (Figure 8.9). Hardware of this type is also used to piggyback cameras and guidescopes onto fork-mounted telescopes. The premier maker of

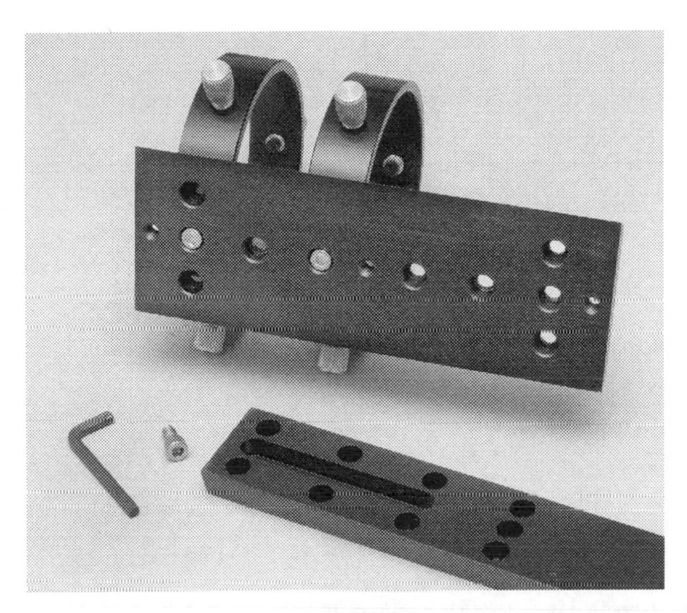

Figure 8.9. Rings and other accessories are attached to dovetail bars with socket-head 1/4″-20 screws whose heads are sunk in holes or slots on the underside.

dovetail hardware is ADM Accessories (www.admaccessories.com), but there are many others.

There are two main sizes of dovetail bars, type D (Losmandy) and type V (Vixen). Type D is larger, 75 mm wide at the base, and the sides slope 60° from the base (that is, 30° from perpendicular). Type V is 44 mm wide at the base and slopes 75° (15° from perpendicular). Widths are not critical, since all dovetail clamps and saddles are adjustable. The thickness of the bar is highly variable, but is typically 5 to 10 mm for Type D and 10 to 15 mm for type V. Type D dovetail bars often have a larger plate on top, which rests on the top of the saddle; type V dovetail bars usually lack this plate and rest on the bottom of the saddle, although both sizes are made both ways.

Other sizes exist. ADM promotes a "mini dovetail system" (MDS), 38 mm wide, 60° slope. The Arca-Swiss dovetail used on camera tripods is 37 mm wide, 45° slope. Smaller dovetails are used to attach finders to telescopes.

Unfortunately, some dovetail saddles make visible marks on the dovetail bar the first time it is put into them, as they secure it with screws (Figure 8.10). This is not serious damage; you can expect to see these marks on telescopes that have been set up in the showroom even if they have not actually been used. A more sophisticated type of saddle avoids the problem by holding the dovetail in a clamp.

One last hint. Sometimes the rings do not hold a guidescope or small refractor high enough above the dovetail bar; the focus knobs may collide with the saddle. If that happens, you can make a spacer out of a flat aluminum bar with two holes drilled in the right places, and install it between the dovetail bar and the rings. That is much sturdier than using washers. A more expensive solution is

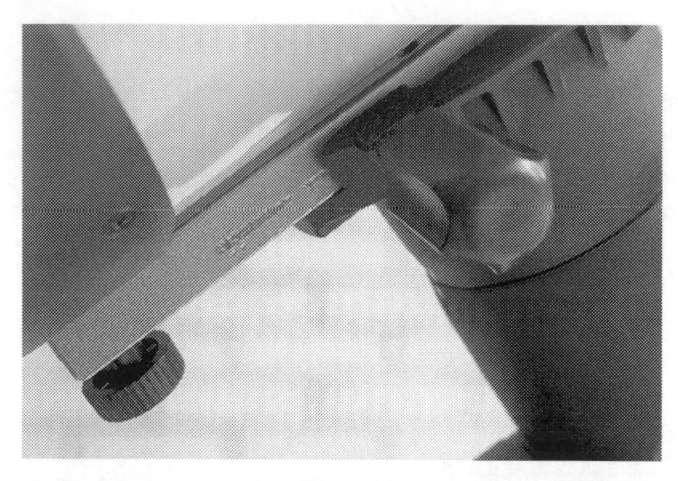

Figure 8.10. Some saddles make marks on the dovetail bar every time they are used. This is not serious damage.

to stack a second dovetail bar, upside down, on top of the first one. Still another way to raise a telescope away from the dovetail bar is to position it off center in its rings, with the top screw backed out and the lower screws farther in.

8.4.2 Counterweights

The job of a counterweight is to be heavy enough, stay in place, and not bump into anything as the telescope moves. Be sure to position counterweight locking screws so they extend in the same direction as the front of the telescope, so they will not protrude toward the mount and bump into it.

Several popular mounts have 20-mm shafts and can take each other's counterweights. This is fortunate, because Celestron does not make a counterweight light enough to balance my AVX mount when it is carrying only a telephoto lens; an iOptron 2-kg counterweight does the job. (Celestron shafts are not 3/4 inch, as some vendors' catalogs indicate; that would be only 19 mm.) Larger mounts from several manufacturers have 1 1/4-inch shafts (32 mm).

Counterweights are expensive to ship cross-country because of their weight. Unfortunately, nobody presently makes a mount that takes standard barbell weights, abundant in every sporting-goods store.

8.5 Setting up a Computerized Equatorial Mount

All equatorial mounts come with detailed instructions, but what I want to do here is address some underlying principles. This section deals with computerized "go-to" mounts;[1] the next, with classic techniques that apply to any mount.

[1] I spell it *go-to*, to avoid confusion with Goto, a distinguished Japanese planetarium company (www.goto.co.jp).

8.5.1 The Difference Between Polar and Go-to Alignment

Polar alignment means making the axis of the mount physically parallel to that of the earth. It can only be done by moving the mount, not by entering information into a computer. Go-to alignment is a separate process that involves sighting stars and telling the mount's computer where they are.

For a visual observing session, it is sufficient to point the polar axis roughly toward the pole and then follow the computer's prompts to sight and center stars. When this is done, the telescope can "go to" objects very accurately and track them moderately well, but it does not track well enough for photography.

During go-to alignment, the telescope measures the error of its polar alignment. It corrects for this known error when finding objects, but not when tracking them. It tracks with just one motor, driving only the right ascension axis for smoothness' sake. However, since it has measured its polar alignment error, it can use that information to help you do a more accurate polar alignment – more about that shortly.

8.5.2 Don't Judge it by the First Star

A common mistake is to judge the quality of a go-to mount by how well it finds the first alignment star. In fact, however, when finding the first go-to alignment star, the mount is just guessing. It has to assume the polar alignment is perfect, the mount head is perfectly level, the date, time, latitude, and longitude are set perfectly, the mount started out exactly in the index position, the index position actually is where the encoders think it is, and the declination and right ascension axes are perfectly perpendicular (perfect orthogonality, no "cone error").

By the time you have shown it two stars, it is only assuming that the axes are perpendicular. After you show it two or three additional stars, it isn't assuming even that. It's relying on its own measurements instead.

The biggest unknown, when finding the first star, is usually the initial position of the telescope. Typically, when you set the mount to the index marks, you don't achieve arc-minute accuracy; it's easy to be a degree or two off. Then, moving a known distance from a guessed initial position, the mount fails to hit the star precisely. No problem; that's why you are going to center the star and tell the computer you have done so.

Some mounts try to learn the actual (slightly inaccurate) positions of their index marks so they can make a better guess next time. Reportedly, some versions of the firmware of some Celestron mounts have had an error that causes this process to malfunction, resulting in substantial right ascension error on the first star the next time the mount is set up. It was corrected in a firmware update.

At no point in this process of go-to alignment have you had to do a precise polar alignment. Your mount can find objects very accurately even if the axis is several degrees off the pole.

8.5.3 Must You Level the Tripod?

A common mistake is to be fanatical about leveling the tripod precisely. A moment's thought will tell you that tracking depends on the polar axis, and go-to accuracy depends on measured star positions, and neither one depends on the levelness of the structures supporting the mount.

When guessing the position of the first star, the mount assumes that the tripod is level, but that's all. As soon as you show it two stars, it drops all assumptions about the tripod. You can proceed to do a very precise polar alignment, and it still doesn't matter whether the tripod is level, although software assumes it is approximately so.

Leveling does matter, at least in an approximate way, when you refine the polar alignment with the aid of the mount's computer or with the drift method (both of which I'm about to describe), because these techniques assume that when you're told to move the mount in altitude or azimuth, that is what you will do, rather than moving at some cockeyed angle, and that, in turn, assumes the mount is level.

Having said that, let me add that I've done the mathematics, and the effect of having the mount one or two degrees away from level is almost imperceptible unless you're starting with a truly enormous polar alignment error. I do have a bubble level on my tripod, but I only check it casually. If your mount looks level, it is level enough.

8.5.4 Hints for Go-to Alignment

The go-to performance of the telescope will be no better than the accuracy with which the stars were centered during setup. Use an eyepiece with crosshairs, and be sure to approach all the stars the same way (with the "up" and "right" buttons on a Celestron hand controller – that implies that just before the final centering, you use the other two buttons to put the star off center). Don't change the orientation of your diagonal between one star and the next.

Star misidentification is the most common cause of alignment problems, and it happens to even the most experienced observer. Just before writing this chapter, I had an observing session that began strangely, with the mount reporting a 4-degree polar alignment error. It turned out I had misidentified Polaris! I saw a star in the polar scope and assumed it was the right one. But it was an exceptionally clear night, all the stars looked brighter than usual, and instead of second-magnitude Polaris, I was actually looking at fourth-magnitude Delta Ursae Minoris.

Think of the perils that await if the telescope slews to a star you don't recognize, says "Center Diphda," and then "Align Diphda" and you assume the star is Diphda without checking. Always look at the map. It can be handy to use binoculars to confirm what you're going to see in the finderscope.

8.5.5 Go-to Alignment with just a Telephoto Lens

You can perform go-to alignment without a telescope. If your mount is carrying nothing but a telephoto lens (about 150 mm or longer), use your DSLR itself, in Live View, to center stars. Magnify the star images as much as possible and center them in the screen, possibly putting a small mark on it (on clear tape) for the purpose. Some cameras show a dot in the middle of the magnified Live View display; it's a great help. Some Canons have the option to display a grid, and I get good results by centering a star exactly on the central grid lines, viewing the screen with a magnifying glass.

8.5.6 Using Go-to Alignment to Refine Polar Alignment

A mount that has been aligned for go-to has measured its polar alignment error. Many mounts can use this information to tell you how to refine the polar alignment.[2]

As an example, here is how Celestron mounts do it. The method is called All-Star Polar Alignment (ASPA). The first step is to set up the mount with its axis roughly pointing toward the pole, and do a full go-to alignment, preferably with four calibration stars as well as the initial two alignment stars.

Then go to a named star that is moderately high in the southern sky (the northern sky if you're in the southern hemisphere). Avoid stars near the pole or low in the east or west. (If you choose an unsuitable star, the adjustment will be hard to make because the two directions in which you move the mount will not be perpendicular; but if you make it successfully, the adjustment will still be accurate.)

Next, press Align, Display Align. You'll be shown the measured alignment error in degrees, minutes, and seconds.

Back up to Align, Align Mount, and step through the procedure. The telescope will again slew to the star you chose and ask you to center and align on it. Next, the mount will move some distance away from the star. By adjusting the polar axis altitude and azimuth, *without* slewing the mount, bring this star back into the center of the field. Then you're done.

If you Display Align again, the error will be shown as all zeroes, which is meaningless; the mount is just *assuming* you've eliminated all the error. To find out how well you actually did, you need to align all over again. Choose Menu, Utilities, Home Position, Go To, then turn the mount off, then turn it on and do the go-to alignment all over again.

If the initial error was large, you may find you need to go through the whole process again. Using Celestron ASPA on either of my two mounts, I get

[2] Early methods for doing this did not work well, and I did not recommend them in *How to Use a Computerized Telescope*. Technology has advanced.

reproducibile alignment accuracy of about 5′ (that is, about 1/12 degree), which is good enough for almost any purpose.

8.6 Classic Methods

8.6.1 Finding the Pole in the Sky

Now for methods that don't require computerized mounts, though they certainly apply to them. The obvious way to align a mount is to find the celestial pole in the sky, then point the polar axis toward it.

Here's how it's done. If your mount has a polar scope, use it. Otherwise, point your telescope exactly parallel to its polar axis and use its finderscope. Don't trust the index mark or the 90° mark on the declination circle; check its accuracy. (I was once frustrated by a telescope that was about 2° off without my knowing it.) Also make sure the finder is lined up with the main telescope.

Then use a star map to find the celestial pole. In the Northern Hemisphere, this means looking for Polaris, but Polaris is 2/3° away from the true pole. The chart in Figure 8.11 makes finding the true pole a snap. When the outer part of the chart is lined up with the constellations in the sky, the inner circle shows how Polaris should look in the finderscope or polar scope. For example, if Cassiopeia is directly above Polaris in the sky, then to center an inverting polar scope on the true pole, put Polaris on the ring, below and a little left of center.

Adjust the mount so that Polaris is right on the crosshairs, and then move it slightly so that Polaris is 2/3° away from the pole in the direction specified by the map. Then you're done – or rather, you're ready to refine the alignment by the drift method.

8.6.2 More about Polar Scopes

If your equatorial mount has a polar scope, you will probably find that you can only see through it when the saddle is turned sideways, to aim a telescope 90° away from the pole. Don't let this mystify you. Not all mounts have this quirk.

The polar scope must of course be aligned so that its reticle is centered on the true direction of the mount's right ascension axis. To do this, aim the polar scope at a distant object in the daytime, such as the side of a building or tree, and twirl the mount around its right ascension axis while you watch. The center of rotation – that is, the point on the building or tree around which the others seem to revolve – should be in the center of the reticle, on the crosshairs. If it is not, adjust the reticle to take up *half* the error, and try again. This process can be difficult but only has to be done once, and maybe checked once a year.

Instead of using Figure 8.11, you can get a very accurate picture of where to put Polaris on your polar scope reticle by using an iPhone app. Three that I recommend are *iOptron Polar Scope, Polar Scope Align,* and *Polar Scope Align Pro,* all in Apple's App Store. The middle one in the list is free, and the last

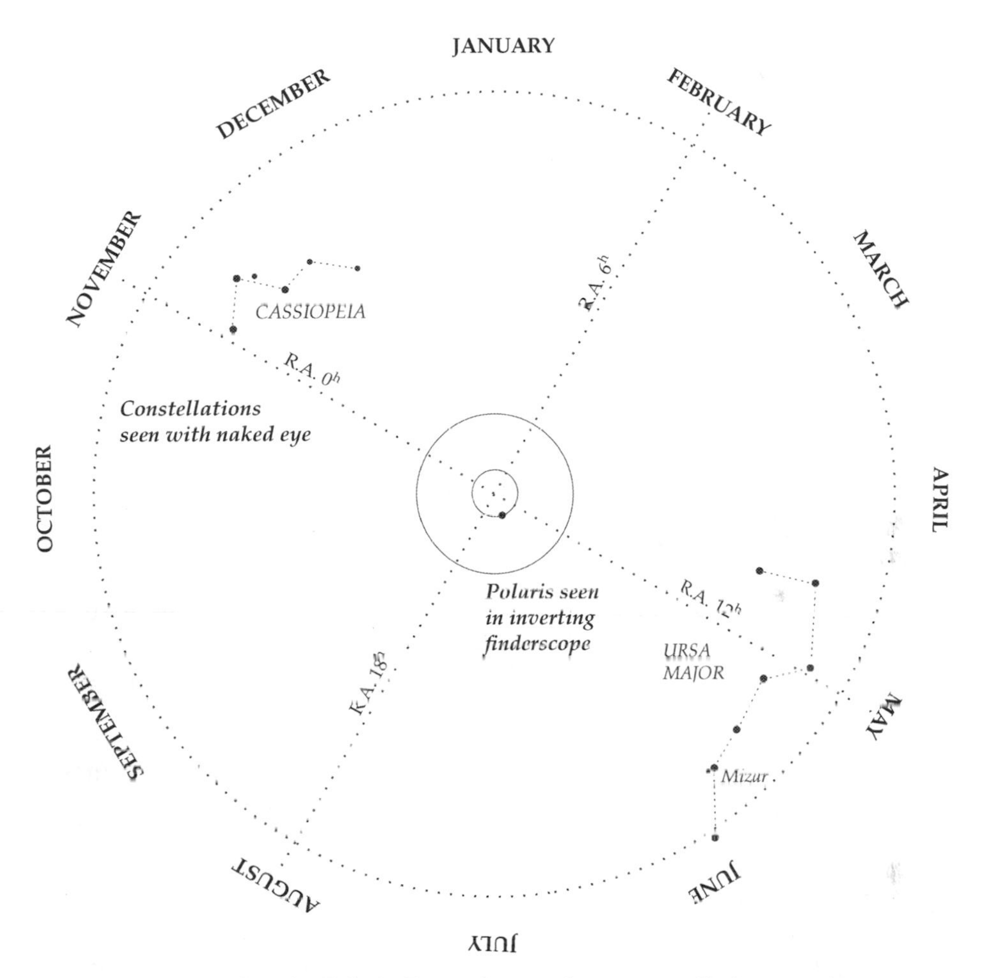

Figure 8.11. Finder chart for Polaris. Rotate chart until outer constellations match your naked-eye view of the sky; center of chart then shows view of Polaris through a finder or polar scope.

one is exceptionally full-featured – it gives you the times of sunset and twilight, weather data, and many other things, as well as accurate pictures of many different brands of polar scope reticles.

With my chart or a good app, I can generally align to within 1/3° with the polar scope. Others can do better. For the ultimate in accuracy, though, you can let a camera do the work. The QHY Polemaster (www.qhyccd.com) is a tiny camera that attaches to your mount (whether or not it has a polar scope) and connects to your computer. Then it sights and identifies stars near the pole. The camera does not have to be aligned with your mount's axis; it determines the axis of rotation as you move the mount in right ascension. Then it tells you how to adjust the mount to make the two axes coincide; you perform the adjustment

Table 8.1 *How to adjust your polar axis using the drift method.*

Drift Method Alignment			
Star high in the SOUTH, near declination 0° In S. Hemisphere, star high in the NORTH, near declination 0°			
If star drifts	NORTH 2″	per minute, *move polar axis*	1/8° RIGHT.
	NORTH 4″		1/4° RIGHT.
	NORTH 8″		1/2° RIGHT.
	NORTH 16″		1° RIGHT.
If star drifts	SOUTH	*do the opposite.*	
Star at 40° N, low in EAST, >60° from meridian In S. Hemisphere, star at 40° S, low in WEST, >60° from meridian			
If star drifts	NORTH 2″	per minute, *move polar axis*	1/8° DOWN.
	NORTH 4″		1/4° DOWN.
	NORTH 8″		1/2° DOWN.
	NORTH 16″		1° DOWN.
If star drifts	SOUTH	*do the opposite.*	

by superimposing images on the computer screen. Alignment accuracies better than *one arc-minute* (1/60 of a degree) are reportedly achievable.

8.6.3 The Drift Method

The sure way to check and refine your polar alignment is called the drift method. What you do is check whether stars in two regions of the sky are drifting northward or southward relative to the telescope's tracking. You can do this without a computer; you can even do it without a motor drive, moving the telescope in right ascension by hand.

You'll need an eyepiece with double crosshairs whose separation corresponds to a known distance in the sky. To measure the separation, compare the crosshairs to a double star of known separation, or point the telescope at a star near the celestial equator and time its motion with your motor drive completely turned off. Such a star will move 15″ per second of time.

Then turn on the drive motor and track a couple of stars. First track a star near declination 0°, high in the south. Center it on your crosshairs and see if it stays there. Measure how fast it seems to drift north or south, and the upper part of Table 8.1 will tell you how to adjust your tripod.

Only northward or southward drift is significant. If the star drifts east or west, bring it back into the center of the field (by moving east or west only) before making the measurement. In fact, you can do drift alignment without running the drive motor at all – just center the star, wait a few minutes, and try to re-center the star by moving the telescope by hand in right ascension only.

Once you've adjusted the polar axis in the left–right direction (azimuth), track a star that is low in the east, around declination $+40°$, and use the lower half of Table 8.1 to adjust the polar axis in altitude. You can continue the process to achieve any desired level of accuracy.

Table 8.1 also gives directions for drift method alignment in the Southern Hemisphere, where the drift method is more necessary because there is no bright pole star.

8.6.4 Automated Drift Method

Instead of an eyepiece with crosshairs, you can use your guidescope for drift-method alignment, if you have one. Automated versions of the drift method are provided in *PHD2* autoguiding software, *PEMPRO* mount-testing software, and other packages. They work whether or not your mount is computerized. Of course, any software that will display an image on the screen will help you perform the drift method manually.

8.6.5 Why the Drift Method is Best

The drift method has a superpower. Recall that what you actually want is not perfect polar alignment, but perfect tracking. If some kind of flexure is gradually shifting the image while you track, you'd like to compensate for it – and that's what the drift method does. Instead of giving you perfect alignment with the earth's axis, the drift method gives you *whatever alignment will keep the stars from drifting*. If your mount is suffering slight overall flexure, the drift method will set up an equal and opposite drift to counteract it.

For this to work, of course, you have to do the drift method in the area of sky where you will be photographing. You may even want to do a second drift-method alignment when you move to another area of the sky.

8.7 How Accurately Must We Polar-align?

An equatorial mount, even a computerized one, tracks the stars with only one motor, driving the right ascension axis. (It uses its declination motor to go to objects and to accept corrections from an autoguider.) Accordingly, the quality of tracking depends on polar aligment.

The effect of polar axis misalignment is to make the image drift. The drift can be in any direction but almost always has a strong north–south component.

The greatest possible drift is about 16 arc-seconds per minute of time per degree of alignment error. In many situations, the actual drift is much less, but to be safe in the worst case, without making guiding corrections, you must align as follows:

- If you can tolerate 1″ of drift per minute of time, you must align to within 1/16 degree (that is, just under 4′);
- If you can tolerate 4″ of drift per minute of time, you must align to within 1/4 degree (that is, 15′).

The first criterion is the strict one; drift of 1″ is likely to be "lost in the noise" because it is comparable to the errors from flexure of portable mounts and from atmospheric effects. The second criterion is more applicable to focal lengths under 500 mm, and it shows why sky trackers usually perform well when aligned with a polar scope. Any drift less than about 5″ is going to be lost in the noise of periodic gear error.

If you *do* make guiding corrections, drift doesn't matter, and your only concern is the other effect of misalignment – field rotation. Suffice it to say that field rotation large enough to affect your pictures is unlikely to occur. To get 0.1° of field rotation in a 10-minute exposure of a star on the celestial equator, you'd have to be 2.5° off the pole. Recall that field rotation does not depend on magnification (since a rotating wheel, viewed through a telescope, still rotates at the same speed), and don't be tricked by lens aberrations that look like field rotation (Figure 7.13).[3]

A better reason to want good polar alignment is so that your guiding corrections don't have to be too large. In particular, corrections in declination involve running a second motor (besides the right ascension motor that is already running), and some astrophotographers prefer not to make them. (With an older, non-computerized mount, you might not even have a declination motor.) In that case, polar-align as accurately as you can, and make corrections only in right ascension.

The effect of polar axis misalignment is described in more detail in *Astrophotography for the Amateur* (1999), pp. 278–282.

[3] Field rotation does become important when you're photographing in the immediate vicinity of the pole; obviously, if the stars are rotating around one pole, you don't want your camera to rotate around a different pole that is not near the true one. It can actually be better to turn the drive motor off completely.

Chapter 9
Precision Tracking and Guiding

No telescope mount tracks the stars perfectly, but some do better than others. Astrophotography at long focal lengths almost always requires guiding corrections. In the past this meant the astrophotographer had to watch a star continuously and press buttons to keep it centered on the crosshairs. This was an agonizing process when film exposures typically lasted an hour.

Nowadays we use autoguiders, which watch the star automatically, or else, increasingly, we make exposures short enough, and tune our mounts carefully enough, that no corrections are needed. This is presently an area of rapid technical progress – and, if you're not careful, of great potential frustration because expectations are rising so rapidly, and because we have the ability to measure imperfections more precisely than ever before.

9.1 Why Telescopes Do not Track Perfectly

There are at least four reasons a telescope doesn't track perfectly. The two most important are polar alignment error, which you can correct with better polar alignment, and irregularities in the gears, which can be addressed by recording and playing back automatic corrections.

The third factor is atmospheric refraction: objects very close to the horizon appear slightly higher in the sky than they ought to, and the extent of the effect depends on humidity, so it's not completely predictable.

The fourth is flexure (bending). Every mount bends a little as the load shifts position during a long exposure.

On a millisecond time scale, there is a fifth factor, "seeing," or atmospheric unsteadiness. Visual observers talk of "good seeing" when the air is steady and fine detail is visible, and "bad seeing" when it isn't. Even when the seeing is relatively good, star images don't hold still; they jump around randomly, sometimes as far as several arc-seconds, and wiggle constantly over distances of one arc-second or so. That is one reason star images in long exposures are always much larger than the smallest images the optics are capable of forming.

9.2 Must We Make Guiding Corrections?

9.2.1 Sometimes, no

Many DSLR enthusiasts make no guiding corrections during the exposure. In the film era, this would have been absurd, but it's possible today for several reasons. Equatorial mounts are better built than they used to be; they track more smoothly. Drift-method alignment has come into wide use, leading to more accurate polar alignment. Periodic-error correction (PEC) greatly smooths out the tracking.

Most importantly, exposures are shorter. It's much easier to get two minutes of acceptable tracking than 20. In fact, periodic gear error being what it is, some 1-minute exposures are bound to be well-tracked even with a cheaply made mount. You can select the best ones and discard the rest. You can even work with 30-second or 15-second exposures if that's all your mount will permit.

Figure 6.3 is an example. With a focal length of 420 mm, I needed no guiding corrections for any of the 50 1-minute subs taken with an AVX mount using PEC. That is a setup I use regularly.

9.2.2 A Futile Quest

At the same time, I want to caution you against undertaking a futile quest. "I can get unguided 3-minute subs," says one enthusiast in an online forum, and immediately someone with the same kind of mount responds with dismay, "I can't." Maybe the second astrophotographer is just more critical of picture quality; maybe he hasn't trained his PEC as well; or maybe it's just luck.

Exceptionally good mounts can sometimes give 10-minute subs, but even the best mount isn't immune to atmospheric refraction or flexure. Observatories with permanently installed equipment make elaborate computer models of their pointing and tracking errors, make corrections in real-time by computer, and often get 5-minute subs without guiding. But with portable or potentially portable equipment, it's generally better to go ahead and use an autoguider when it's needed.

9.3 Mount Performance

9.3.1 How Tracking Error is Measured

Unlike our predecessors in the film era, today we can easily measure how well a mount is tracking. Any of various software packages – which I'll get to – can take repeated short exposures of a star through an astronomical video camera and

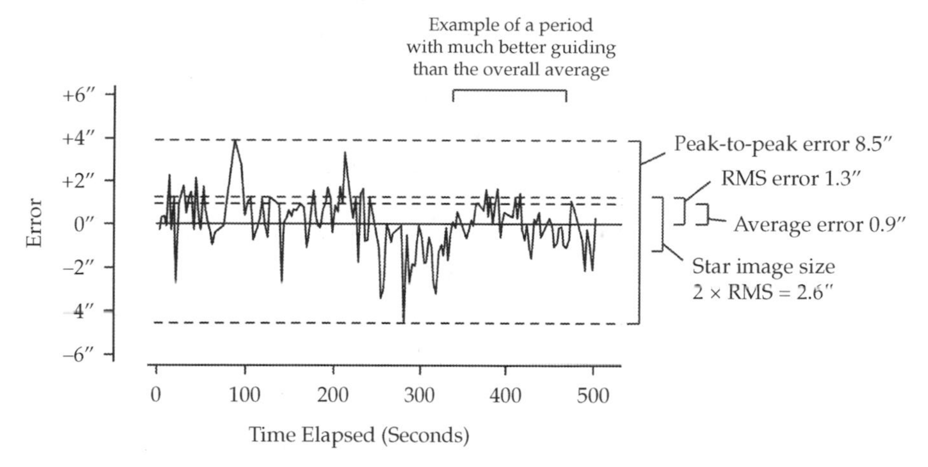

Figure 9.1. Tracking error is measured by taking frequent short exposures of a star and plotting its apparent movement. See text for mathematics.

plot its apparent movement. Of course, with perfect tracking, the star would stand still. In real life, you get a graph like Figure 9.1, plus any of several numerical measurements of how much the star image moves around.

The error is measured in arc-seconds, and the small random fluctuations (up to 2″ and occasionally more) are due to atmospheric unsteadiness.

The amount of error can be measured several ways. Consider a sequence of n consecutive error measurements; call them $x_1, x_2, x_3 \cdots x_n$. Then the *peak-to-peak* error is the maximum minus the minimum:

$$\text{Peak-to-peak error} = x_{\text{max}} - x_{\text{min}}.$$

The trouble with peak-to-peak error is that it measures only the extremes of the error, which may be momentary flukes. We'd rather have some kind of average that shows how much error is constantly present as we go along. That leads to two more measures:

$$\text{Average error} = (|x_1| + |x_2| + |x_3| + \cdots + |x_n|)/n,$$

$$\text{RMS error} = \sqrt{(x_1{}^2 + x_2{}^2 + x_3{}^2 + \cdots + x_n{}^2)/n}.$$

That is, for the average error, simply treat all the values as positive, and average them. For the RMS error, square them all (which makes them positive), average them, and take the square root. (*RMS* stands for "root-mean-square," root of the mean of the squares.)

These same statistics are also used to measure other fluctuating quantities, such as the differences between pixels; the RMS error is the standard deviation, the average error is also known as the average absolute deviation (AAD), and you will sometimes see the median used instead of the average (median absolute deviation, MAD). Guiding errors are measured relative to zero, but in other situations fluctuations are referred to the average value of whatever is being measured.

There are no formulae to interconvert peak-to-peak, average, and RMS error because the relation between them depends on the shape of the waveform. As a rule of thumb, though, RMS error is much less than peak-to-peak error.

Figure 9.1 consists of real data from one of my mounts, and it also shows another important fact: there are periods of a minute or two when tracking is much better than overall, when the gears hit a smooth spot. If you have lots of time but not a very good mount, you can simply take lots of subs and throw out two thirds of them, retaining the ones that happened to track well.

9.3.2 Periodic Gear Error

The drive mechanism of any modern mount consists of a motor; a gearbox containing small spur gears to reduce the speed; a worm, which is a screw-shaped component that engages the teeth of a gear and turns it; and the driven gear itself, which has slanted teeth to match the worm.[1]

Most of the tracking error of a typical mount recurs every time the worm turns. No worm is perfectly smooth, regular, and centered. Accordingly, at some points in the worm's rotation, the movement of the driven gear is a bit too fast, and at other points, too slow. This is called *periodic error* (PE). It can be corrected by recording the errors and automatically varying the motor speed to make up for them (*periodic-error correction*, PEC).

Depending on the design of the mount, each turn of the worm takes 4, 8, or 10 minutes (technically minutes of sidereal time, which are 99.73% the length of normal minutes). Many, though not all, mounts are designed with integer ratios of the smaller gears nearer the motor. That makes them mesh on the same teeth every time they go around, so that they do not introduce periodic errors recurring at intervals longer than the rotation of the worm. That makes PEC more effective.

Figure 9.2 shows the measured tracking error of my Celestron AVX, plotted with *PEMPRO* software, as the worm went through slightly more than two cycles. Fortunately, a small drift (probably due to polar alignment error) kept the curves from overlapping, so you can see them. The overall periodic error has the shape of a sine wave with an amplitude of about 24″. Even the smallest fluctuations, though, are mostly recurrent.

The figure also shows that although the total error is 24″ peak-to-peak, *the error during any given one- or two-minute period is much less*, which is what makes unguided subs practical, even if periodic error is not corrected.

---

[1] Some people use the term *worm gear* to denote the worm, but it more properly refers to the gear that the worm drives.

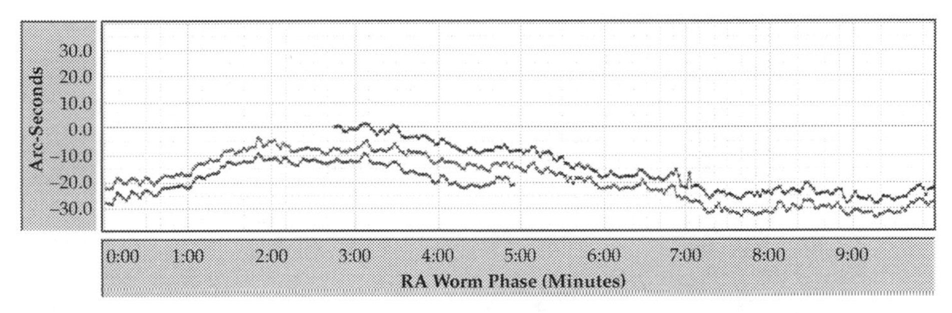

Figure 9.2. Measured tracking error during slightly more than two cycles of the worm of a Celestron AVX. Note how much of the error is recurrent, including small fluctuations.

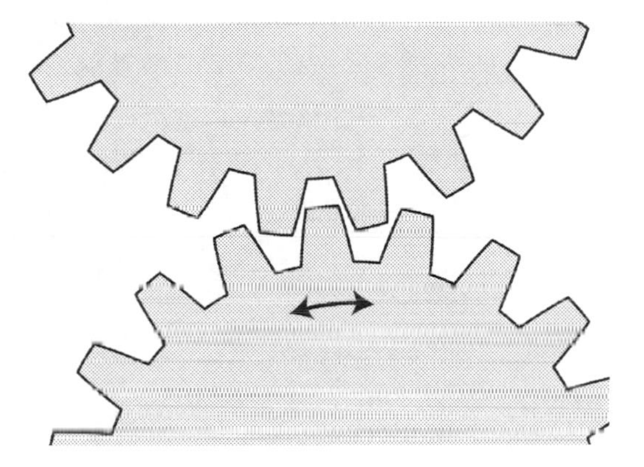

Figure 9.3. All gears have some backlash (space to move freely), or else they would jam.

9.3.3 Backlash

Backlash (Figure 9.3) is free movement of gears (including gears driven by worms) due to space between them. It normally does not affect tracking because one gear is driving another from one side, and the free space on the other side does not matter.

However, an excessively well-balanced mount can wobble back and forth across its backlash, producing strange guiding problems. Mounts actually track better when they are slightly out of balance – not enough to create a heavy load, but enough to take up the slack in the gears. That is often achieved by moving a counterweight perhaps an inch from the position of best balance. The mount should be just on the verge of moving when the brakes are released. In right ascension, the mount should be east-heavy; that is, the counterweight should be to the east of the perfectly balanced position. Whether this means moving it up or down the shaft depends on which side of the meridian the mount is on.

All gears have to have some backlash, or they will jam. Mounts can be adjusted internally to reduce backlash, but trying to completely eliminate it is not a good idea. It is not uncommon for the backlash of a portable GEM to be as large as 120″.

Backlash is most noticeable when you are using the telescope visually and slewing by pressing buttons. Specifically, if you slew in one direction and then the opposite direction, the second movement will not start immediately; there will be a pause of a second or two (or more) to take up backlash. Slewing in right ascension, you will see another effect: when you slew east (opposite the direction of tracking), the movement continues for a few seconds after you let go of the button. That is because tracking does not resume until the backlash is taken up.

Computerized mounts include electronic backlash correction. You can tell the motors to make an extra, fast "jump" to take up backlash when the slewing reverses direction. This is very handy for visual observers but can wreak havoc with autoguiders, and I usually leave it set to zero. However, some mounts with large backlash may benefit from partial electronic backlash correction, especially in declination.

9.3.4 Flexure

All equipment bends noticeably as you move it around. Recall that an error of one arc-second is big enough to notice, and 1″ is 1/206 265 of a radian, which means it is about 5 parts per million. That is, it is equivalent to moving the end of a 1-meter-long rod by a distance of 0.005 mm, much less than the diameter of a grain of sand. An unwanted arc-second of flexure is *very* easy to get.

There is *overall flexure* of the entire mount and instrument package, and *differential flexure* of the guidescope versus the main telescope. The former is just another source of tracking inaccuracy; the latter is a serious problem, and we do our best to eliminate it by mounting the guidescope sturdily or using an off-axis or on-axis guider.

One unobvious source of differential flexure is movement of the mirror in a Schmidt–Cassegrain or Maksutov–Cassegrain telescope ("mirror flop"), usually not a serious problem with short exposures. The mirror has to be movable for focusing, but some telescopes, including the EdgeHD series, include a provision to lock down the mirror during an exposure.

A typical rate of overall flexure with a portable mount is on the order of 1″ per minute of time. All flexure, overall and differential, including mirror flop, tends to be worse – often markedly worse – when tracking across the meridian. That is when the direction of the load on every structural component is changing fastest. Many of us get much better results tracking objects that are about 30° away from the meridian in either direction.

9.4 Periodic-error Correction (PEC)

Perhaps the single most important advance in guiding technology since my early days is periodic-error correction (PEC). The corrections needed to

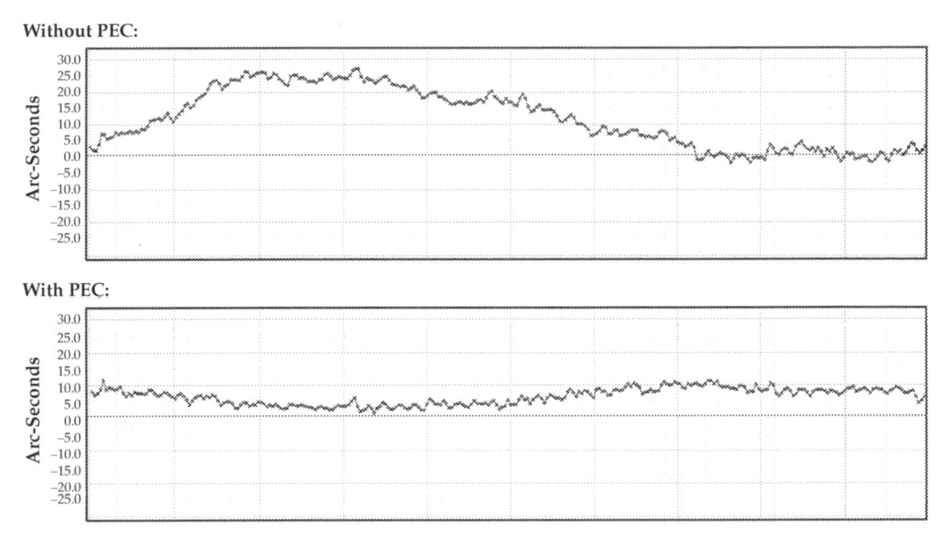

Figure 9.4. The benefit of periodic-error correction (PEC). Note slight overcorrection because amplitude of periodic error varies with mount position and load; when recorded, periodic error was slightly greater than when played back.

overcome periodic error are recorded by a microprocessor and played back as corrections to the motor speed every time the worm turns, greatly reducing the tracking error (Figure 9.4). Even imperfect PEC is a lot better than none.

When PEC is to be used, the mount's computer needs to know the position of the worm. Accordingly, some mounts slew as much as 2 degrees, to find an index mark, when the PEC is turned on. Other mounts perform this initialization when they are first powered on.

How do you train PEC? Any of several ways. The simplest is to use an eyepiece with crosshairs, watch a star, and make guiding corrections by pressing buttons. Do this for the entire length of the worm cycle (commonly 8 minutes) while the PEC is recording. Then switch to PEC playback, and your corrections will repeat automatically.

Better ways to train PEC are based on autoguiding, but you don't actually need a guidescope – you can put a video camera in the main telescope and use autoguiding software to send corrections to the mount. Any autoguiding software will do, but *PEMPRO* does the job very thoroughly and elegantly.

Do not cross the meridian while training PEC. That is, do the entire training session with the telescope aimed either east or west of the meridian. That way, a shift in flexure will not happen in the middle of the training session.

Most mounts have "permanent PEC," which means that once you've done the training, your corrections will be stored and can be used again at the next session. With permanent PEC, I usually re-train every couple of years to account for mechanical wear on the gears. A few older mounts have PEC that is not permanent – you have to train it at the beginning of every session.

PEC works better if you can average the results of several training sessions. With Celestron mounts, this can be done with Celestron's PECtool (free from www.celestron.com/pages/drivers-and-software); other mounts may provide a similar feature built-in.

The software package *PEMPRO* (from www.ccdware.com) is a powerful tool for training PEC and analyzing mount performance. It was used to acquire the graphs in Figure 9.2 and can also do Fourier transforms (to measure gear periods), combine many worm cycles for more accurate PEC training, and upload corrections to the mount. It works with a wide variety of astronomical video cameras and is compatible with practically all mounts – if it cannot upload correction files to your mount, it can simply play back the corrections while the mount records them its own way. It also measures backlash and helps you perform drift-method polar alignment.

9.5 Autoguiding

9.5.1 The Concept

An autoguider watches a star and sends guiding corrections to the mount in order to maintain good tracking. It consists of:

- An astronomical video camera or modified webcam that can take repeated short exposures of a star;
- Either a guidescope through which the camera watches the guide star, or an off-axis or on-axis guider that enables it to track a star in the image from the main telescope; and
- A computer that calculates the corrections and sends them to the mount.

A *standalone autoguider* combines the camera and computer in one package. More commonly nowadays, though, the camera is connected to a laptop computer running special software and also connected to the mount.

9.5.2 Subpixel Accuracy

Crucially, the autoguider – whether a standalone device or software in a laptop – analyzes star images with *subpixel accuracy*. It doesn't just figure out which pixel in the guiding camera corresponds to the star image. It analyzes how the light is spread over several pixels and finds the centroid of the image relative to several pixels: "15% of the way from this one to that one," for instance.

That means the guidescope and guide camera do not need to have the same resolution as the main telescope. If the pixels in the guide camera and main camera are about the same size, the guidescope can be as short as 1/5 the focal length of the main telescope. Even smaller guidescopes, as in Figure 8.3, are sometimes successful.

9.5.3 Communication with the Mount

There are two ways the autoguider can issue slewing commands to the telescope. Older autoguiders use relays that simulate the buttons on a traditional control box connected to a 6-pin modular socket (sometimes called an SBIG-type or ST-4-type port); for a wiring diagram, see *How to Use a Computerized Telescope* (2002), p. 158.

The newer approach, favored when the autoguiding is controlled by a laptop computer, is to issue slewing commands to the telescope through its serial or USB port. This is preferable for two reasons. First, corrections can be made more quickly and more accurately, without allowing time for relay contacts to close and open. Second, the autoguider can find out the declination at which a computerized mount is pointed and can adjust the calibration appropriately (a minute of right ascension is a different distance at different declinations). The autoguider can even find out that a meridian flip has occurred and reverse the directions of the movements accordingly.

Autoguiding through a computer interface generally requires the computer to have the ASCOM platform installed (https://ascom-standards.org); it provides device drivers for Windows software to use. The same drivers can be used by more than one software package at the same time, such as an autoguider and also a sky map program, both controlling the same mount. The newer INDI platform (http://indilib.org) performs a similar function under Linux and macOS as well as Windows.

Even the newest mounts still have ST-4 interfaces for use with older autoguiders, and old mounts that only have ST-4 interfaces are still in use (such a mount need not even be computerized, as long as it has motors on both axes). A computer, however, cannot communicate with an ST-4 interface unless an adapter is used. USB to ST-4 adapters are available from www.shoestringastronomy.com and also are often built into autoguiding cameras. If a camera has an ST-4 port on it, and is not a standalone autoguider, that is why.

9.5.4 Autoguiding Software

From here on, when I say "autoguider" I mean either a standalone autoguider or a computer with appropriate software connected to an appropriate camera and interfaced to a mount. Usually I mean the latter. Standalone autoguiders are now uncommon, and the few that are still on the market have computer interfaces so you can view the image and make the settings. For years I used an SBIG ST-V, a standalone autoguider with a control box bigger than a laptop, with its own video screen.

All autoguiders work very much the same way, but my specific examples will be from *PHD2* (https://openphdguiding.org), software for Windows and macOS, which was developed as an open-source successor to Stark Labs' *PHD Guiding* (www.stark-labs.com), which remains available and works well. *PHD* stands for "push here, dummy!" and the intent of the software is to make everything as easy as possible. Both *PHD* and *PHD2* come with copious, clear instructions.

Autoguiding is also included in *MaxIm DL* and other software packages; INDI-based autoguiding is planned for *PixInsight*. In general, these use some of the same algorithms as *PHD2* and the same instructions will apply.

One package takes a different approach. Instead of the usual one- or two-second time exposures, *Metaguide*, free from www.astrogeeks.com/Bliss/MetaGuide, uses much shorter exposures taken in rapid succession and then combines and sharpens them using the same methods as planetary video astronomy. This enables corrections to be more prompt and more precise. Also, because it reconstructs a diffraction-limited star image, *Metaguide* can tell you how well your telescope is collimated and will even re-center the collimation star as you adjust it.

9.6 Cameras, Guidescopes, and Off-axis Guiders

9.6.1 The Guide Camera

The camera used for autoguiding needs to take short exposures (up to 3 seconds or so) and deliver them quickly to the computer. Apart from that, the requirements are not exacting. I use an ImagingSource DMK video camera (Figure 1.6), but many others work well, and it is best to start by looking at the list of cameras supported by your software. The ZWO ASI120MM is especially popular, has a built-in USB to ST-4 interface, and is versatile enough that you can also take 1.2-megapixel deep-sky images with it (which of course begs the question of how you will autoguide while doing so!).

Monochrome cameras are preferred, but color cameras will do; use what you have. You can even make a guide camera by removing the lens from a cheap webcam and installing a tube that will fit the telescope. The guide camera needs to be small and light, but other things being equal, a larger sensor is better because it includes more stars in the field.

9.6.2 Guidescopes

To autoguide, you need either a guidescope or an off-axis or on-axis guider so that a camera can monitor the tracking. Figure 9.5 shows my present setup. Because autoguiders have subpixel accuracy, the guidescope can be considerably smaller than the main telescope, as small as 1/5 the focal length and aperture. It need not have tip-top optical quality; a simple achromatic refractor is good enough. Slightly blurry star images may actually give more accurate centroids.

The advantage of using a guidescope rather than an off-axis guider is that it is easy to find guide stars; with an $f/4$ objective and a typical camera, there is

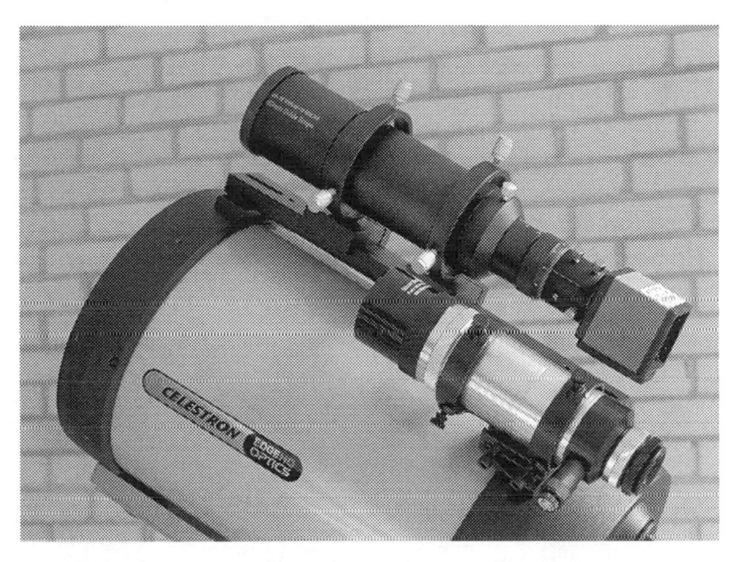

Figure 9.5. 6-cm $f/4$ guidescope and DMK video camera mounted on dovetail atop 20-cm telescope. (Cables omitted for clarity.) For greater stability, the camera could also be supported.

at least one suitable star in the field almost anywhere in the sky. It helps if the camera sensor is not too small.

The disadvantage of the guidescope is that eliminating differential flexure can be hard. After trying a number of alternatives, I chose a guidescope with a special focuser that locks down very sturdily – the focuser is the main point of flexure. The guidescope is mounted in rings on a dovetail on top of the telescope, and for extra sturdiness, I am considering adding a support for the camera. I have also been known to zip-tie the camera cable to the dovetail to keep it from wiggling the camera.

Of course, the guidescope must be focused accurately. If it is out of focus, no stars will be seen, and even a small focusing error makes faint stars disappear. I often do the initial focusing of a new setup by taking aim at the moon, which is too big to miss, and then carefully mark the focus setting. The focus can then be refined on the stars, preferably in a rich cluster such as the Pleiades or M44.

The guidescope does not have to be aimed in precisely the same direction as the main telescope, although it should be close. Sometimes you will have to guide on a star a couple of degrees away from the object you're photographing. This affects accuracy only if you have a huge polar alignment error.

9.6.3 Off-axis Guiders

To avoid the risk of one telescope flexing relative to the other, you can guide through the main telescope itself with an off-axis guider (Figure 9.6), which

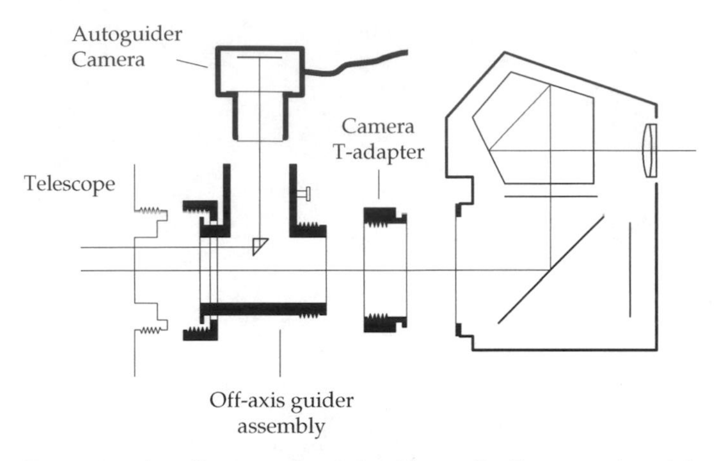

Figure 9.6. An off-axis guider picks off a small off-axis portion of the image and feeds it to the camera. It can be very hard to find a suitable guide star.

picks off a small off-axis portion of the image (which wouldn't fall on the sensor anyway) and uses it for guiding. The off-axis guider even follows along if the mirror of a Schmidt–Cassegrain is gradually shifting position.

As you can imagine, off-axis guiders can be tricky. Simply getting the main camera and guide camera both into focus can be difficult (again, the moon is a good first target). It can also be very hard to find a star that is bright enough to guide on and can be placed in exactly the right position. Especially when you are photographing galaxies, guide stars are scarce, and the star clusters of Sagittarius or the Orion Nebula are a better place to start.

The optical quality of the star image picked up by the autoguider is likely to be poor, since it is outside the normal field. That is not a problem; the autoguider will follow the brightest part of it.

9.6.4 On-axis Guiding

To avoid having to find a guide star at the right place just outside the field, you can even guide on-axis, by using a mirror that only reflects infrared. The visible-light image goes to the camera unobstructed, but infrared light from the same object goes to the autoguider. (CCD and CMOS cameras are generally quite sensitive in the near infrared, just outside the visible spectrum.) On-axis guiding (ONAG) devices are made by Innovations Foresight (www.innovationsforesight.com) using high-quality dichroic beam splitters; the guide camera works at about 750 nm, a wavelength well within the reach of any sensor that does not have an infrared-blocking filter on it.

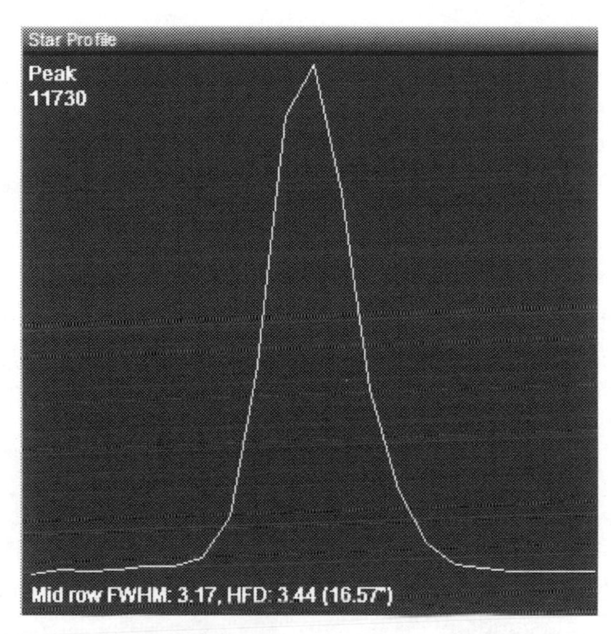

Figure 9.7. Profile of a well-focused star in *PHD2*. Compare Figure 3.11

9.7 Using an Autoguider

9.7.1 Choosing a Guide Star

The first step in using an autoguider is to view the image from the camera, touch up the focus, and choose a star to guide on. The guide star need not be in the center of the field as long as it is far enough from the edges that it won't go out of view during calibration or guiding.

Best guiding occurs when the star is neither too bright nor too dim. Every autoguider is different, but in general, the best centroids are measured on stars that are moderately bright. (You'll have to figure out what magnitude is "moderately bright" with your guidescope and camera.) Many autoguiders show you a graphical representation of the star image (like Figure 9.7), and you want a sharp peak in the middle, which indicates correct focus and appropriate brightness. Besides choosing a different star, you control brightness to some extent by adjusting the camera gain and the exposure time.

9.7.2 Hot Pixels and Dark Frames

Like DSLRs, guide cameras can have hot pixels. *PHD2* provides a way for you to take a library of dark frames, store them, and apply the right one depending on the exposure time. That eliminates hot pixels completely.

The alternative is to watch out for hot pixels when choosing the guide star. A single pixel does not look quite like a star image, and if you try to track it, you will get seemingly *perfect* tracking while the telescope wanders all over the place and takes terrible pictures.

9.7.3 Calibration

Calibration of an autoguider means making it determine how far and which way the telescope moves in response to control signals. Without that knowledge, autoguiding is impossible. Calibration has to be done before you start autoguiding. Most autoguiders do it automatically, but three things can go wrong.

First, obviously, the star may go completely out of the field as the autoguider is trying out its movements. The cure is to start out with a star nearer the center.

Second, the autoguider may not be able to make the star image move. This happens if the mount has a lot of backlash and the calibration movements are not long enough to overcome it. The cure is to specify longer calibration movements, as long as 45 seconds.

Third, the autoguider may report inconsistent results – it simply can't move in all four directions, or the directions in which it moves are not perpendicular. In that case, if the problem isn't backlash, look for severe flexure (something loose or snagged on a cable) and loose electrical connections.

If calibration is difficult, check the guiding speed settings for your mount. In all four directions, the guiding speed should normally be about 0.5× to 1× sidereal-rate tracking. This setting determines how fast the mount moves when making a guiding correction.

Also, turn off backlash compensation in the mount, or set it to a low value. Many autoguiders, including *PHD2*, compensate for backlash themselves. Too much backlash compensation is much worse than too little.

9.7.4 Autoguider Settings

The most important autoguider setting is the exposure time – typically 1 second in very steady air, 2.5 seconds under average conditions, and as long as 5 seconds when the air is rough. If the exposures are too short, the autoguider will "chase seeing" (try to track atmospheric fluctuations), and if they are too long, it will be too slow to make needed corrections.

If you need long exposures but have to make short ones because the star is too bright, you can tell the autoguider to combine successive exposures. Also consider adding a red filter; the air is steadier in red light than in blue, and you can get more accurate guiding that way.

The next setting is aggressiveness. Since part of the movement seen by the camera is atmospheric and not reproducible, the autoguider should not try to correct all the observed error, but only about half or three quarters of it. That means the aggressiveness is typically set to 50% or 75%. There are separate settings for right ascension and declination.

If your software offers you still more settings (and *PHD2* offers plenty), the most important of them is minimum move, the minimum error that can trigger a correction. By setting a higher minimum move, you tell the autoguider to ignore low-level fluctuations.

9.7.5 Algorithms

Mathematically, autoguiding is predictive time-series analysis. The goal is to respond to trends that would persist if uncorrected, but not to random fluctuations that are going to reverse themselves immediately anyhow.

It follows that two successive corrections in opposite directions are probably wrong. Most autoguiders include some *hysteresis* (reluctance to reverse direction) to keep such things from happening, and you can adjust the hysteresis.

That's only the beginning. *PHD2* and other software packages offer a variety of guiding algorithms, and new ones are still being invented. The mathematical challenge is basically the same as predicting the stock market for short-term trading, or predicting weather, or many other things that involve separation of signal from noise. When a new algorithm comes out, it may take some time before the community reaches a consensus about whether it is a good one, and if so, under what conditions and with what settings it works best.

9.7.6 Quality of Guiding

Figure 9.8 shows how *PHD2* tells you how well it is guiding. The continuous lines on the guiding graph show the actual error, exposure by exposure, blue for right ascension, red for declination. Corrections sent to the mount are plotted as rectangular bars. At the left you see the accumulated RMS error, but be careful – this is the RMS error since the last time you pressed "Clear," and it may include periods when you were not actually taking pictures. It is shown as error in right ascension, error in declination, and error in the two combined.

Here is how to interpret RMS error:

- 2.0″ is acceptable in unsteady air or with low-end equipment;
- 1.5″ is good;
- 1.0″ is very good;
- 0.5″ is exceptional.

Remember that your goal is good pictures, not low RMS error for its own sake.

Also at the bottom left of *PHD2*'s display is a number reporting RA oscillation. That is the fraction of right ascension corrections that are in the opposite direction than the previous one. If you're not "chasing seeing," it should be low, 0.33 or less. Here it is 0.22, which indicates good guiding.

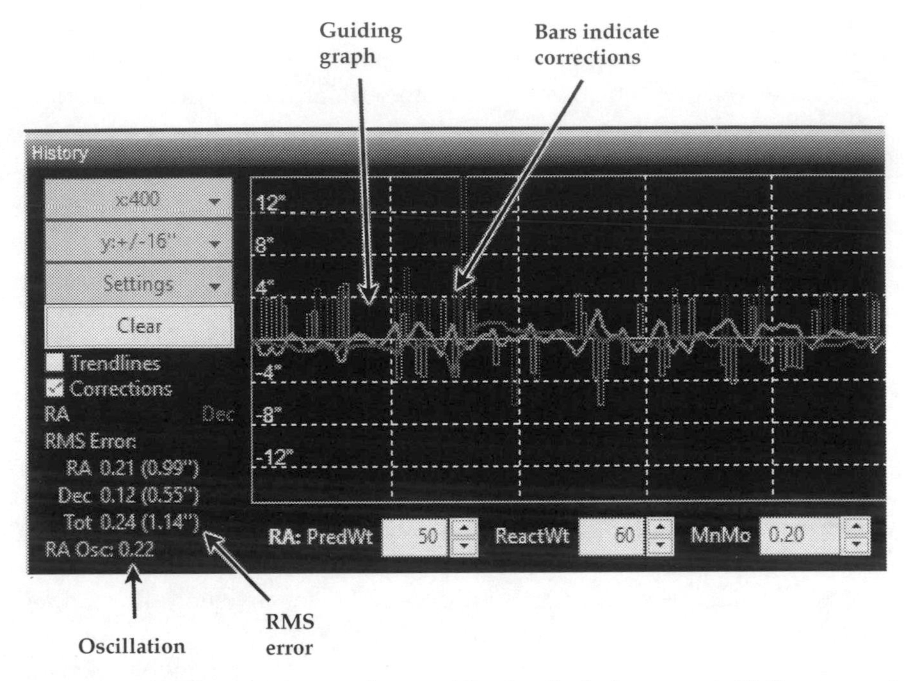

Figure 9.8. *PHD2* reports how well it is guiding by displaying a graph, RMS errors, and oscillation. Other autoguiders are similar. RA is shown lighter here, and declination darker; on the screen they are blue and red respectively.

9.7.7 Interpreting Guiding Graphs

The guiding graph tells you not only how well you're guiding, but also how to improve it. Figure 9.9 shows examples.

The first graph is an example of good guiding. Here is how to respond to the others:

Fast and choppy guiding indicates that the autoguider is "chasing seeing," overreacting to atmospheric turbulence. Lengthen the exposure, turn down the aggressiveness, and/or turn up the minimum movement.

A *single disrupting event* tells you that something happened only once. Maybe you stepped on the concrete pad close to the telescope; maybe a cable briefly snagged something, then got loose; maybe your cat decided to brush against a tripod leg. Such events usually cause a jerk on both axes, right ascension and declination.

Or the problem may be deeper. One of the symptoms of an overloaded mount is a sudden jerk like this every few minutes. That can also indicate a mount that needs lubricating. Think of the latter two problems if the jerk occurs in right ascension but not declination.

Drift raises the suspicion that corrections are not being made. A number of factors can cause drift – poor polar alignment, flexure, refraction, or the wrong

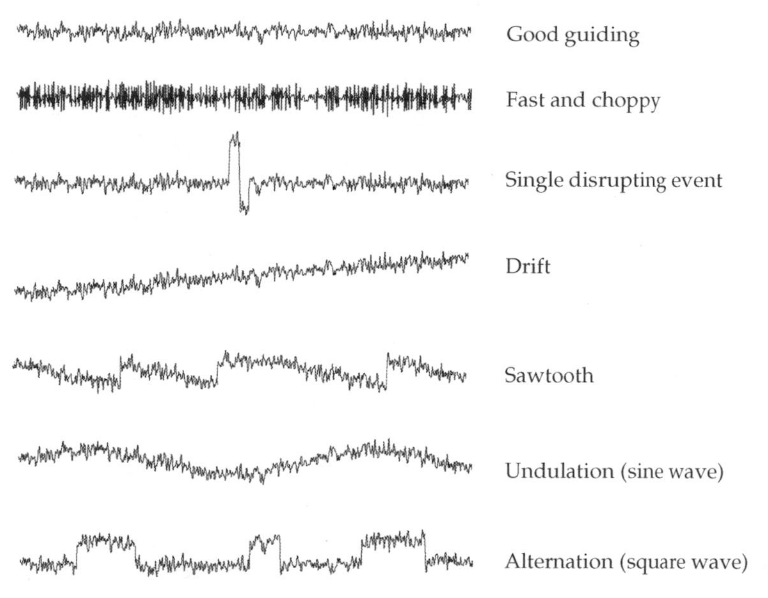

Figure 9.9. Examples of guiding graphs. See text for interpretation.

tracking rate – but the autoguider should be correcting it. Check that corrections are turned on, the minimum movement is not set to an absurdly high value, and the aggressiveness is not too low.

Sawtooth guiding is a symptom of minimum movement set way too high, or else "stiction" (static friction) or backlash. A large correction is made, followed by a long period when corrections are needed but not made, and then another large correction.

The question is whether, in between the large jumps, other corrections are being sent to the mount and having no effect. If so, the problem is stiction (in declination) or backlash. If no corrections are being sent except the big ones that you observe, then the problem is an autoguider setting, probably minimum motion, possibly hysteresis or aggressiveness. The problem could also be excessive electronic backlash compensation in the mount.

Undulation could be uncorrected periodic error (corrections not being made in right ascension) or any situation where corrections are too small, so that it takes them a long time to catch up with the actual error, which by then has reversed itself. Make sure corrections are actually being sent, and if so, turn up the aggressiveness. If you have periodic-error correction (PEC), use it so that the autoguider has less work to do.

Alternation, a square-wave pattern, has the same causes as sawtooth and undulation, but is more likely to come from backlash. The mount may be wobbling across its own backlash. The cure is to put it deliberately slightly out of balance, preferably east-heavy. Alternation can also come from excessive electronic backlash compensation.

9.7.8 Right Ascension and Declination are Different

One of the most important facts about guiding is that you have a motor continuously running in right ascension but not in declination. Guiding corrections in right ascension are made by varying the speed of that motor, which never reverses direction.

That means backlash in right ascension normally has no effect; the gear is driven from the same side all the time. Any backlash is taken up when you first start to track an object. However, if the mount is too well balanced, it can still wobble across its own backlash. To prevent that, make the mount slightly east-heavy, so that the motor is always working against gravity. To do that, move the counterweight an inch or so east of the balanced position. East may mean toward or away from the mount head, depending on which side of the meridian you're on.

Backlash in declination is a problem. Guiding corrections are made by running the motor in either direction, which means it often needs to take up backlash. You can prevent or reduce this problem by not balancing too well; make your telescope slightly camera-heavy.

Many autoguiders can measure and correct declination backlash. With less sophisticated autoguiders, you may need to set electronic backlash correction in the mount, but set it to a lower value than seems correct for visual use. Too little is always better than too much.

Another challenge in declination is *stiction* (static friction). Since it rarely moves, the declination axis tends to stick in place. Then a correction is sent to it and does not overcome the stiction. The autoguider sends another correction, and another, and suddenly the stuck axis breaks free and moves too far. The result is a sawtooth or square wave guiding graph in declination only.

There are work-arounds. One is to do an excellent drift-method polar alignment and then make no declination corrections at all. For exposures of just a few minutes, this approach works surprisingly well. Another is to offset the polar alignment deliberately so that there will be constant drift in declination; all the corrections are then in the same direction, backlash is taken up, and corrections are frequent enough that stiction is not a problem. It's a kluge, but it works.

9.7.9 PEC while Autoguiding?

If you have PEC, should you turn it on while you autoguide? In theory, it shouldn't make a difference. In practice, the answer is usually yes.

If the PEC is very accurately trained (with *PEMPRO* or the like), it will know what has happened on multiple worm cycles and will correct low-level mechanical fluctuations that the autoguider could not distinguish from noise.

Bad PEC training, on the other hand, could make the autoguider's job harder. So if you find that your PEC does not get along with your autoguider, turn it off, and at first opportunity, re-train it.

9.7.10 Good Autoguiding, Bad Pictures

What do you do when the autoguider reports good, or at least decent, guiding but your pictures show elongated stars? Look for flexure in any of three places:

- Between guide camera and guidescope;
- Between guidescope and main telescope;
- Between main telescope and main camera.

A good tactic is to beef up the supports of any part of the system – even crudely – and see if it makes a difference. If so, come back and reinforce that part of the system more thoroughly.

Flexure is proportional to the duration of the exposure. If your star images are not twice as elongated in a 2-minute exposure as in a 1-minute exposure, the problem isn't flexure and may not be related to guiding at all. It's an optical problem, maybe a lens defect, or maybe even image stabilization (IS or VR) accidentally turned on. Any genuine guiding problem will show up in the guiding graphs.

One notorious cause of flexure is cables. The weight of a cable doesn't pull in the same direction all the time as the mount moves around. Cables do not slide smoothly across other objects as they move. One remedy is to secure cables very tightly to dovetails so that they cannot pull on cameras.

Another point of vulnerability is where the guide camera attaches to the guidescope. Unless your guidescope has a focuser that locks down with unusual tightness, and your guide camera is very light, you will need to support the camera. Preferably, the guide camera, guidescope, and guide camera cable should all be secured to the same dovetail.

Yet another source of flexure is shifting of the mirror in a Schmidt–Cassegrain or Maksutov–Cassegrain. I have never had problems with this, even with telescopes that provide no mirror lock, but others have.

9.8 The Challenge of Round Star Images

9.8.1 What Should a Star Image Look Like?

We judge guiding by how round and small the star images are. Random guiding errors in all directions can produce an image that is round but larger than it should be. Elongated stars show that you are guiding better in one direction (almost always declination) than the other (right ascension).

Even though stars are perfect point sources, and the diffraction limit of most of our telescopes is smaller than one arc-second, the atmosphere prevents us from getting images smaller than a few arc-seconds in diameter. A reasonable expectation is that in a very well-guided image, the diameter of star images will be about 3″ or 3 pixels, whichever is larger. Many professional observatory images have 3″ to 5″ star images.

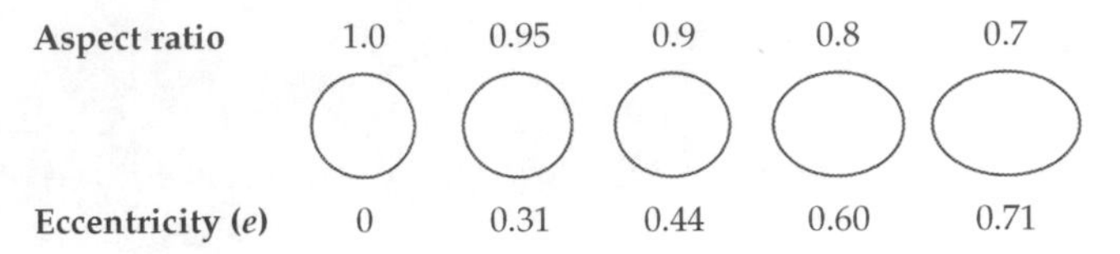

Figure 9.10. Two ways of measuring roundness.

9.8.2 How Roundness is Measured

PixInsight and other software packages can measure the roundness of star images. In *PixInsight* puts this is under Script, Batch Preprocessing, SubframeSelector, and also under Script, Image Analysis, FWHMEccentricity. These scripts also give you the full-width-half-maximum (FWHM) size of the star images in pixels.

Figure 9.10 shows two ways of measuring roundness. *Aspect ratio* is height divided by width; it should be the same whether the image is or is not gamma-corrected, since contrast stretching does not change the shape of objects. The other measurement, *eccentricity* or e, is the same quantity used in Kepler's theory of elliptical planetary orbits; denoting aspect ratio as a, the two are related by the formulae:

$$e = \sqrt{1 - a^2},$$
$$a = \sqrt{1 - e^2}.$$

It is correct that these two formulae look alike; the function is its own inverse, and $a = e$ at the value 0.707.

The usual recommendation is to keep e below 0.42 for round-looking star images. Note, however, that if the star images are not very large, they don't need to be perfectly round. The stars in Figure 9.11 have $e = 0.63$, and until you read this, you probably didn't consider them unsatisfactory.

The elongation in Figure 9.11 is systematic, but if the star images are small enough, there will also be random elongation because star images do not hit groups of pixels symmetrically. In that case you can get relatively high e values, up to 0.7 or so, from perfectly guided images. All the stars in such an image are slightly non-round, but not all in the same direction.

9.8.3 Some Practical Tips

Understand that numerous factors affect the shape of your star images, and it is unlikely that their total effect will be symmetrical. An image that seems to be bad in one direction may actually be exceptionally good in one direction and normal in the other.

Understand also that with the tiny pixels of modern DSLRs, you are seeing imperfections you never could have seen before. Round star images are not sharp; sharp star images are not round.

Figure 9.11. The galaxy M51 autoguided with the setup in Figure 9.5 using *PHD2*. Main telescope is 20-cm (8-inch) Celestron EdgeHD with focal reducer giving $f/7$, focal length 1400 mm, Celestron CGEM mount. Stack of 6 3-minute exposures (selected from 10) with a Nikon D5300 at ISO 400.

Understand that portable equipment is different every time you set it up – maybe even every time you slew to a different area of the sky. Don't judge tracking by the first 2 or even 5 or 10 minutes; keep tracking, and the mechanism will often settle down. In my experience, after any slewing, the second turn of the worm is often smoother than the first, perhaps because lubricants have smoothed out.

The rated load capacity of your mount is for visual and planetary work. For long-exposure photography, a good rule of thumb is to allow only half as much. Random "noise" in right ascension can be a symptom of overloading. So can alternation.

Manage your cords and cables. Be mechanically fastidious. Distrust tube rings, which are often flexure opportunities.

Pay attention to the seeing (steadiness). When the stars are twinkling strongly, the air is unsteady and it's time to take pictures through medium telephoto lenses rather than an autoguided telescope. Your best guiding will happen on nights that seem slightly hazy. Weather forecasts that predict seeing are available from Clear Sky Chart (www.cleardarksky.com/csk), Environment Canada (for all of North America, https://weather.gc.ca/astro), Meteoblue (www.meteoblue.com), and others. To judge actual conditions, just look at a star through your telescope at a high power.

Watch out for *terra non firma*, vibration of the ground from people walking near the telescope, especially on concrete pads, or from other sources. I once came across someone who had a strange optical problem, causing elongated images in various directions, which turned out to be vibration from an air conditioner some distance away. There is also the possibility of *lens non firma*, a lens whose mount is loose or whose image stabilization is accidentally turned on.

Stack a lot of exposures; the star images in a stacked image are rounder than those in the individual subs, especially if you use median or kappa-sigma stacking.

Go for percentage. Be willing to take 20 or 30 subs to get 10 good ones (Figure 9.12). That is normal procedure when working near the limits of your equipment.

As a last resort, remember that the quality of mass-produced mounts is variable; if yours is unduly recalcitrant, there's a chance you got a lemon. That is one reason to test a mount thoroughly (perhaps with the help of an experienced astrophotographer) while it is still under warranty. Don't hesitate to contact the manufacturer if there are persistent problems. Mounts can also be overhauled by third parties such as Deep Space Products (www.deepspaceproducts.com).

Finally, remember that people don't publish the images that didn't turn out well. You don't know how much luck and how many tries it took someone else to get an exceptionally good image.

9.8.4 Downsampling

One reason getting round star images is such a challenge is that our cameras now have more resolution than ever before, indeed more resolution than our optics. Every imperfection is clearly visible when you view the image at maximum magnification.

Accordingly, one of the easiest ways to get better guiding is to resample the image and make it smaller. Many guiding problems and optical flaws simply go away if you downsample the image to 1/4 or 1/5 its original resolution – putting it into the same resolution range as film.

Figure 9.12. This sharp image of M64 was achieved by selecting the best 10 of 20 3-minute subs. Same equipment as Figure 9.11. Star image diameter (FWHM) is 6.2 pixels = 3.5″, roundness, $e = 0.47$.

You can go further. It smacks of fakery, but there is a simple way to get very round star images – just reduce your image to perhaps 1/10 of its original size, then enlarge it again. With the right reduction and enlargement algorithms, you can produce an image that seems to be very well guided, though it lacks resolution.

9.8.5 Deconvolution

The last trick for improving badly guided images is a surprising one – motion-blur deconvolution. Elongation of star images, and of everything else in the picture, can be undone digitally. "Deconvolution" means undoing a change to an image, in this case smearing.

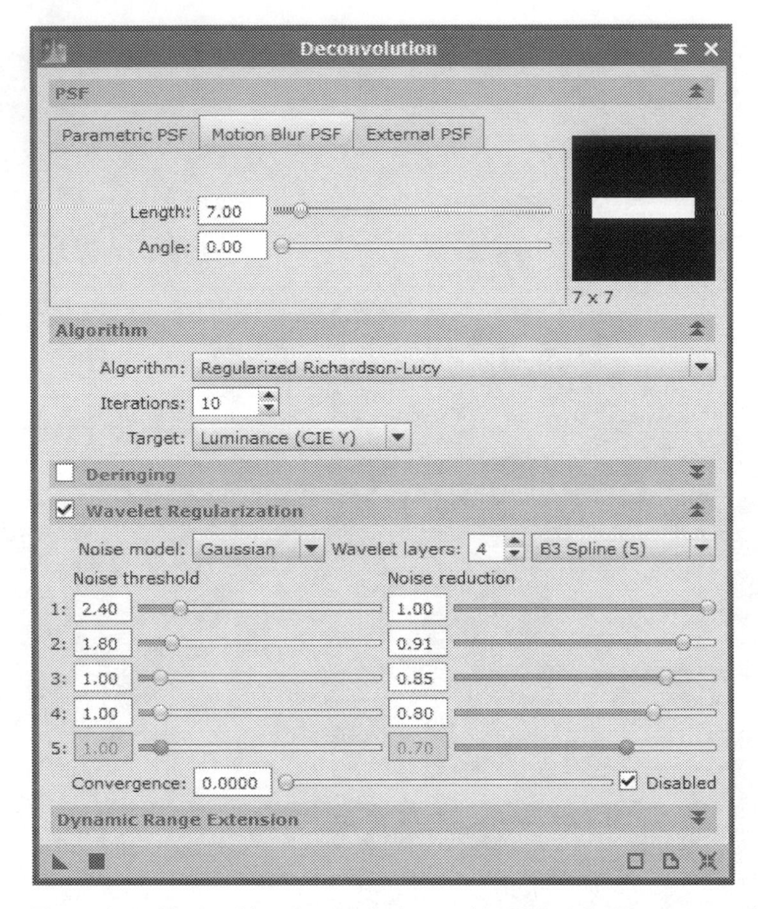

Figure 9.13. Motion-blur deconvolution tool in *PixInsight*. Here more noise reduction is chosen than the default.

The key idea is that any blur that is mathematically understood can be mathematically reversed. Unsharp masking is a special case; it undoes a Gaussian blur, approximately. Undoing any other kind of blur is a slow, iterative process, but it can be done. Only small guiding errors can normally be corrected, but that is often enough.

This is a method that needs to be further developed. Figure 9.13 shows the motion-blur deconvolution tool in *PixInsight*, and Figure 9.14 shows what it can do. You select the elongation that best matches the star images in length and orientation, then apply the filter. A great deal of trial and error is required. Somewhat to my surprise, it does not seem to make much difference whether the image is linear or gamma-stretched, but in either case the noise level increases, and there are likely to be artifacts around bright stars. A more sophisticated approach would use a mask to apply the deconvolution mostly or entirely to the bright stars, not the fainter surface of the galaxy or the background.

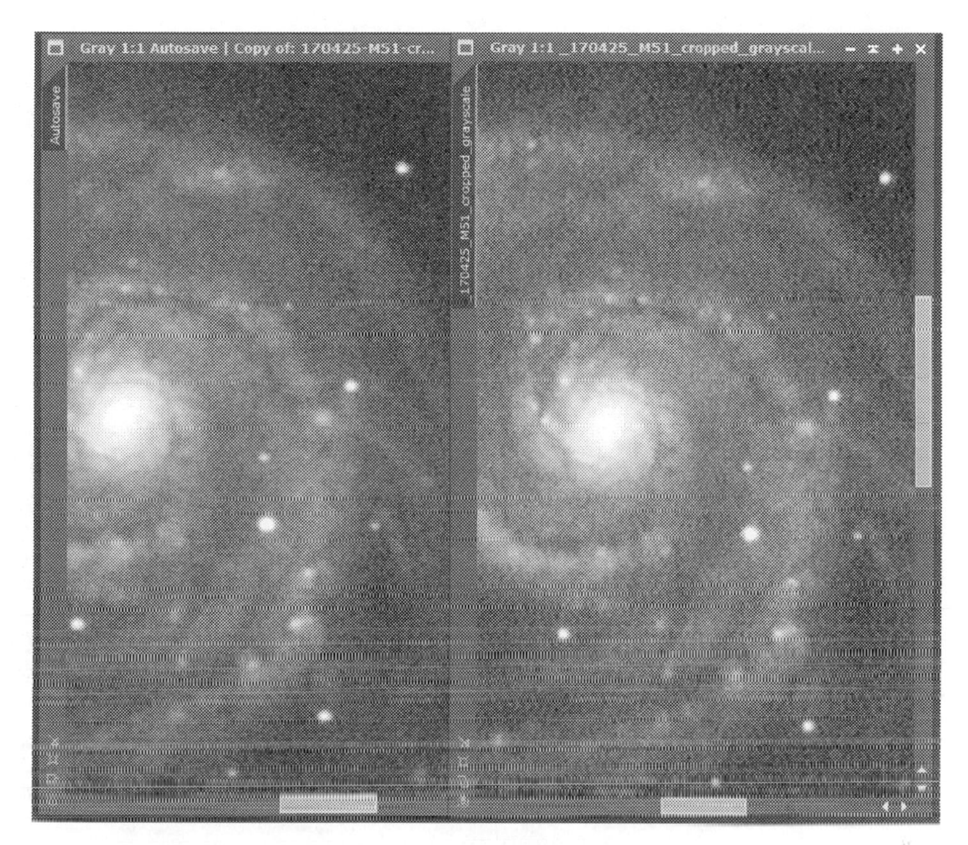

Figure 9.14. Effect of motion-blur deconvolution on an image. Elongated stars become rounder but noise increases.

Normally, deconvolution is used for a different purpose, as an alternative to unsharp masking to bring out fine detail. Both *PixInsight* and *MaxIm DL* offer it for that purpose, but only *PixInsight* offers it neatly packaged to undo motion blur. (*PixInsight* also offers a simpler matrix-based filter to counteract motion blur, and something similar is in *Photoshop* under Smart Sharpen.) I would like to see people develop deconvolution tools aimed specifically at correcting imperfect guiding.

Chapter 10
Power and Camera Control in the Field

Although some observing sites are blessed with AC power, most astrophotographers rely on batteries in the field. In fact I often use battery power even at home, both to make sure my field-trip equipment is working and for greater safety. Many of us also bring computers into the field, both for autoguiding and for camera control.

Since this is a book about image acquisition and processing, not primarily about interfacing electronic equipment, this chapter will be brief; I'll just give you the basics. One important change since the previous edition is that we no longer have to build our own gadgets; all the components needed for an elaborate system are available off the shelf.

A word of caution: Keep it simple. As your power and control system grows, make sure you understand each component (and are sure it works correctly) before adding more. It can be very frustrating to string together five or ten pieces of equipment and find the combination not working, but not know where to look for the problem.

10.1 Portable Electric Power

10.1.1 Power for the Telescope

Most telescopes operate on 12 volts DC and can be powered from the battery of your parked car. In fact, telescopes usually come with a lighter socket adapter for doing this, but I don't recommend it. What if you drain the battery and the car won't start when you're ready to go home?

It's much better to use portable lead-acid batteries, either gel cells or deep-cycle boat batteries. Check the battery voltage periodically during use and don't let it get below 12.0. Recharge the battery before storing it. Store it with a float charger connected. Float chargers that can also recharge a battery overnight are cheap and widely available.

Figure 10.1. The author's battery pack, made from a "jump starter" containing a 12-volt gel-cell battery.

Figure 10.1 shows my main battery pack, made out of a 12-volt "jump starter" from an auto parts store. I removed the big, heavy jumper cables and added more connectors and a digital voltmeter. It holds an 18-amp-hour gel-cell battery which I replace every few years. I carry a second battery pack as a backup or when a laptop is to be used.

The capacity of a rechargeable battery is rated in ampere-hours (amp-hours, AH), thus:

$$\text{Hours of battery life} = \frac{\text{Battery capacity in ampere-hours}}{\text{Current drain in amperes}},$$

or sometimes in milliampere hours (mAH), where 1 AH = 1000 mAH.

Battery packs that perform voltage conversion are often rated in watt-hours (WH) or milliwatt-hours (mWH), where:

$$\text{Ampere-hours} = \frac{\text{Watt-hours}}{\text{Voltage}},$$

and again 1 WH = 1000 mWH. So a pack rated at 48 000 mWH = 48 WH, when delivering 12 volts, has a capacity of 48/12 = 4 AH, not especially big.

If possible, you should measure the current drawn by your equipment using an ammeter. If you have to guess, a telescope mount draws about 0.5 ampere (amp) (with peaks of 3 or 4 amps when slewing fast), a dew heater system draws

1 to 3 amps, and an autoguider or laptop computer may draw 2 to 5 amps. That's enough to run down a commonly available 17-AH battery pack very quickly.

In such a situation, a 60-AH deep-cycle battery is worth paying for. Better yet, put the telescope on one battery, the laptop on another, and the dew heater on a third. This will prevent ground loops and will ensure that a mishap with one piece of equipment doesn't disrupt everything.

Don't use a car battery; it is not designed for deep discharge, and if you use it within its rated limits, it won't do you any more good than a much smaller, lighter gel cell pack. If you discharge it deeply, it will wear out prematurely.

Recently, lithium-ion battery packs have become available that deliver 12 volts. They are delightfully lightweight and maintenance-free (you can charge one, put it away for six months, and use it, with no float charging in the meantime), but, at present, expensive. The cost may fall rapidly.

10.1.2 DC Power Connectors

The power connector to the telescope mount is usually a DC coaxial plug (Figure 10.2). These come in many sizes, several of which look alike. In particular, Meade equipment uses a plug 5.5 mm in outside diameter that accepts a 2.5-mm pin in the middle. Celestron's plug is the same except that it expects a 2.0-mm pin (sometimes listed as 2.1 mm). If you plug a Meade plug into a Celestron mount, you'll get a poor connection; the mount may be completely dead, or it may work most of the time but fail momentarily when the telescope makes a sudden movement. Other sizes to look out for are 5.0 × 2.0 mm and 5.0 × 2.5 mm; these go into 5.5-mm sockets but often fail to make a good connection. The barrel length also varies but is not the crucial dimension.

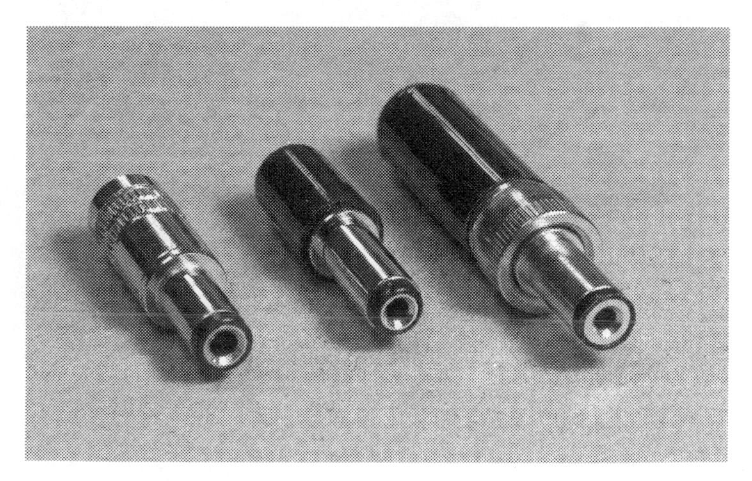

Figure 10.2. Coaxial plugs: 5.5 × 2.5 mm (Meade), 5.5 × 2.0 mm (Celestron and iOptron), and 5.5 × 2.0 with locking nut (Switchcraft S761K, used by Celestron).

Polarity is *extremely important.* Meade, Celestron, and iOptron DC plugs have positive in the center. Many pieces of equipment are not protected against reverse polarity and will suffer severe damage if positive and negative are reversed. Several years ago, I used an SBIG autoguider whose plug was just like that of my Meade telescope except that it was negative in the center. One false move could have damaged the telescope severely. I ended up modifying the telescope to add a protective diode. Note that conventional rectifier diodes cost you 0.7 to 1.0 volt; Schottky diodes are better, and it's better yet to use a fuse followed by a diode that would conduct only if polarity is reversed, blowing the fuse.

At the other end of the cable, lighter socket connectors are bulky and somewhat unreliable. Many portable equipment operators in the ham radio community have switched to the Anderson Powerpole system (www.andersonpower.com). In fact, ham radio operators have a lot of experience with portable 12-volt power systems; you can learn a lot by making contact with them.

For about 20 years, I have used RCA phono plugs for 12-volt power, always positive in the center, and always using plugs for devices that consume power, and sockets for sources of power (analogous to AC line plugs and sockets); this ensures that plugs are never "live" and will not short out if they touch metal. I've been meaning to convert to Anderson Powerpoles, but it hasn't happened.

Powering a mount that may draw 3 or 4 amperes requires a thick cable. Thin wires designed for smaller equipment may have enough resistance to drop the voltage appreciably. I have found nothing better than ordinary 2-wire lamp cord (Figure 10.3).

10.1.3 Voltage

A fully charged 12-volt lead-acid battery actually delivers 12.6 volts, or briefly, even more (up to 13.3 or so) when freshly charged. These voltages do not harm 12-volt equipment, which can generally tolerate up to 13.8 volts, the nominal voltage of an automotive 12-volt system with the engine running. Avoid higher voltages unless you are sure they are safe for the equipment.

On the other end of the scale, 12-volt telescope mounts generally work well at voltages down to 11.0 and sometimes a good bit lower, provided the voltage does not dip precipitously when the telescope slews. Discharging below 11 volts is not good for a lead-acid battery, though.

For readers unfamiliar with electrical theory, I should explain that volts are what the power supply delivers, and amperes are what the load takes. A 12-volt, 4-ampere power supply will not harm a 12-volt, 1-ampere mount. The voltage of the power source needs to be correct, and the amperage needs to be equal to or higher than what the load demands. Part of the appeal of batteries is that they can deliver heavy current momentarily (such as when the telescope slews) even if their typical average load is much less.

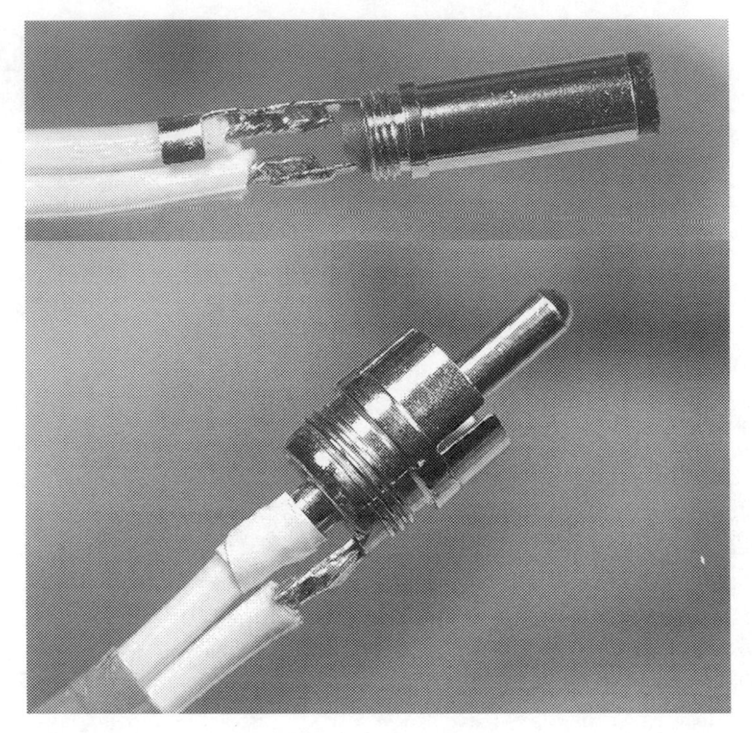

Figure 10.3. Soldering phono plug to 18-gauge lamp cord. Note use of tape or heat-shrink tubing as extra insulation. Empty space is the best insulator, if you can be sure it will stay empty.

10.1.4 Powering the Computer and Camera

The most convenient way to power a laptop computer in the field is to use its own battery, which is good for two or three hours; you can bring a fully charged spare or two. To save power, turn down the LCD brightness, reduce the CPU speed if you have that option, and turn off the network adapter.

Second choice is to equip the laptop with a 12-volt power supply, such as you would use in a car or aircraft, and connect it to the main 12-volt battery. To reduce drain on the battery, start with the laptop fully charged so that it does not have to charge its battery.

Third choice, distinctly less efficient, is to use an inverter to convert 12 volts DC to 120 or 240 volts AC, and feed that to the laptop's AC line power supply. The problem is that every voltage conversion wastes energy; even the best switchmode converters are not 100% efficient. But a big advantage of DC-to-AC inverters is that they provide transformer isolation, eliminating ground loops (see next section).

The same goes for the camera. My own approach is to bring three or four fully charged camera batteries and swap them as needed. I can keep them in my pocket (in insulated wrappers, of course) so that they don't get too cold.

The alternative is to use an external power supply for the camera, if available. There are also "car chargers" for DSLR batteries, so that you can recharge one camera battery from your big 12-volt battery while the camera uses another one.

10.1.5 Care of Li-ion Batteries

The lithium-ion (Li-ion) batteries commonly used in cameras and laptop computers have a big advantage over earlier NiCd and NiMH types: they retain much more of their charge while unused. You can charge a battery, put it away, and expect it to work without recharging a month later.

Li-ion batteries always require a "smart charger" rather than just a source of regulated voltage or current. The smart charger senses the state of the battery to charge it rapidly without overcharging.

This means no harm will result if you "top up" your already-charged Li-ion batteries before an observing session. In fact, I usually do so. This is done by putting them in the charger in the usual way. In a few minutes, it will report that the battery is fully charged, but keep the battery in the charger anyway; it continues to gain charge for as much as an hour after the full-charge indicator comes on.

As batteries age, they lose capacity. There is no quick trick for restoring old Li-ion batteries, but sometimes, all they need is another charging cycle. One of my Canon batteries became almost unusable, but I ran it through the charging routine twice, and since then, it has worked fine. Perhaps one of the measurements used by the smart charger had come out of calibration.

10.1.6 Ground Loop Problems

If you use one battery for several pieces of equipment, you may have problems with *ground loops*. A ground loop is what happens when two devices are tied together through "circuit ground" (typically at USB or serial ports), and also share a power supply, but the voltage of circuit ground relative to the power supply is not the same in both devices.

Here's an example. If you power a laptop through a 12-volt power supply, its circuit ground may be tied to the negative side of the battery. But circuit ground in a classic (non-GPS) Meade LX200 telescope is not connected to the negative side of the battery; the current indicator (LED ammeter) is between them. If you connect the laptop to the telescope, serial port to serial port, you'll have a ground loop. Fortunately, in this case only the LED ammeter will be disrupted.

Fortunately, ground loops are not common, and well-designed equipment is not vulnerable to them. (I think Meade made a wrong decision with the classic LX200, not anticipating the complex setups it would be connected to.) You should suspect a ground loop if you have equipment that malfunctions only when it shares the battery of another piece of equipment to which it is connected.

One way to eliminate ground loops is to power each device from a separate battery. Another is to use an AC inverter followed by the AC power supply of the device, which is required by safety regulations to provide true transformer isolation. In fact, one inverter can power several accessories, all of which are isolated by their own AC power supplies.

10.1.7 Safety

Think of any rechargeable battery as something like a fuel canister. It contains a tremendous amount of energy, and if short-circuited, it can produce intense heat, explode, and start a fire.

It's obvious that a large lead-acid battery deserves respect; it's big, heavy, and full of sulfuric acid. There should always be a fuse between a lead-acid battery and whatever it is powering, preferably a separate fuse for each piece of equipment, as in the electrical system of a car or boat.

Do not connect an AC line power supply directly to a rechargeable battery unless the power supply is designed for use as a float charger. The reason is that some power supplies contain a "crowbar" circuit that shorts the output if the voltage goes too high, the theory being that it is much better to deliver 0 volts (and blow the fuse) than deliver an excessive voltage. That's fine as long as no battery is present, but shorting across a battery can make it explode. To use both a line power supply and a battery, connect them with a battery isolator that uses diodes to keep current from flowing backward into the power supply. It is also not a good idea to connect rechargeable batteries in parallel, even if they are well matched, because a shorted cell in one of them would cause current to flow backward, discharging and overheating all the remaining cells in both batteries.

Small rechargeable batteries for cameras or laptops are also potentially perilous. *Always keep camera batteries wrapped in insulating material.* If you drop an uncovered DSLR battery into your pocket along with your keys, you may soon find your clothing on fire. I keep camera batteries in thick plastic bags, and I only put them in my pocket if the pocket doesn't contain anything else.

The AC line at your observing site, if there is one, must be protected by a ground-fault circuit interrupter (GFCI). This does not totally prevent electric shock, but it does protect you in situations where electricity tries to flow through your body from the power lines to the earth. Remember that there will be heavy dew during the night; keep high-voltage connections dry. Also remember that AC from an inverter is almost as dangerous as AC from the mains; the only difference is that the output of the inverter is isolated from the earth.

10.2 Camera Control

10.2.1 How Camera Control is Done

Software that can control DSLRs is abundant. Most camera manufacturers bundle something of the sort with the camera or offer it as a free download, although the astronomy-specific software recommended later in this section is much better. Camera control is also built into *Nebulosity*, *MaxIm DL*, and other general-purpose astrophotography software packages. It offers you the ability to do everything from the computer – including viewing the screen – that you could do from the camera itself.

Camera control relies on device drivers supplied by the camera maker. You may need to install the camera maker's control software, even if it's not adequate for astronomy, in order to get the device drivers. In general, only Windows is supported, and some older DSLRs only have drivers for 32-bit editions of Windows.

Most cameras communicate entirely through the USB cable, but a few older DSLRs cannot make long exposures unless a second connection is made from the computer to the camera's cable release socket through an interface circuit. Suitable interfaces are available from www.shoestringastronomy.com, and plans to enable you to make your own are published by some software makers. The Canon interfaces described in the previous edition of this book work with the cameras for which they were designed, but not necessarily others.

10.2.2 Choosing a Laptop

There are two reasons not to use your best laptop for telescope control in the field. One is to avoid exposing it to hazards such as dew and accidental damage. The other is because you may need a different operating system. Your best laptop might be a MacBook or Linux system, but camera control is almost confined to Windows. Even if you are a Windows user, you might need a 32-bit edition of Windows to control an older DSLR. Or you may have to roll back operating system updates in order to keep your equipment interfaces working (as happened with Windows 10 in 2016 and 2018), and this may leave the computer not secure enough for online banking and regular web surfing.

A secondhand laptop can be very inexpensive, and you don't need great computer power to autoguide and control a camera. (Image processing is a different matter.) Choose something designed for long battery life, not high CPU speed, and look for an illuminated keyboard if possible.

10.2.3 Cables

You will need a USB cable that fits your camera, probably a longer one than usual, and it may need a special connector on the camera end. Long, reliable camera USB cables – bright yellow so you can see them on the ground – are made by Tether Tools (www.tethertools.com) and used widely by professional photographers.

10.2.4 Camera Control Software

Two camera control packages specially designed for DSLR astrophotography are *BackyardEOS* and *BackyardNIKON*, both from O'Telescope Corporation of Ottawa (www.otelescope.com). Besides making available all the camera functions you need, including Live View on the computer screen, they also help with drift alignment, analyze focus, can focus telescopes with electric focusers through ASCOM drivers, and even download and log current weather information for your site. Figure 10.4 shows *BackyardNIKON* in action.

They also perform automatic dithering by communicating with *PHD2* or *Metaguide* or directly through ASCOM mount drivers. Dithering means the telescope is aimed slightly differently for successive exposures so that the same features don't fall on the same pixels.

Sequence Generator Pro, from Main Sequence Software (http://MainSequenceSoftware.com), approaches the same goal in a less DSLR-specific way; it also works with dedicated astrocameras. The functionality offered is largely the same, but with more emphasis on planning and automating an entire

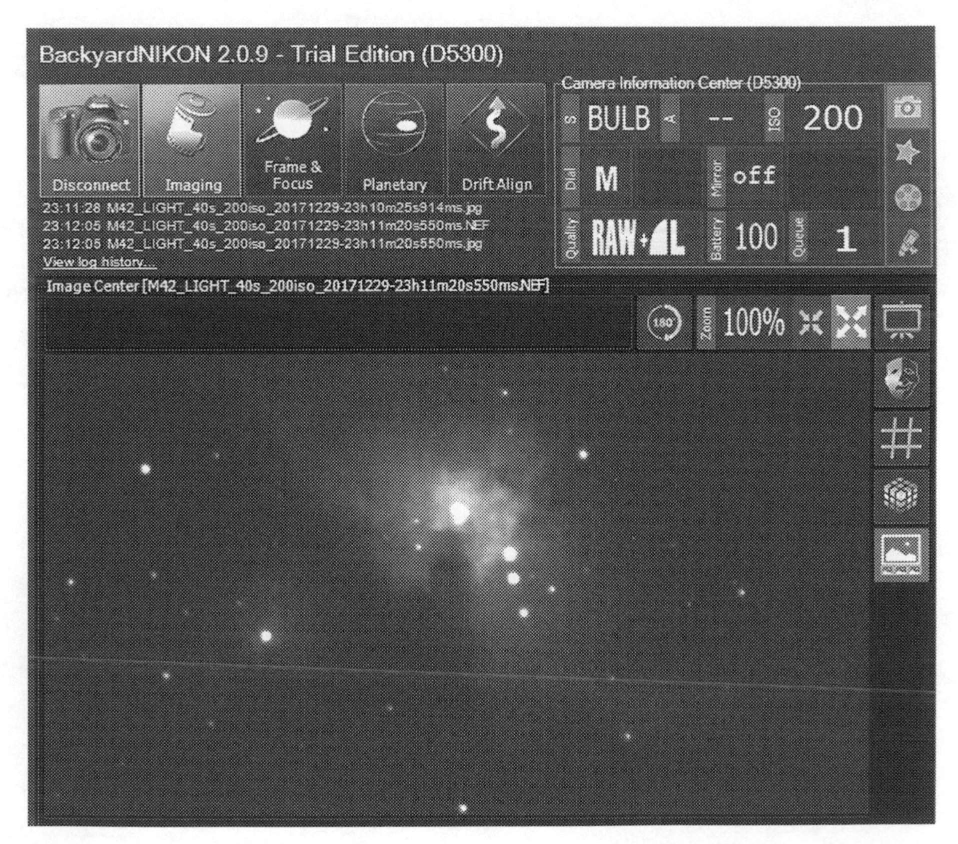

Figure 10.4. *BackyardNIKON* in action, controlling a camera, with Live View on the computer screen.

astrophotography session, so that you set the computer going, make sure it's autoguiding, and walk away (which is also possible with *BackyardEOS* and *BackyardNikon*).

10.3 Networking Everything Together

Now it's time to sum up. The typical modern DSLR astrophotographer uses a laptop computer for control, a video camera for autoguiding, and a computerized telescope mount. Here are all the connections that have to be taken care of:

- Power for the telescope, the computer, and the camera (preferably all separate);
- USB or serial connection from the computer to the telescope for guiding, and possibly also for telescope control (finding objects) and electric focusing;
- USB connection from the computer to the autoguiding camera (this also supplies power to the autoguiding camera);
- USB connection from the computer to the DSLR to control settings and download images;
- Possibly a serial- or parallel-port cable from the computer to the DSLR cable release socket to control long exposures.

If the telescope is an older one with a 6-pin modular autoguider port, and there is no 6-pin interface on the autoguiding camera, you can use cables available from www.shoestringastronomy.com that connect a 6-pin port to the computer's parallel or USB port. And if you need serial or parallel ports that you don't have, use USB-to-serial or USB-to-parallel converters.

Now count the connections. You may need as many as five USB ports, in which case a USB hub enters the picture. Small USB hubs, powered from the USB port itself, are to be preferred if they work adequately.

In a few years, we will be using short-range wireless communication instead of a rat's nest of USB cables. Already, some equipment has Wi-Fi or Bluetooth communication, and some experimenters have gone so far as to mount a single-board Windows computer on the telescope mount and access it via Remote Desktop Protocol over Wi-Fi. As wireless technology progresses, the only cables we will have to deal with will be for power.

10.4 Operating at Very Low Temperatures

The most obvious effect of low temperatures on DSLRs is that there is a lot less dark current. This is a good thing and is why astronomical CCDs are internally cooled. It's also why we take dark frames at the same temperature as the exposures from which they are to be subtracted.

A second effect, not so benign, is that batteries lose some of their capacity. If you remove a "dead" battery from a camera and warm it up in your hand, it will often come back to life. LCD displays may also lose contrast in the cold, but the effect is temporary.

There are unconfirmed reports of laptop screens and laptop batteries being harmed by freezing. Of course, a running laptop gives off a fair bit of heat, so the greater concern is about equipment left outdoors for a long time unused.

At temperatures down to about 0° F (−18° C), those are probably the only effects you will notice. At lower temperatures, though, electronic components may begin to malfunction. Those most likely to be affected are flash memory devices and data communication chips because they rely on voltage thresholds that can shift with temperature. This means your camera memory card may have problems, as well as the flash ROM, USB port, and serial port of a microprocessor-controlled camera or telescope. Disk drives are typically not rated to work below 32° F (0° C).

Do not bring a very cold camera or laptop computer directly into warm, humid air, or moisture will condense on it. Instead, put it in a plastic bag, or at least a closed case, and let it warm up gradually. If condensation forms, do not turn it on until it has dried.

Part III
Image Processing

Chapter 11
Deep-sky Image Processing

The purpose of this chapter is to tell you what deep-sky image processing consists of; the next chapter will tell you how to do the basic steps with three major software packages.

11.1 Processing Workflow

Figure 11.1 shows my preferred workflow for image processing. There are several variations on it, which we'll get to. Some people may be disconcerted that this version of the process does not include bias frames, but it is correct as shown, as we shall see.

One thing that makes image processing seem complicated is that there are many names for almost everything in it. We've already mentioned that flat darks are the same thing as dark flats. Color images with a Bayer matrix are also called CFA (color filter array) or OSC (one-shot color) images. Decoding them is called deBayering or demosaicing. Then we align (register) and stack (combine, integrate) them. Through this step, the image is linear (not gamma-corrected). Then it undergoes nonlinear stretching, or gamma correction, or histogram adjustment (all three are the same thing) and becomes viewable. The process is not nearly as complex as the vocabulary suggests!

11.2 Calibration

11.2.1 Image Arithmetic

Chapter 5 covered how to make calibration frames; this chapter explains how they are used. Calibration is based on the concept of image arithmetic or pixel arithmetic. From here on, an equation like

$$\text{Image 1} = \text{Image 2} + \text{Image 3}$$

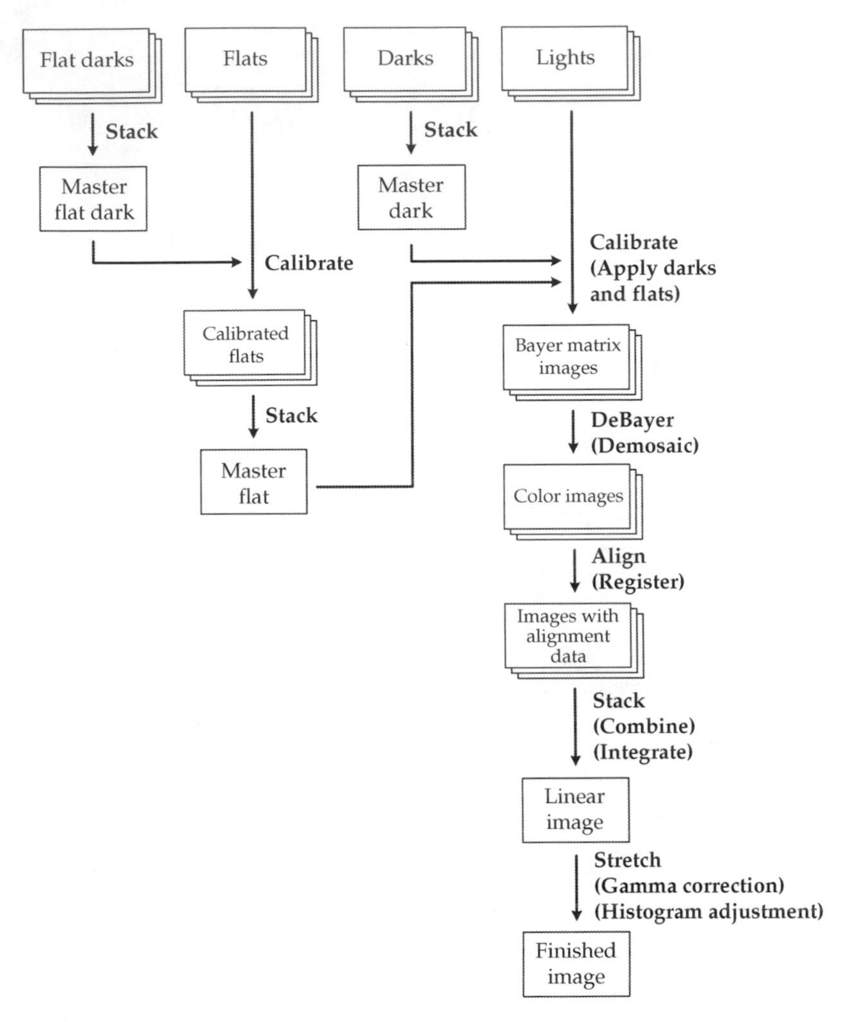

Figure 11.1. Workflow for image processing. Fortunately, most of this can be automated.

means that every pixel value in Image 1 is obtained by adding the pixel value at that position in Image 2 to the pixel value in the same position in Image 3.

That holds for all operations, including multiplication and division. For example,

$$\text{Image 1} = \frac{\text{Image 2}}{\text{Correction}}$$

means to divide every pixel in Image 2 by the value in the same position in the correction image (which is not exactly an image, but does have numbers in the positions of all the pixels). This is not matrix division; pixel arithmetic always applies to individual pixels in matching positions.

11.2.2 Components of a Raw Image

Now calibration can be defined more precisely. Dark frames and flat fields are used to overcome:

- **Dark current**, leakage within sensor pixels that makes them act as if they're receiving some light even when they're not. This comprises "hot pixels" that always read as very bright, plus ordinary pixels that are a bit leaky. Dark current is proportional to exposure time and ISO setting and increases with temperature. It is *added* to the original image in the camera.
- **Bias**, the fact that pixel values don't start at zero. Even with no light and no exposure, each pixel typically has a nonzero value. Like dark current, this nonzero starting point is *added* to the original image in the camera. In fact, DSLRs add bias to ensure that they do not need negative numbers.

  Bias frames are strictly necessary only if the software wants to convert a dark frame to a different exposure time than it was taken with. Otherwise, you do not need bias frames because dark frames and flat darks include bias.
- **Uneven response**, which comprises two things: differences in sensitivity from one pixel to another, and differences in the amount of light reaching the pixels because of vignetting, dust, and other obstacles.

  Uneven response is counteracted with flat fields, from which we extract a mathematical model that tells how much each pixel responds to light. For example, a sensor pixel that responds to only three quarters of the usual amount of light is modeled with a factor of 0.75. We treat the pixels in the image as if their light levels had been *multiplied* by this factor, and we correct it by dividing.

Table 11.1 describes the kind of images that come out of the camera. Translating the formulae into English:

- A bias frame contains only the bias. The exposure is too short for it to record anything else.
- A flat dark contains bias plus the dark current from a very short exposure. *If the exposure is about 1/10 second or less, the dark current is indistinguishable from zero, and flat darks are the same thing as bias frames.*

Table 11.1 *Components of five kinds of images.*

Bias frame =	Bias
Flat dark =	Short-exposure dark current + Bias
Flat field =	(Constant × Uneven response) + Short-exposure dark current + Bias
Dark frame =	Long-exposure dark current + Bias
Light frame =	(Astro image × Uneven response) + Long-exposure dark current + Bias

- A flat field contains what a flat dark contains, plus an image of a constant amount of light (a uniform bright surface) affected by uneven light transmission and pixel sensitivity. We model this as a constant exposure multiplied by the uneven response.
- A dark frame is like a flat dark except that the exposure is longer, to match the light frames.
- A light frame is like a flat field except for two things: the light comes from an astronomical object rather than a uniform source, and the dark current comes from a longer exposure, because light frames are typically exposed for seconds or minutes, while flat fields are only exposed for a small fraction of a second.

What we want is to recover what I called "Astro image," initially as a correct linear image (what would have come out of a perfect sensor, or as near as we can get), and then gamma-correct it to account for the fact that pixel values on a computer screen have a radically different meaning than pixel values on a sensor.

To perform whichever of the following calculations your software supports, generally all you have to do is tell the software where to find the files, and set it going. But it is useful to know how the calculations are done.

11.2.3 Master Darks, Flats, Flat Darks, and Bias Frames

Normally, each calibration frame is actually a stack of many, averaged together to reduce random noise. Typically you might stack 10 to 100 lights, the same or a greater number of darks, at least 10 flats, and at least 10 flat darks. As Figure 11.1 shows, calibration frames are stacked early, because they do not need alignment, but light frames are stacked after most of the other processing, because they need to be aligned to star image positions rather than leaving all the pixels where they were.

11.2.4 Should Flats Be Binned or Smoothed?

It used to be common practice to bin the pixels of a flat-field image 2×2 or blur them on a larger scale. The idea was that flats represent overall gradients due to vignetting and the like, and differences between individual pixels should be smoothed out. Binning is still default in some software, such as *Nebulosity*.

That is no longer recommended. The imaging community has realized that one of the functions of flat fields is to correct for differences in response between individual pixels. For that to work, the pixels must be handled individually.

Binning flat fields 2×2 does have one benefit – it makes the Bayer matrix colorless. The flats can no longer introduce (or correct) a color cast.

11.2.5 Method 0: Just Lights and Darks

The simplest way to calibrate images is to leave flat fields out of the picture (literally) and just subtract dark frames from lights. Then:

$$\text{Astro image} \approx \text{Light frame} - \text{Dark frame}.$$

As you can see by comparing to the formulae in Table 11.1, this gets rid of dark current and bias but does not correct for uneven response. It is what cameras do automatically when long exposure noise reduction is turned on.

In Method 0, the lights and darks must have the same exposure time and ISO setting, and, as far as possible, be taken at the same sensor temperature.

11.2.6 Method 1: Lights, Darks, Flats, and Flat Darks

Perfect calibration results from these calculations:

$$\text{Uneven response} = \frac{\text{Flat field} - \text{Flat dark}}{\text{Average pixel value}},$$

$$\text{Astro image} = \frac{(\text{Light frame} - \text{Dark frame})}{\text{Uneven response}}.$$

That is just like the previous example except that correction for uneven response, using flat fields, is included.

The first formula requires some explaining. We want to model uneven response such that normal pixels are 1.0, pixels with half the normal response are 0.5, and so on. Then, by dividing by that model, we can correct the image to look as if every pixel responded to light equally well. Pixels that responded to only half the light will be doubled, and so on.

In Table 11.1, the flat fields include a constant that we cannot recover. That doesn't matter. To model uneven response, we don't need to know the original exposure, only the proportional response of each pixel. That is achieved by dividing by the average pixel value.

What is much more important is that the flat fields have to be dark-subtracted using flat darks. If that is not done, we are not working with values truly proportional to the response to light.

Here is a worked example to show why that is so. Suppose one pixel has a true response of 0.5, and another, 1.0. Suppose further that the camera's pixel values are on a scale from 0 to 16 383 and the flat fields were given an exposure corresponding to 10 000 pixel value units (DNs, as they are termed in Chapter 15). Suppose further that the camera adds a bias of 1024 to each pixel. Then the two pixels we are considering will have values of:

$$(0.5 \times 10\,000) + 1024 = 6024,$$

$$(1.0 \times 10\,000) + 1024 = 11\,024.$$

The ratio between them will be 6024/11 024 = 0.546, which is not correct. To model the first pixel correctly as 0.5 of the second one, we have to subtract the bias first (plus any dark current that might be present, having the same effect). Failure to do this causes undercorrection; flat-fielding does not do quite as much good as it ought to.

One advantage of Method 1 is that flats and lights do not have to be taken at the same ISO setting. They have to be taken with exactly the same optics, of course, but because there are no bias frames, nothing has to match the ISO setting of the bias frames. The dark frames have to match the ISO setting of the lights, and the flat darks have to match the ISO setting of the flats. Taking flats at a low ISO setting (such as 200) is convenient and leads to lower noise.

11.2.7 Method 2: Lights, Darks, Flats, and Bias

An older way of doing calibration includes bias frames but not flat darks. The advantage of this method is that, with bias known, the dark frames can be (approximately) scaled to different exposure times, if necessary (a point to which we shall return).

Note that, with reasonably short exposures (under 1/10 second or so), bias frames are the same thing as flat darks anyhow, so this method is what you get with older software if you simply put in the flat darks as bias frames.

The main difference from the previous method is that, because bias frames are used with both flats and lights, the flats and flat darks have to be taken at the same ISO as the lights. The calculation is:

$$\text{Long-exposure dark current} = \text{Dark frame} - \text{Bias frame},$$

$$\text{Uneven response} \approx \frac{\text{Flat field} - \text{Bias frame}}{\text{Average pixel value}},$$

$$\text{Astro image} = \frac{\text{Light frame} - \text{Long exposure dark current} - \text{Bias frame}}{\text{Uneven response}}.$$

The last formula is the same as in Method 1, but calculated in a more roundabout way. A bias frame was subtracted from long exposure dark current, which is now being subtracted from the light frame; that means the bias is being added back in and needs to be subtracted again.

Note also the $\approx$ sign in the second formula. Theoretically, it is more accurate to calibrate a flat field against a flat dark than a bias frame, but if bias frames and flat darks are indistinguishable, the point is moot.

Method 2 requires all the calibration frames to be taken at the same ISO setting because lights, darks, and flats are all calibrated using the same bias frames.

11.2.8 Method 3: Lights, Darks, Flats, Flat Darks, and Bias

The most thorough kind of calibration requires all four kinds of calibration frames, all taken at the same ISO setting as the lights. It is as accurate as Method 1 but allows scaling the dark frames to different exposure times if your software supports it. Here's how it's done:

$$\text{Long-exposure dark current} = \text{Dark frame} - \text{Bias frame},$$

$$\text{Short-exposure dark current} = \text{Flat dark} - \text{Bias frame},$$

$$\text{Uneven response} = \frac{\text{Flat field} - \text{Short-exposure dark current} - \text{Bias frame}}{\text{Average pixel value}},$$

$$\text{Astro image} = \frac{\text{Light frame} - \text{Long-exposure dark current} - \text{Bias frame}}{\text{Uneven response}}.$$

The fourth formula is roundabout in the same way as in Method 2. The third formula is also roundabout; it is actually equivalent to the formula for uneven response in Method 1. In both of these cases we are subtracting from the image a dark current from which bias has been subtracted. Because the bias was left out, it then has to be subtracted as a separate step. Method 3 makes the theory clearest but doesn't minimize the arithmetic.

Like Method 2, Method 3 requires all the calibration frames to be taken at the same ISO setting as the lights because the same bias frames are used on all of them.

11.2.9 Scaling the Dark Frames

What if your dark frames were taken at a different exposure time, or temperature than your astronomical images? Can you still use them?

Maybe. Scaling of dark frames was a widespread practice in early CCD astronomy, but many people found it unsatisfactory, and normal practice now is to match the darks to the lights whenever possible. But the need for it may be coming back as we use DSLRs that record their sensor temperatures but do not control them. It is often painfully obvious that the darks do not match the temperatures of the lights.

Of course, we only want to scale the dark current, not the bias. That implies we have done the calculation

$$\text{Long-exposure dark current} = \text{Dark frame} - \text{Bias frame}$$

in Methods 2 and 3. Any algorithm to scale the dark frames has to have bias frames, or flat darks that can serve as bias frames.

What we want is:

$$\text{Scaled long-exposure dark current} = \text{Long-exposure dark current} \times \text{Factor},$$

where the factor allows us to apply the old dark frames to the new lights taken under different conditions.

How do we get the factor? Part of the answer is easy: The effect of dark current is (or should be) proportional to exposure time and ISO setting, so:

$$\text{Factor} = \frac{\text{New exposure time}}{\text{Old exposure time}} \times \frac{\text{New ISO setting}}{\text{Old ISO setting}}.$$

If the temperatures are also mismatched, you can also multiply the factor by

$$\exp[0.09 \times (\text{New temperature in }^\circ\text{C} - \text{Old temperature in }^\circ\text{C})]$$

which *may* be a reasonable model of the effect of temperature on dark current.

There is a better way. In *PixInsight*, dark frames can be scaled automatically by analysis of their actual noise levels. Measuring the change in dark current (which is irregularly distributed across pixels, hence measurable as noise) is much better than trying to estimate it. The option is called "Optimize dark frames" in the ImageCalibration and BatchPreprocessing tools.

11.3 Cosmetic Correction

Dark-frame subtraction is not the only way to get rid of hot pixels. It is not even necessarily the best way. If the dark frames are adjusted in any way (such as the way *PixInsight* adjusts them to minimize final noise level), subtraction no longer takes out hot pixels perfectly.

Enter *cosmetic correction*, the automatic detection and removal of deviant pixels in the image, whether or not a dark frame is available. The key idea is that our sensors are sharper than our optics, so no genuine object in the image will occupy a single pixel. (Images of stars, which are point sources, are typically 3 pixels in diameter.) Accordingly, any pixel wildly different from its neighbors is incorrect and should be removed. In its place is put the average of the adjacent pixels.

Cosmetic correction in major software packages is covered in the next chapter. When included in the process, it is an extra step in Figure 11.1, right after calibration.

11.4 DeBayerization

After images have been calibrated, but before they are aligned and stacked, the Bayer color matrix (the CFA, color filter array) must be decoded (a process also called demosaicing).

The traditional way to deBayerize images was to fill in the unknown color layers of each pixel (e.g., the R and G layers of a B pixel) by averaging its neighbors (Figure 2.2). It was quickly discovered that this is not enough for images containing high-contrast fine detail – which is what star images are. Accordingly, newer algorithms detect brightness gradients in the image also, and software commonly provides several deBayerization algorithms. Adaptive homogeneity-directed (AHD) deBayerization is highly recommended, with the variable number of gradients (VNG) algorithm as second choice. The smaller the pixels are, the less this matters.

One approach to deBayerization that I definitely do not recommend is *superpixel mode*, which means binning the pixels 2×2 in order to deBayerize. You simply take a square group of R, G, G, and B pixels, average the two G values, and take the R and B values as given. Every four pixels in the original image (R, G, G, B) become a single "superpixel."

The trouble with superpixel mode is that the edge of a star image can easily intersect one of these quartets of pixels such that it hits only the R, or only the B, or only a G. The result is a brightly colored pixel when there was no bright color present in the celestial object. All deBayerization algorithms run some risk of color fringing, as it is called, but this one particularly so. It is much better to perform deBayerization using information from all the neighbors of every original pixel, and then downsample the image afterward.

11.5 Stacking

11.5.1 The Concept

Why do we stack (combine) images? To build up the signal while rejecting the noise. The key idea is that the random noise is different in each image and therefore will partly cancel out when they are stacked (Figure 11.2). To be precise, random noise in a stack of N images is only $1/\sqrt{N}$ as strong as in one image by itself.[1] Stacking even two images helps; stacking a hundred can almost work miracles.

Note that stacked images don't have to be raw; you can align and stack JPEG images if you want to. This is a handy way for a beginner to reap the benefits of stacking without mastering image processing *in toto*. It also comes in handy if you have forgotten to set your camera to record raw image files.

11.5.2 Confusing Term: *Integration*

In *PixInsight* and some other software packages, stacking is called integration. In digital imaging generally, though, *integration* often means exposure time. You can integrate 100 frames each of which has a 60-second integration time. To avoid confusion, I no longer use the term *integration* in either sense, except when referring to *PixInsight* menus.

[1] This follows from the root-sum-square law of addition for Gaussian noise sources. If noise sources $N_1, N_2, \ldots N_n$ are uncorrelated, then the resulting noise amplitude of their sum is not $N_1 + N_2 + \cdots + N_n$ but rather $\sqrt{N_1 + N_2 + \cdots + N_n}$.

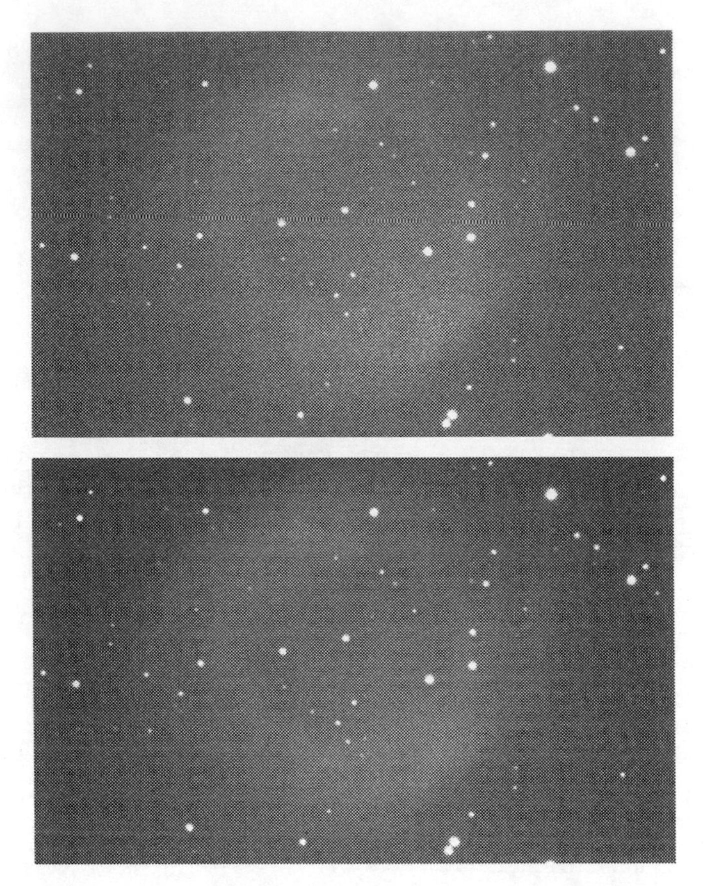

Figure 11.2. Combining images reduces noise. *Top:* Single 6-minute exposure of the Helix Nebula (NGC 7293) through an 8-inch (20-cm) telescope with $f/6.3$ compressor, with Canon Digital Rebel (300D) at ISO 400. *Bottom:* Average of three such exposures. Random noise is reduced by a factor of $\sqrt{3} = 1.7$.

11.5.3 How Images Are Combined

Sum

The most obvious way to combine corresponding pixel values is to add them. This is like making a multiple exposure in a camera; every image contributes something to the finished product.

The problem with adding (summing) is that the resulting pixel values may be too high. If you are working with 16-bit pixels, then the maximum pixel value is 65 535; clearly, if you add two images that have 40 000 in that pixel, the result, 80 000, will be out of range. For that reason, instead of the sum, we use the average.

Average (Mean)

The average (the mean) is the sum divided by the number of images. The resulting pixel values are always in the same range as the originals. As with summing, every image contributes equally to the finished product. In fact, taking the average is exactly equivalent to summing the images and then scaling the pixel values back to the original range.

Median and Sigma Clipping

The trouble with the average is that every exposure makes a contribution. Sometimes some of the pixels of one of the exposures are so deviant that they ought to be ignored. Maybe an airplane flew through the picture (Figure 11.3).

We'd like to average the normal-looking pixels but throw out the ones that are too different. One simple way to do this is to use the median.

The median of a set of numbers is the one that is in the middle when they are lined up in order. For example, if a particular pixel in five images has pixel values of, respectively, 40, 50, 53, 72, and 80, then the median is 53. If the highest value changes from 80 to something much higher, such as 8000, the median is still 53.

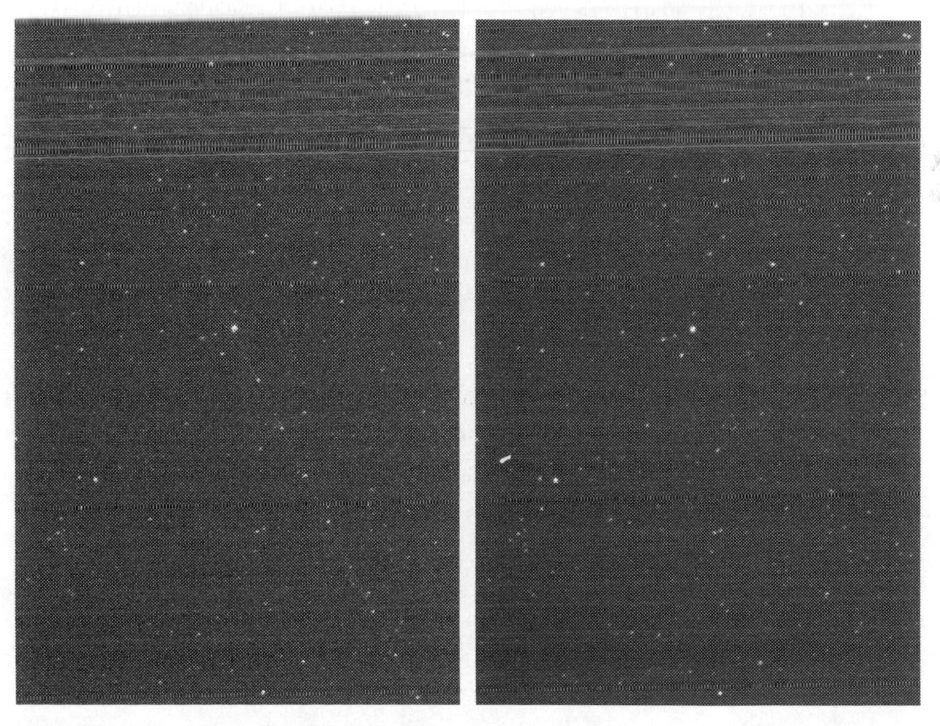

Figure 11.3. Airplane trail in a single exposure (left) disappears completely when this exposure is stacked with six more, using sigma clipping (right). Median combining would give same result.

But the median is not ideal either; it does not produce as smooth an image as the average because it does not get rid of low-level fluctuation. Is there a way to combine the two?

That is the idea behind *sigma clipping* – averaging with abnormal values excluded from the mix. To judge which values are abnormal, the program computes the average and standard deviation (σ, sigma), then rejects any pixels that differ from the mean by more than $\kappa\sigma$ where κ (kappa) is a factor you specify, and computes a new average omitting the rejected pixels. This is done iteratively so that after rejecting one deviant pixel, the updated average can be used to evaluate the others.

I recommend sigma clipping rather than median combining whenever possible. Another alternative is *percentile clipping*, rejecting pixels whose values are not within the middle 90% (or some other percentage) of the distribution.

11.6 Before We Stack, We Align

Master darks and flats are made by just stacking the images that come from the camera. Images that show celestial objects, however, need to be aligned to superimpose the images rather than leaving the pixels in their original positions. The images in successive exposures may shift because of imperfect tracking, deliberate dithering, or even taking images on more than one night.

Fortunately, our images contain lots of stars. The stars provide reference points for alignment and stacking. Some software aligns the images and stacks them immediately; other software outputs registration data (numbers indicating how to shift images), or else registered images (already shifted to match each other), and stacking is a separate step.

Alignment is nowadays a sophisticated process that includes rotating the images as needed (handy for observers with altazimuth mounts) and stretching to accommodate lens distortion. The software has to be able to find stars, of course, usually hundreds of them, and normally you can adjust thresholds for doing so. If the threshold is wrong, either no stars will be found, or grain will be mistaken for stars and alignment will fail.

11.7 Nonlinear Stretching (Gamma Correction)

11.7.1 The Concept

Until you perform nonlinear stretching or gamma correction on your astronomical image, it will probably look very dark or even black (Figure 11.4). If you were dealing with a correctly exposed daytime photograph, you could apply gamma correction using the formula in Section 2.5.4. Your camera does that automatically when it produces JPEG images.

Figure 11.4. A daytime photograph before and after gamma correction. Astronomical images often do not contain anything as light as the lightest parts of this picture.

But in astrophotography you usually have an underexposed picture with overexposed individual stars, a non-black background, and so forth, so gamma correction has to be customized to each picture. What you do is perform extreme stretching as described in Section 4.3. Moving the middle slider to the left is what makes the image nonlinear. That is, the pixel values are no longer linearly proportional to light levels.

Once the image is no longer linear, the pixel arithmetic used in calibration is no longer valid; that's why calibration is done first.

11.7.2 Digital Development Processing (DDP)

Though most of us do our nonlinear stretching by hand, there are ways to automate it, at least approximately, for astronomical images.

One popular approach is Digital Development Processing (DDP), an algorithm invented by astrophotographer Kunihiko Okano that combines gamma stretching with unsharp masking.[2] It is particularly good at preserving the visibility of stars against bright nebulae.

Some care is needed setting the parameters because implementations of digital development that are optimized for smaller CCDs can bring out grain in DSLR images. In that situation, the unsharp masking radius needs to be much larger, or the amount of edge enhancement needs to be less.

[2] Web: www.asahi-net.or.jp/~rt6k-okn/its98/ddp1.htm.

11.8 Postprocessing

Postprocessing is everything else you do to make the picture look good – adjusting the size and color balance, making final tweaks "by eye," and so forth – and may be done with *Photoshop* or *GIMP* rather than your astronomical software.

There are also additional processing steps that can be inserted at various stages in the process, such as noise reduction, background flattening, and sharpening. Some of them are covered in Chapter 13. Others are still being invented.

Chapter 12
Workflow with Specific Software

This chapter explains, as concisely as possible, how to process DSLR images with *DeepSkyStacker Nebulosity, MaxIm DL,* and *PixInsight*. These are not the only good software packages, but they are representative. This chapter is not a complete guide to them, but it will get you started.

Most image processing packages have roots in CCD astronomy with much smaller sensors. Their default settings are not necessarily the best for DSLR work. In these examples, I tell you how to use them with multi-megapixel DSLR images.

Throughout, I assume that you are working from lights, darks, flats, and flat darks. Obviously, if any of these are missing, they can be left out, except that if possible, flats should not be used without flat darks.

12.1 Before We Start

12.1.1 Screen Stretch

One of the first things to ask about any image processing software is how to turn on screen stretch. We will tackle this very early on with each of the software packages discussed.

Screen stretch is a feature that enables you to preview an image with strong stretching (brightening) even though it hasn't really been stretched (Figure 12.1). It's often the only way you can see what you are doing if you need to edit or preview an image that is still linear.

The screen stretch in your software is likely to be heavy-handed – its purpose is to show you what's there, not to produce a fine picture – and may or may not be related to the adjustments for actually stretching the image.

12.1.2 Methods and ISO Settings

If all your calibration frames were taken at the same ISO setting as the lights, you can use any of the methods described in this chapter. If you used one ISO

Figure 12.1. Before gamma correction, many images look completely black (top) unless you view them with screen stretch (bottom).

setting for the flats and flat darks, and another for the lights and darks, you can only use the methods that are descibed as Method 1 (Section 11.2.6) and do not involve bias frames or anything that the software thinks are bias frames.

12.2 *DeepSkyStacker*

DeepSkyStacker has three great advantages: it is free, it is easy to use (with lots of safeguards to keep sleepy astronomers from making mistakes), and it is *fast* (as well as accurate and reliable).[1] Its disadvantage is that it is not a full-featured image-processing system; although you can adjust brightness and contrast, making a finished picture generally requires postprocessing with other software (which can be *Photoshop* or *GIMP*).

Development of *DeepSkyStacker* has recently resumed after a hiatus; the original product had no major updates after 2009 and had started running into the memory limits of Windows 32-bit mode while processing images from newer, larger sensors. In 2018, *DeepSkyStacker* was made open-source and ported to 64-bit mode, and development resumed apace.

12.2.1 User Interface

DeepSkyStacker's user interface is simple, with a menu running down the left side of the window. When files have been chosen, they are displayed in a list at the lower right, and the selected image is displayed above the list (Figure 12.2).

12.2.2 Setting up

Figure 12.3 shows important settings that should be made the first time you run *DeepSkyStacker* so that it will decode DSLR raw files correctly. The settings will stick if you make them and then use them in a processing run, but not if you just make them and then exit *DeepSkyStacker* without processing any images.

12.2.3 Calibrating and Stacking Images

Calibration and stacking in *DeepSkyStacker* are very easy. Just click, in sequence, the six items marked in Figure 12.4. Each of the first four items opens a file dialog box in which you pick the files of the appropriate kind. If you don't have all four kinds of frames (light, dark, flat, and dark flat), just click on the ones that you do have. Ordinarily, with a DSLR, you will not have bias (offset) frames, and *DeepSkyStacker* will display a warning when about to stack your images, but you can ignore the warning; *DeepSkyStacker* chooses the right algorithm to process the types of files you give it.

You must mark the checkboxes on the files you want to use. "Check all" selects all of them in one step.

When you click "Register checked pictures" you can choose "Stack after registering" in order to save a step, or you can register and stack in separate steps. Look at the stacking options to make sure a reasonable algorithm is chosen (normally kappa-sigma). Also, look at the output options; normally you will want *DeepSkyStacker* to write out the linear image file automatically (as

[1] *DeepSkyStacker* is often abbreviated *DSS*, but is not to be confused with the *Digitized Sky Survey*, which is also abbreviated *DSS*.

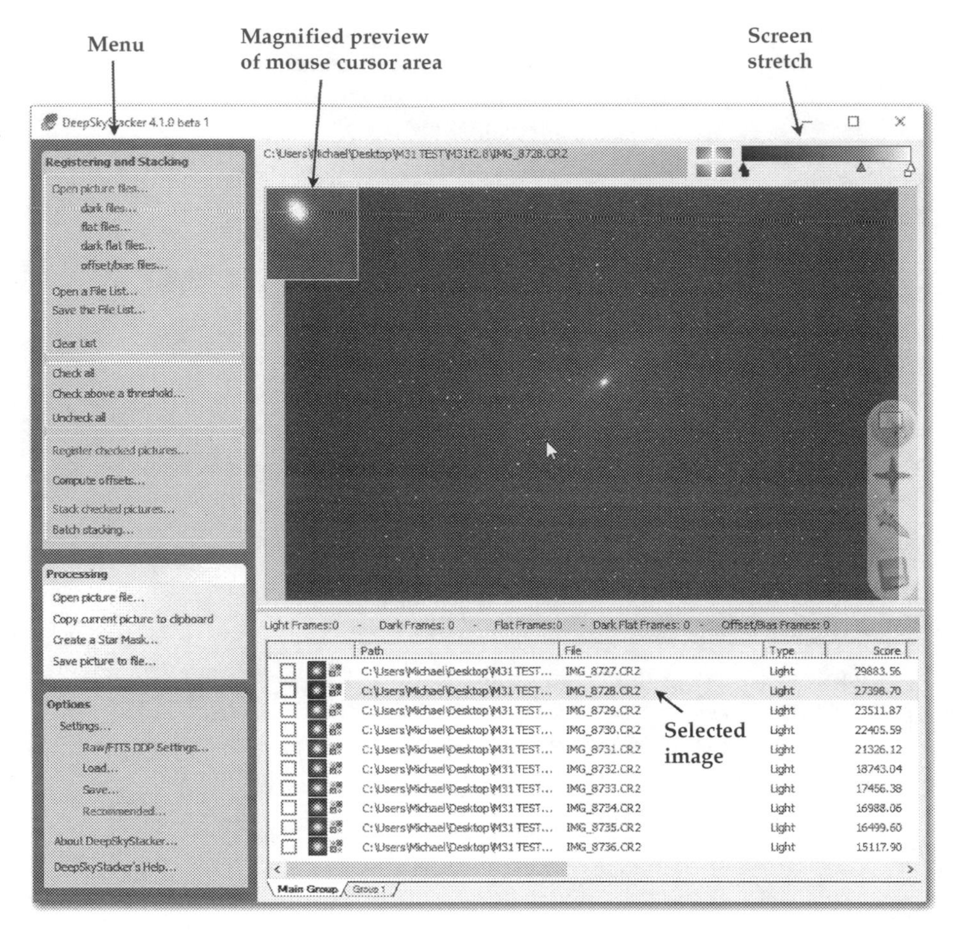

Figure 12.2. *DeepSkyStacker*'s user interface is straightforward, with the menu always visible at the left.

Autosave.tif, in the folder with the light frames) and also to generate an HTML file describing the processing run. These settings will be remembered for the next processing run.

The first time it reads each set of calibration frames, *DeepSkyStacker* creates a master frame (master dark, master flat, etc.) and stores it in the same folder. On subsequent runs, it uses the master frame to save time.

12.2.4 Viewing and Selecting Images to Stack

You don't have to check all the images, of course; you can stack only the good ones. To preview an image, click on it in the list of files, wait a moment for it to load, and look at it on the screen (Figure 12.2). Screen stretch is at the upper right. To see a star magnified, move the mouse cursor over it and look at the upper left of the picture.

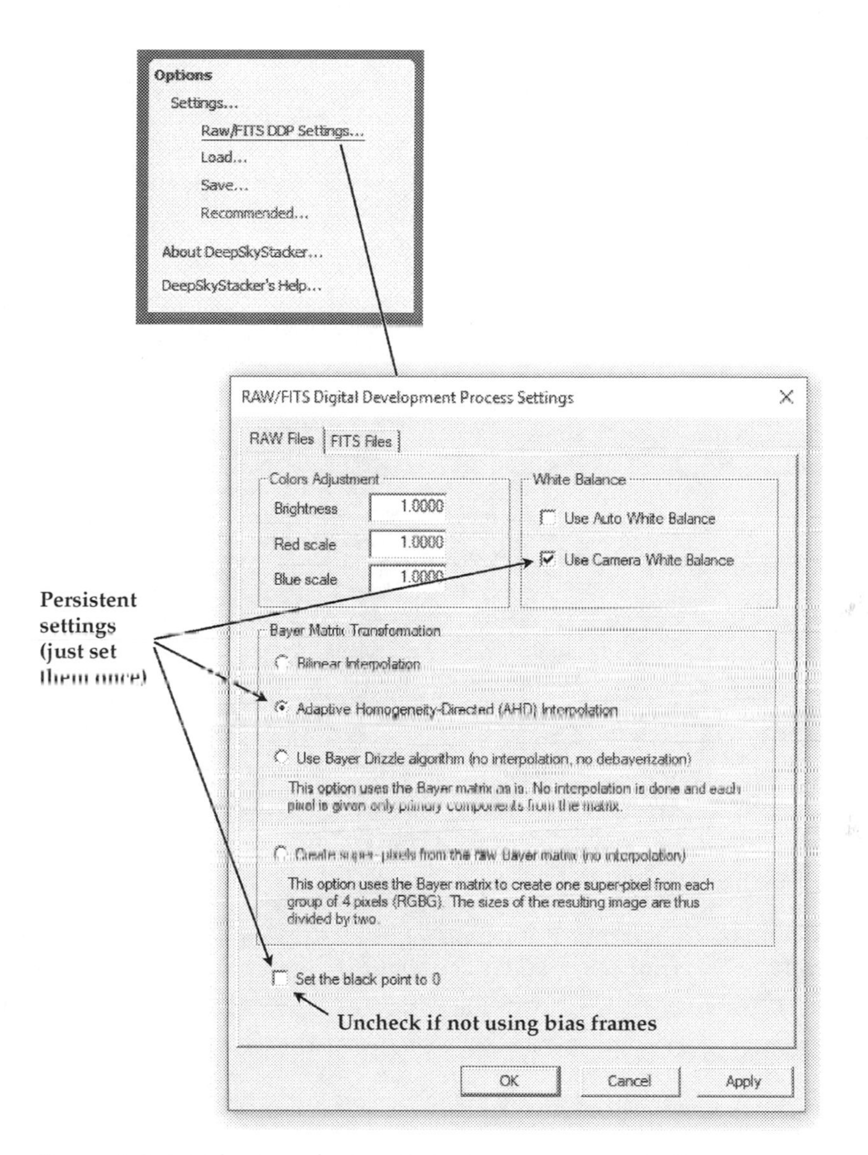

Figure 12.3. Initial settings for *DeepSkyStacker* to process DSLR files.

Once the images have been registered (whether or not they have been stacked), they are assigned scores, shown in the file list. Higher scores indicate images that are better tracked and better exposed. Unchecking images with especially low scores is a good strategy.

12.2.5 Stretching

The Autosave.tif image file generated by *DeepSkyStacker* will look dark and colorless. Your best move is to open it with one of the other software packages and start at the "Stretching" step as described in several places later in this chapter.

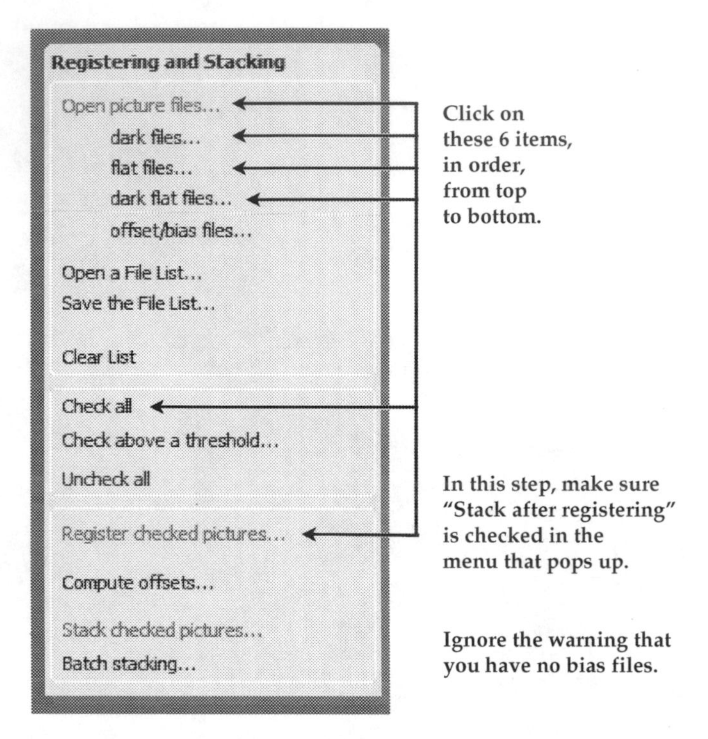

Figure 12.4. Calibration and stacking in *DeepSkyStacker* are very straightforward.

However, you can perform some stretching with *DeepSkyStacker* as shown in Figure 12.5. This feature is likely to be changed and improved in later versions of the software.

12.3 *Nebulosity*

Nebulosity is a product of Stark Labs (www.stark-labs.com), the same company that gave us *PHD Guiding*. It can be downloaded and used in free trial mode, in which you can try out all the features but the saved files have a pattern of short lines superimposed. Some camera control is included, and it is especially important to try out the camera control before you buy, if you plan to rely on it. This chapter does not cover camera control.

The intent of *Nebulosity* is to deliver everything you really need in a small package. Its selling points are ease of use and low price. It is particularly popular with beginners but can definitely be used for advanced work, though it is less full-featured than competitors. The processing steps are straightforward, and it is hard to get lost.

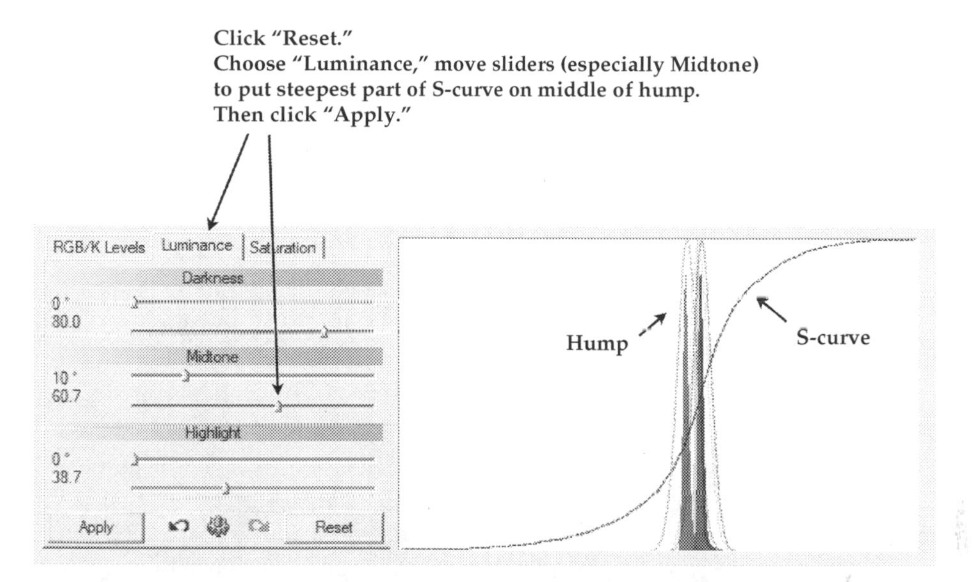

Figure 12.5. *DeepSkyStacker* stretching tool is displayed when an Autosave.tif file has been generated or when a file is loaded for processing. This tool may change in later versions.

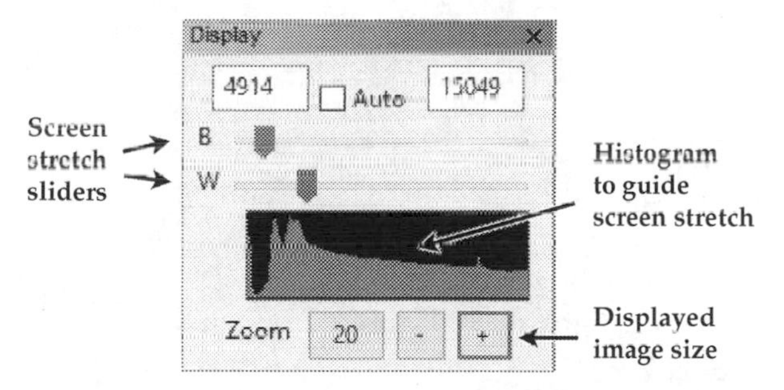

Figure 12.6. *Nebulosity* screen stretch and image size controls are at the upper right of the main window.

12.3.1 User Interface

Nebulosity's user interface is simple, with five well-organized menus. The Help menu contains a link to the complete manual.

"Undo" and "Redo" are under the Edit menu. By default, you can go back and forward three steps, but this can be changed to unlimited (i.e., record all the steps in every session) from the menu under Edit, Preferences.

Screen stretch in *Nebulosity* (Figure 12.6) is automatic and particularly easy to use; it resets automatically whenever the brightness of the image is changed. You can uncheck "Auto" and move the sliders yourself to change the amount of stretch. To disable screen stretch, move them all the way to left and right (0 and 65 535).

The size of the image is controlled by the "Zoom" buttons just below the screen stretch controls. It defaults to 100% (that is, one image pixel per screen pixel), which means that you will only see a small part of the upper left corner of a typical DSLR image until you reduce the zoom.

12.3.2 Basic File Editing

Nebulosity's preferred file format is FITS (extension *.fit*), but it also supports TIFF, JPG, PNG, and others, including virtually all DSLR raw file formats, but not *PixInsight*'s XISF. To open a file, simply choose File, Open File.

When you open a DSLR raw file, it comes in un-deBayered. To deBayer it, choose Image, De-mosaic RAW. Of course it will still not be gamma-corrected, and you may need to increase the screen stretch to see what is in it.

Tools for resizing, rotating (by right angles only), and cropping the image are on the Image menu. To crop, you have to type in how many pixels to take off each edge. In general, file editing operations like these might be better accomplished by saving the finished image as a TIFF and taking it into *Photoshop.*

12.3.3 Calibration

In *Nebulosity,* calibration, deBayering, and stacking are three separate steps. To calibrate images, choose Batch, Pre-process image sets, and proceed as shown in Figure 12.7.

Click on buttons to load sets of darks, flat darks, flats, and lights. You can have several groups of each kind, and normal practice is to put the long-exposure darks in one group (Dark 1) and the flat darks in another (Dark 2). If you wish, you can add additional sets of lights to be calibrated against the same darks and flats.

The settings in the figure correspond to Method 1 described in Section 11.2.6. When you click OK, the calibrated images are written into the same folder as the existing light frames, as FITS files with names prefixed by *pproc_*.

If you have no flats and flat darks, simply leave them out.

12.3.4 DeBayering

Next, the *pproc_* files need to be deBayered. From the menu choose Batch, Batch Demosaic + Square RAW Color. ("Square" refers to an adjustment DSLRs don't need, but which is done at this step for some other cameras – resizing the image in one dimension to make up for pixels that are wider than they are high.)

You don't have to make any decisions; just choose the files and let the process run. The output is another set of files, with names beginning with *recon_*, written into the same folder.

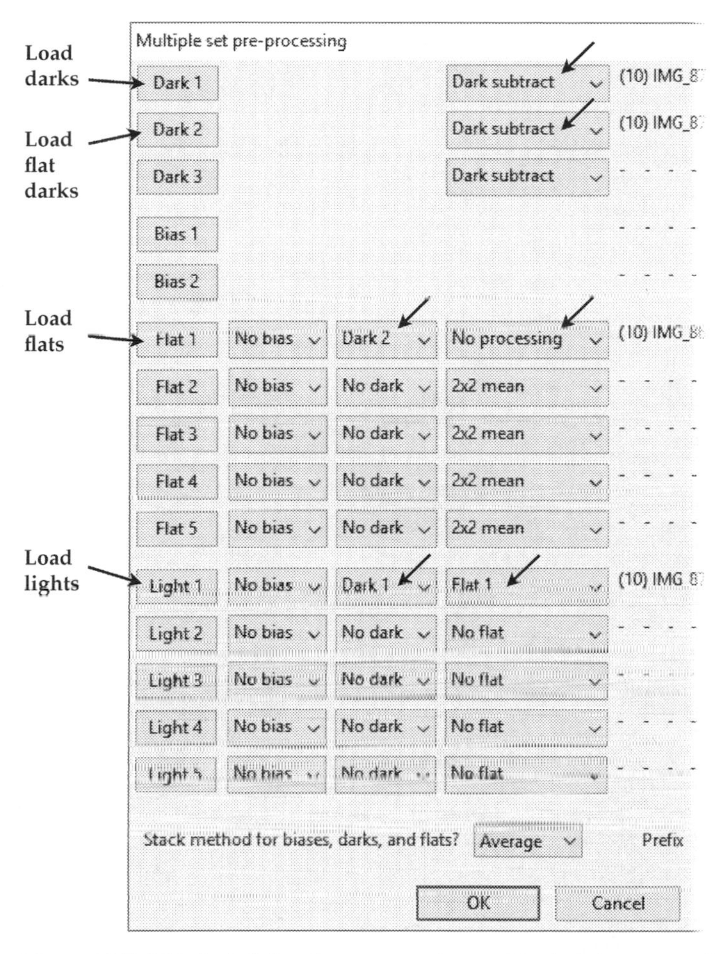

Figure 12.7. *Nebulosity*'s batch preprocessing tool (only left portion of window is shown). Note that flats calibrate against Dark 2 (flat darks), and are not binned, while lights calibrate against Dark 1 and Flat 1.

12.3.5 Choosing Images to Stack

Choose File, Preview Files, in order to load and view a number of images in succession. You can go back and forth through the series of images and delete or rename images as you encounter them.

It is convenient to do that, at this step, with the *recon_* files and rename or delete any that you do not want to use. Actually, you can open and preview any set of files this way. You may need to adjust the screen stretch to see more of what you have.

The files are loaded one by one, so you cannot "blink" or flip through files instantly; loading a DSLR image typically takes two or three seconds. That is generally good enough.

You can also measure the quality of images. Choose Batch, Grade Image Quality, and *Nebulosity* will prompt you to open the image files you want to evaluate. It will then make copies of all the files with designations such as *Q250_* and *Q273_* prefixed to their names. The number is the star image radius (as full-width-half-maximum, FWHM) in hundredths of pixels, so *Q273* means 2.73 pixels. Smaller numbers are better. When alphabetized in a file dialog box, the file names of the better images come first.

12.3.6 Aligning and Stacking Images

In *Nebulosity,* alignment and stacking can be done together or in two separate steps. If you do alignment and stacking together, you can only average the frames, which is not ideal, but it is a quick way to get a preliminary result.

Figure 12.8 shows the Batch, Align and Combine Images tool window. To align your images for later stacking, choose "Save each file," or if you want an averaged image immediately, choose "Save stack." Then choose the type of alignment you want – translation (shifting), or translation plus rotation, or

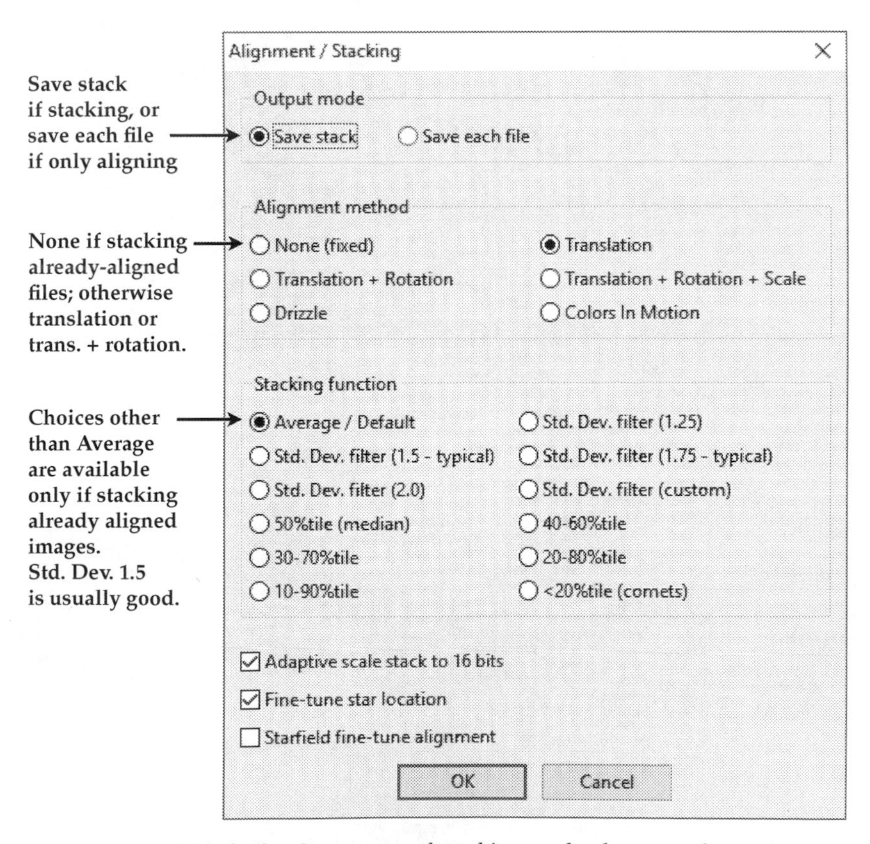

Figure 12.8. In *Nebulosity,* alignment and stacking can be done together or in two separate steps.

even translation plus rotation plus scaling. The first is usually good enough for well-tracked images taken with an equatorial mount.

When you click OK, you'll be prompted to open the files you're going to stack (the *recon_* files from the deBayerization step, or a subset of them). Then, on each image, *Nebulosity* requires you to find the alignment stars manually, clicking on the same star in each image – one star if rotation is not involved, two stars if it is. This is not as difficult as it sounds; Figure 12.9 shows the process in action. You must also check one of the images as the reference image, indicated by the checkbox next to one file name.

If you've saved the stack as one image, you're done. If you've saved separate files, the next step is to open the same tool again (Batch, Align and Combine

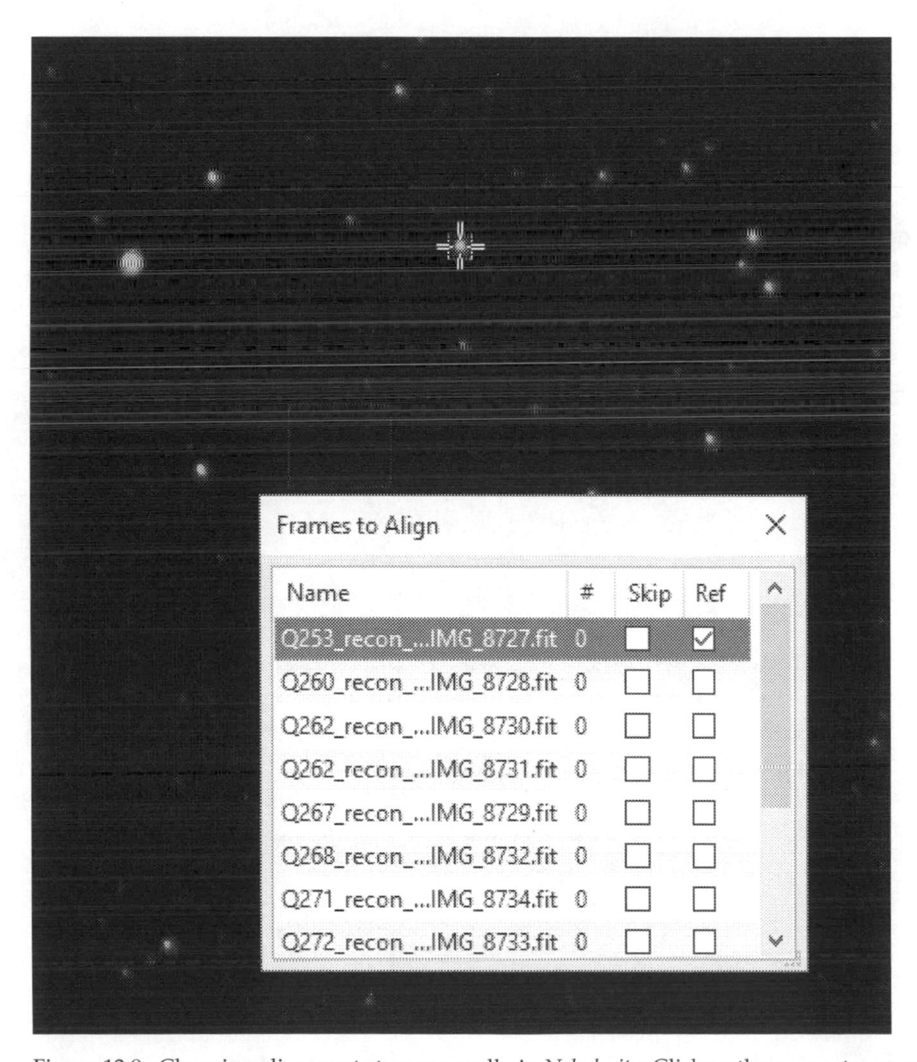

Figure 12.9. Choosing alignment stars manually in *Nebulosity*. Click on the same star as each image is presented. Note that one image has been chosen as reference ("Ref").

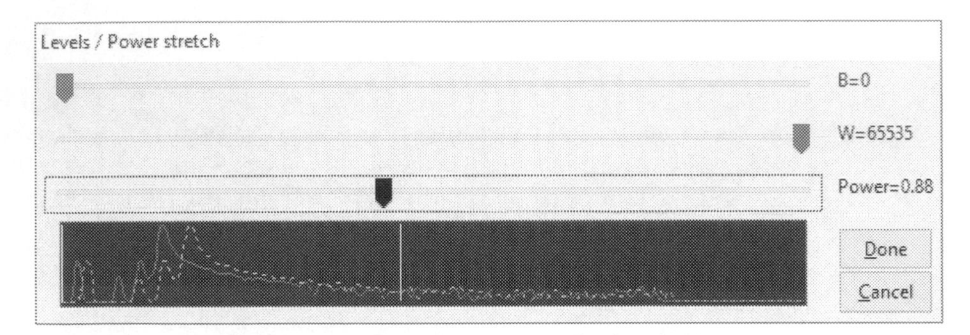

Figure 12.10. *Nebulosity's* Levels tool has the "middle" slider on the bottom. Move it to the left to gamma-correct an image.

Images) and stack the aligned files that you just created (their names begin with *align_*). Choose "None" as the type of alignment, since they are already aligned, and choose a suitable rejection criterion (I suggest "Std. Dev. 1.5") to reject deviant pixels. Click OK, and the images will be stacked.

12.3.7 Stretching

To perform stretching in *Nebulosity,* choose Image, Levels/Power Stretch. This tool works very much like the Levels tool in *Photoshop* (Section 4.3) except that the "middle" slider is actually displayed at the bottom of the tool window (Figure 12.10). Before using it, you should turn screen stretch off.

What you will do is move the middle (i.e., bottom) slider a long way to the left; move the other sliders as necessary.

Alternatively, semi-automatic stretching is provided under Image, Digital Development Processing (DDP) (see Section 11.7.2). It is controlled by three sliders whose effect is best understood by trying them out.

You cannot stretch the three color channels (R, G, and B) separately. Instead, there is another tool, Image, Adjust Color Scaling, which lets you scale the colors separately; it shows you histograms for the three channels, and you move sliders to make them match. After doing this, you can return to Levels for further adjustment. There is also a useful Auto Color Balance tool.

The result is a finished image, which you should save as FITS (to preserve all information) and can also save as TIFF, JPEG, or PNG for viewing and printing.

12.4 *MaxIm DL*

MaxIm DL, from Diffraction Limited (www.diffractionlimited.com), is a time-honored, moderately full-featured image processing system with an emphasis on scientific rigor. It also includes elaborate camera control, telescope control, and autoguiding (not covered here). A lot of knowledge about cameras is built in so that correct settings are made automatically. Differently priced editions give different levels of functionality.

Extensive documentation for *MaxIm DL* is available at the manufacturer's web site, including video tutorials and a 700-page online manual.

12.4.1 User Interface

Overall Approach

The user interface of *MaxIm DL* is designed to guide you to the tools you actually need without bewildering you with too many choices. The menus are well organized and resemble other graphic arts software. Though of completely separate origin (and older), *MaxIm DL's* user interface is similar enough to *Nebulosity* that moving from one to the other should be easy.

Undo and Redo

Figure 12.11 shows the toolbar at the upper left of the screen, including the undo and redo buttons, as well as the buttons that toggle visibility of the Screen Stretch and Information windows. The undo feature only goes one step back, followed by the ability to redo what was undone.

Screen Stretch

Screen stretch in *MaxIm DL* (Figure 12.12) is normally automatic; you can choose low, medium, or high levels of stretching, as well as several other options. It switches to manual automatically when you move either of the sliders. To disable screen stretch, set it to Manual and pull the sliders to the extreme left and right.

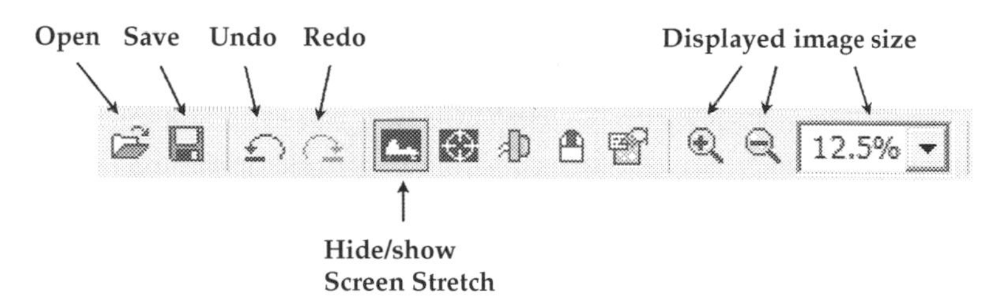

Figure 12.11. Important toolbar items in *MaxIm DL*.

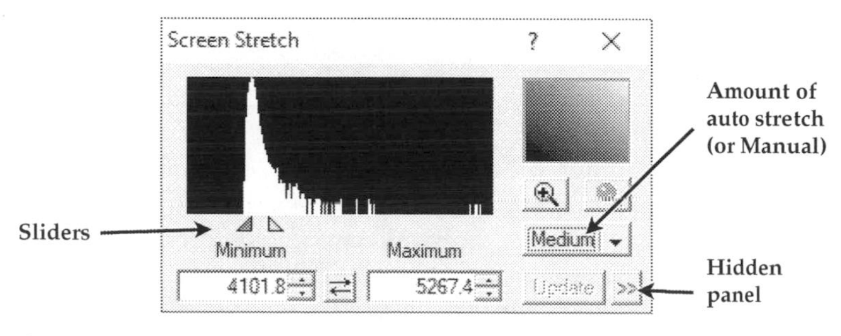

Figure 12.12. *MaxIm DL* Screen Stretch window.

Hidden Panels

Many *MaxIm DL* windows have a button marked ≫ that reveals an additional panel. There is an example in Figure 12.12; there, the ≫ button displays settings to let you define the screen stretch in terms of percentiles.

12.4.2 Basic File Editing

Opening Existing Files

MaxIm DL's preferred file format is FITS (extension *.fit*). It also opens and saves TIFF, PNG, BMP, and JPEG files, as well as many camera raw formats (but not XISF). Just choose File, Open on the menu.

Opening DSLR Raw Files

MaxIm DL opens DSLR raw files un-deBayered. To deBayer a file after opening it, use Color, Convert Color, and choose the make and model of camera. You can specify scale factors for red, green, and blue if the default ones are not satisfactory.

The default image size is 100% (one screen pixel per image pixel), so you will see only part of the picture until you reduce it with the minus zoom button on the toolbar.

Cropping and Sizing

Image resizing is under Process, Resize (with Half Size and Double Size as shortcuts). See also the Bin 2 × 2 and Bin 3 × 3 buttons under Edit.

Cropping is under Edit, Crop, and allows you to draw a rectangle on the picture by dragging the mouse cursor, although this isn't obvious.

12.4.3 Choosing Images to Stack

MaxIm DL does not have a Blink tool for flipping rapidly through images, but you can easily open a number of images at the same time and view them with the same screen stretch. You can judge tracking and overall image quality without deBayering them.

You can also measure the quality of images using the Quality tab in the Stack tool (described below).

12.4.4 Calibration and Stacking

Setting Up the Calibration

Calibration and stacking in *MaxIm DL* are done in two steps that do not correspond to the steps in other software. The first step is to set up the calibration by choosing the calibration frames and other parameters. The second step is to perform the stacking, automatically preceded by calibration and deBayering.

What we're going to use is Method 1 (Section 11.2.6), with one group of flats and two groups of darks (the regular darks and the flat darks).

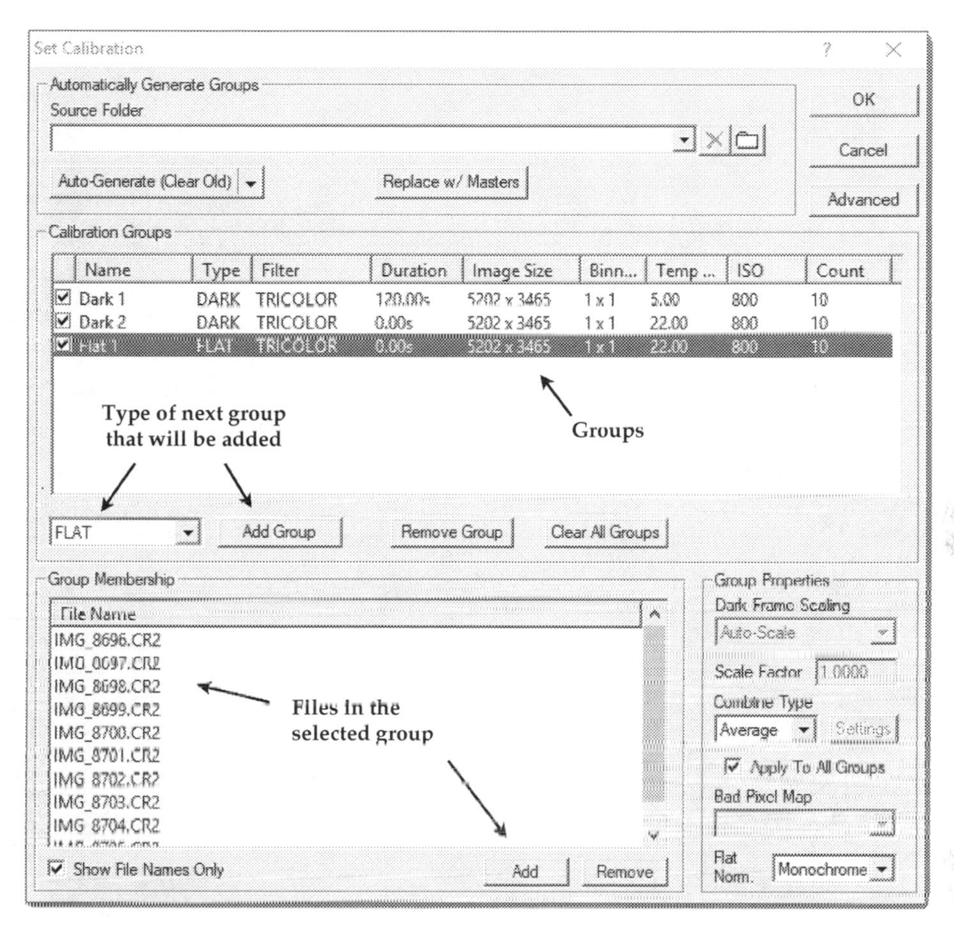

Figure 12.13. *MaxIm DL* Set Calibration window is where you create groups of flats and darks.

To set up the calibration, open Process, Set Calibration (Figure 12.13). Create two groups of darks and one group of flats. Then you put the regular darks into one group of darks, the flat darks into the other, and the flats into (of course) the group of flats. Since there are no bias frames, dark frame scaling will be unavailable and grayed out. Notice that there is no group for lights.

Click OK, and you've finished this step. No computation happens yet because all you've done is make settings. Note that Set Calibration will retain its settings after you close it and even the next time you run *MaxIm DL* – they are saved until you change them.

Stacking Images

Now choose Process, Stack, and in the window that opens, click Add Files and choose your light frames (Figure 12.14). Don't panic when *MaxIm DL* reports

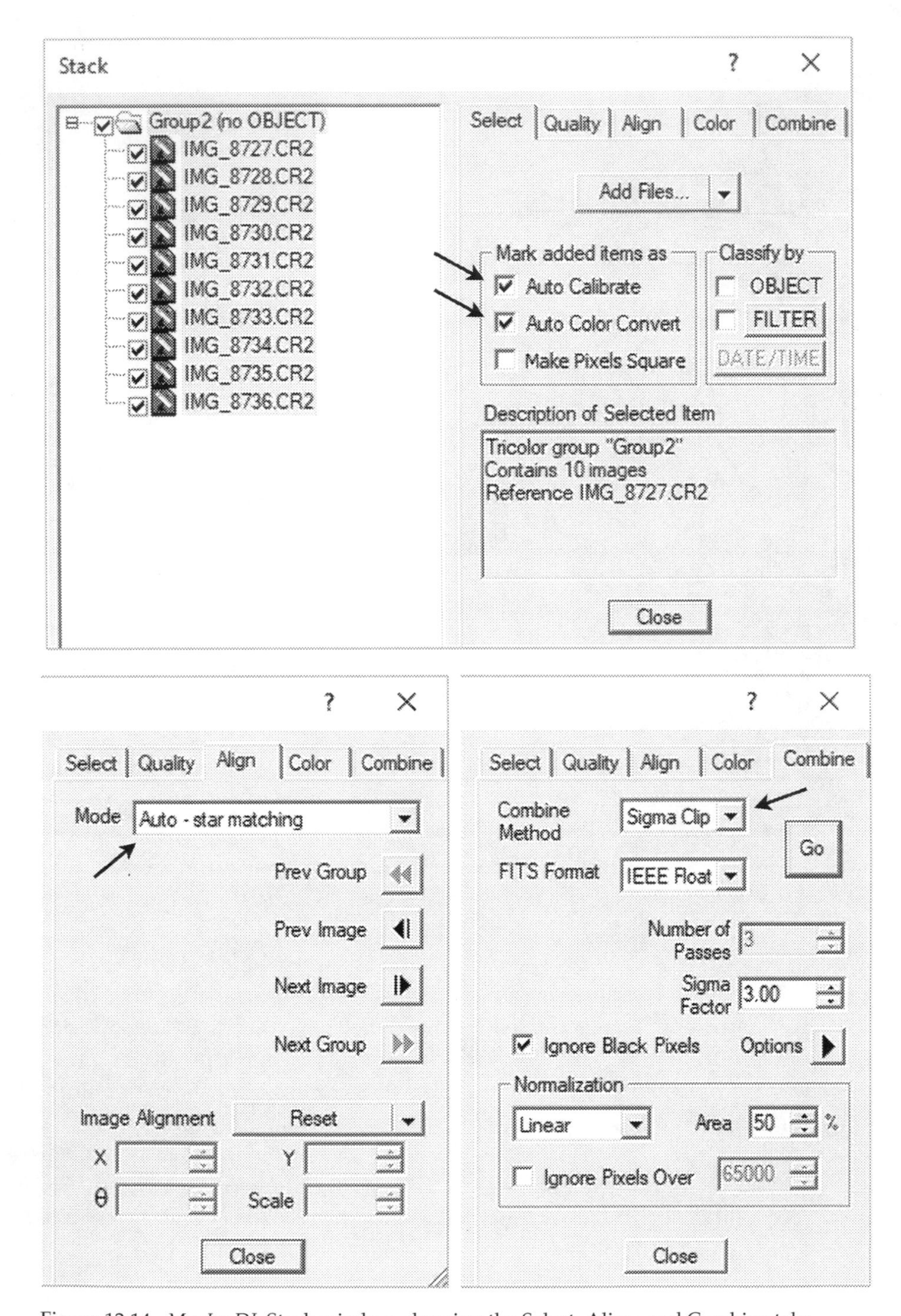

Figure 12.14. *MaxIm DL* Stack window, showing the Select, Align, and Combine tabs.

that it is "processing" them – what it's doing is scanning them and making a group. The group will probably be labeled "No OBJECT," which only means that your raw files do not say what celestial object you were photographing (FITS files from a camera control program might include this information).

If you want to choose a reference image, left-click on one of the image files to select it, and then right-click and check Reference Image. Otherwise the first image in the list will be used as the reference image.

Then check some menu settings:

- On the Select tab, make sure Auto Calibrate and Auto Convert Color are checked;
- On the Align tab, select how you want to match star images. This is normally Auto – Star Matching, but you can match stars manually if you want (which would require clicking on the same star in every image);
- On the Combine tab, choose Sigma Clip as the method of combining; for the other settings, the defaults are generally correct.

Finally, on the Combine tab, click Go, and calibration and stacking will begin. The result will be displayed as an open image in *MaxIm DL*.

12.4.5 Stretching

The easiest way to do stretching and gamma correction in *MaxIm DL* is to use Process, Levels (Figure 12.15). The Levels tool works very much like the classic one in *Photoshop* (Section 4.3) and is previewed on the screen as you

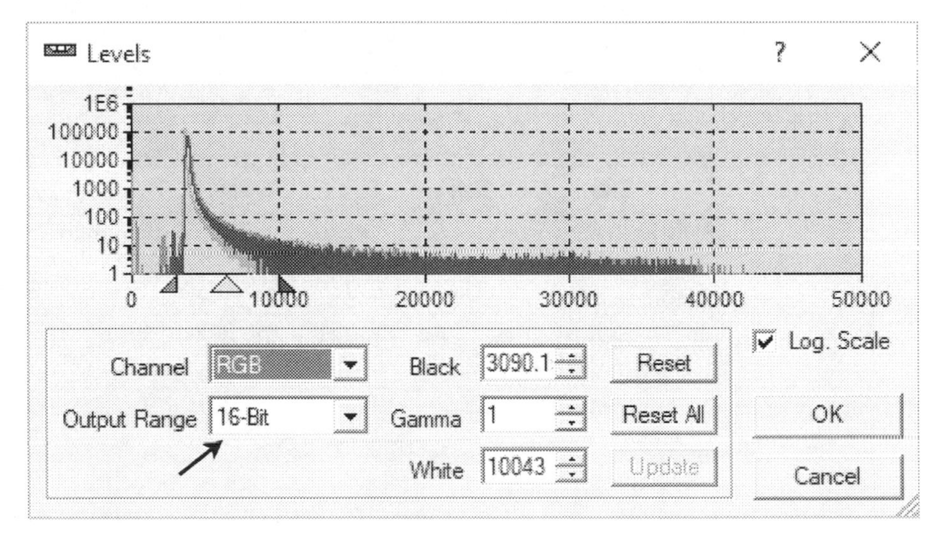

Figure 12.15. *MaxIm DL* Levels tool. Move middle slider to left to gamma-correct. Be sure to choose 16-bit output.

make adjustments. Be sure to select 16-bit output. Screen stretch automatically switches off when you use this tool so that you can see the effect of your corrections.

Gamma correction is done mostly by moving the middle slider to the left; it is best to do this in several small operations rather than all at once. You can adjust red, green, and blue separately or together.

12.5 *PixInsight*

PixInsight (www.pixinsight.com) is designed for serious experimenters and researchers. It is unusually full-featured; its Process menu contains more than 100 tools, there are a couple of dozen more tools under Scripts, and you can create your own by programming in Javascript. Quite a few of the tools reflect alternative theories of how to do the same thing. The intent is to include every image processing tool that could possibly be useful, even if there is controversy about whether it should be used. Sometimes the same tool is accessible from menus as many as four different ways.

The drawback of *PixInsight* is that, looking at those dozens of menu choices, it is far from obvious how to process an image. To get started, don't try to learn everything at once. The gaps in the explanation that I'm about to give are deliberate; I'll ignore much of what is there in order to get you to the most-used features quickly. Further tutorials on *PixInsight* are abundant, on line and even in the form of seminars and training courses. I particularly recommend the book *Inside PixInsight*, by Warren A. Keller (Springer, 2016); arguably its 376 pages still only scratch the surface, but they scratch it much more deeply than I do.

12.5.1 User Interface

Overall Approach

All the features of *PixInsight* are on menus, some of them reachable from more than one place. Many features are also on tool bars across the top and down the left side of the screen. Using View, Tool Bars, you can control which tool bars are visible.

To give you more screen space, *PixInsight* provides four "workspaces" (desktops), which are chosen by clicking on squares at the left bottom of the screen (often not visible when *PixInsight* is maximized).

Undo and Redo

PixInsight can undo every step of your processing, then redo it again if you choose. The undo and redo icons are on the Image menu and are also on the tool bar at the upper left corner of the screen (Figure 12.16) and also on the Image menu.

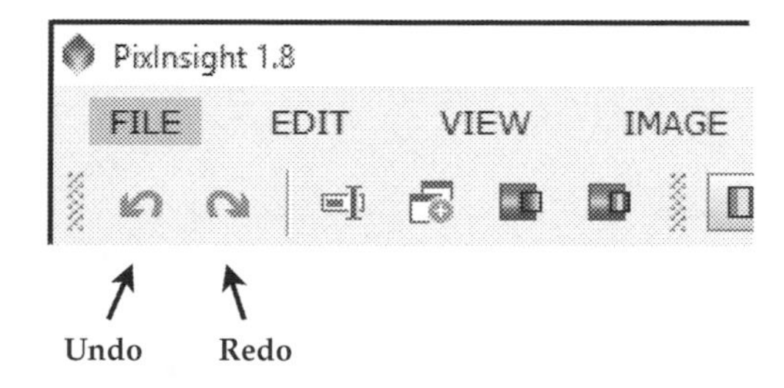

Figure 12.16. *PixInsight* can undo every step of a processing session, then redo the steps in order if you choose.

Figure 12.17. *PixInsight* screen stretch icons are on a tool bar and also the Image menu.

Screen Stretch

Screen stretch in *PixInsight* is called Screen Transfer Function (STF) and is turned on and off with the icons shown in Figure 12.17. If the Screen Transfer Functions tool bar is not visible at the top of your screen, go to View, Tool Bars, and activate it. These icons are also reachable on the Image menu (some of them under Screen Transfer Functions) and through Process, IntensityTransformations, ScreenTransferFunction.

Using Tool Windows

Across the bottom of many of *PixInsight*'s tool windows are the components shown in Figure 12.18. The apply and reset buttons are of course indispensable. The apply button is square if it works on the currently selected image, or round if it works on a set of images chosen within the tool.

The arrow at the right side of the figure shows a panel that can be collapsed (hidden) or expanded. Many windows consist of several such panels, and the screen is not big enough to expand all the panels at once. You will often have to collapse some panels in order to see others.

The instance triangle is one of *PixInsight*'s most unusual features. Basically, it is a representation of the tool window (with its current settings) which you can drag around. If you drag the instance triangle onto an open image, the tool will be applied to that image. If you drag it onto the desktop, you create an

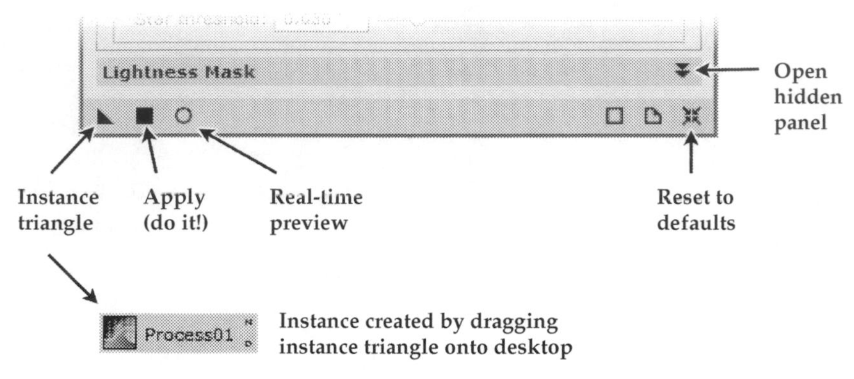

Figure 12.18. *PixInsight* tool windows often include these components.

"instance," which can be dragged onto an image, or opened up as another copy of the tool window, or used to transfer settings into another tool in various ways.

12.5.2 Basic File Editing

Opening Existing Files

Although it prefers XISF, *PixInsight* accepts files in a great variety of file formats, often with a greater variety of bit depths than other software packages support.

If you open a file with File, Open, and then save it in the same format, your original file will be overwritten. If, however, you use File, Open a Copy, then you can only Save As (that is, give it a new name), and that is what I recommend, to eliminate the risk of destroying the original.

Projects

Besides opening and saving files, you can save and load *projects*, which are records of everything you've done in a *PixInsight* session, including all versions of the processed images. The operations are recorded on an XML file with extension *.xosm*, and the images are stored (in an internal format) in a folder with extension *.data*. You can open a project and resume working right where you left off, including the ability to undo previous steps.

Opening DSLR Raw Files

Decoding of DSLR raw files is controlled by the setting in View, Explorer Windows, Format Explorer (usually also accessible from a tab on the left edge of the screen). Double-click on DSLR_RAW to make settings. Usually, either "Pure Raw" or "de-Bayer RGB" is what you want. The first of these omits deBayerization; the second performs it using *dcraw* (Section 2.2.4). If you don't like the color balance you are getting, you can scale red, green, and blue individually by factors other than the default 1.0. These settings are remembered from session to session.

Cropping and Sizing

Tools to zoom in and out on the image are in the View menu and on the Zoom tool bar, which is activated by default.

To crop an image, use either Process, Geometry, Crop (to type in the number of pixels you want taken off each edge) or Process, Geometry, DynamicCrop (to crop the image by dragging the corners of a box); I prefer the latter.

Tools for resizing and rotation are also under Process, Geometry. The IntegerResample reduces or enlarges an image by a whole-number factor; it is fast and accurate. The regular Resample tool resizes by any factor. Analogously, the FastRotation tool only rotates 90° or 180°, by rearranging pixels without any interpolation, and the regular Rotation tool can rotate through any angle at some risk of losing sharpness.

12.5.3 Choosing Images to Stack

To preview your images quickly, use Process, ImageInspection, Blink. This tool lets you open a number of raw files and view them in a single image window, switching instantly from one to another. They are displayed with automatic screen stretch, and you can zoom the window in the usual way to get a close look at the star images.

To evaluate the star images quantitatively, choose Script, Batch Processing, SubframeSelector, choose the image files, click Measure, and you will see a table of the star image size and roundness for each image, along with other measurements, including weights that can be used to give the better images more influence when stacking.

From there, you can either manually take note of which images you want to use, or let SubframeSelector help. Uncheck the images you don't want to include in the stack and open the Output tab. There you will find that you can copy the selected images to a folder, and the copies will be annotated with weights using an XISF or FITS keyword that you supply (I suggest "weight"). The stacking (ImageIntegration) tool can read these weights and use them. Once you've made your choices, click "Output subframes" and the copying will be done.

12.5.4 Raw or FITS?

PixInsight can stack FITS files incrementally, saving memory, but due to the additional decoding needed, it has to hold raw files in memory separately and will run out of memory when stacking large sets. You can convert raw files to FITS in bulk using Script, Batch Processing, BatchFormatConversion.

12.5.5 Calibration and Stacking

The BatchPreprocessing Script

Calibration and stacking in *PixInsight* can be an intricate, highly customized process, but the quickest way to do it is to choose Script, Batch Processing,

BatchPreprocessing and use the tool shown in Figure 12.19. It involves clicking four buttons in succession and selecting four types of files.

On each of the tabs (Bias, Darks, Flats, Lights) you have further options, such as how to combine the images for master calibration frames (usually sigma clipping is appropriate) and what to do with the lights (in the example, they are to be deBayered, aligned, and stacked). Master bias frames, darks, and flats are made automatically and used as soon as they become available.

How all of this applies to a DSLR is not always obvious. There are several approaches.

Lights, Darks, Flats, and Flat Darks

The simplest way to proceed is shown in Figure 12.19. Put in the lights and flats in the usual way. Put both kinds of darks – regular darks and flat darks – as darks; *PixInsight* will keep them separate by exposure time and will dark-subtract lights and flats each with the right set of darks. Make sure "Optimize dark frames" is not checked.

This is Method 1 from Section 11.2.6. If you have no flats and flat darks, simply leave them out, and you're doing Method 0 (Section 11.2.5).

You will be surprised how much time and disk space are required for calibration. *PixInsight*'s documentation suggests that after clicking Run, you should go get a cup of tea. An order-of-magnitude estimate is to expect to process 4 frames per minute, counting the lights and all the calibration frames. Fast CPUs, lots of RAM, and solid-state disks speed this up.

You will get a warning message that image integration (stacking) should not be done in this script and that your results will only be preliminary. Actually, though, the results are quite good; you can ignore the warning and check the box that says not to show it again.

If you get warnings about the methods of combining images (e.g., that you do not have enough images for sigma clipping), read them and heed their advice, but do not fear catastrophe. *PixInsight* works well even when settings are not optimal.

When the process is finished, the output folder will contain a lot of files. There will be four directories, one for logs, one for master files, one for registered (aligned) images, and one for calibrated images.

The finished, calibrated, and stacked picture is the "master light" in the "master" directory. It still needs stretching and other processing, of course. You can delete the "registered" and "calibrated" folders unless you are going to experiment with re-doing the stacking manually.

Using Bias and Scaling

PixInsight can automatically scale dark frames in an intelligent and useful way. If you check "Optimize dark frames," the darks will be scaled to match the actual noise level of the corresponding lights. This is more accurate than

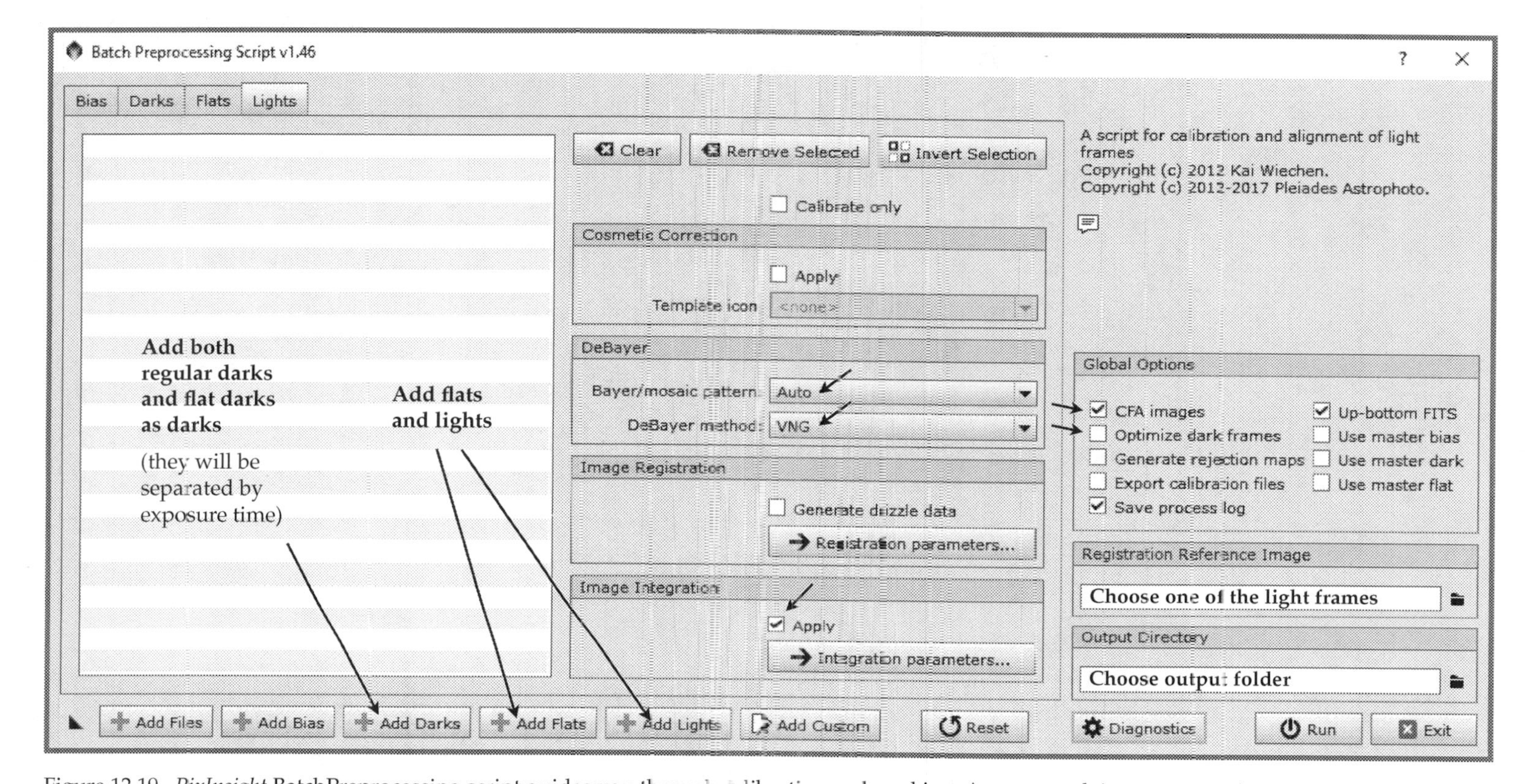

Figure 12.19. *PixInsight* BatchPreprocessing script guides you through calibration and stacking. Arrows mark important settings.

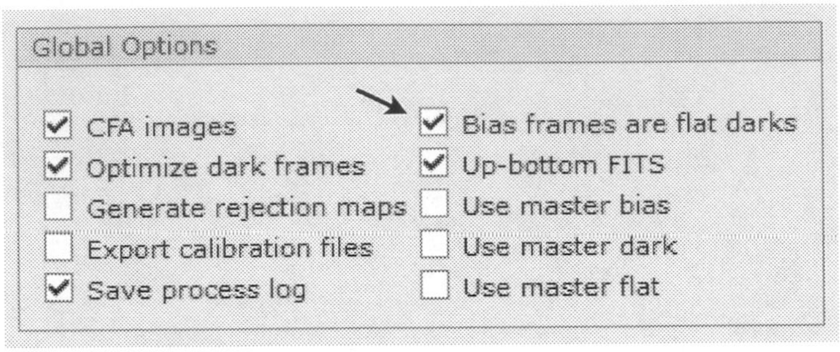

Global Options

- [x] CFA images
- [x] Optimize dark frames
- [ ] Generate rejection maps
- [ ] Export calibration files
- [x] Save process log
- [x] Bias frames are flat darks
- [x] Up-bottom FITS
- [ ] Use master bias
- [ ] Use master dark
- [ ] Use master flat

Figure 12.20. *PixInsight* script BatchPreprocessingFD, from www.dslrbook.com, adds the ability to use the same frames as bias and flat darks without an extra step.

trying to predict the effect of changes in exposure time and temperature. Since the temperature of a DSLR sensor can change appreciably during the course of a session, I recommend doing this.

Scaling the darks requires that bias frames be present. Since DSLR flat darks are indistinguishable from bias frames, all we have to do is tell *PixInsight* to use them (Method 3 in Section 11.2.8; note that all the calibration frames must be taken at the same ISO setting as the lights).

There are two ways to proceed. One is to make a second copy of the flat darks, with different file names or in a different folder, and put in one copy as darks and the other copy as bias frames. *PixInsight* won't let you put the same file in more than one group, hence the need for the second copy.

The other is to download the BatchPreprocessingFD script from this book's web site (www.dslrbook.com) or possibly from other sources. It is a modified version of BatchPreprocessing that gives you one more checkbox (Figure 12.20) so that you can tell it that you have put in the flat darks as bias frames. Then the (so-called) bias frames are used not only for bias but also for dark-subtracting the flats.

Adding Cosmetic Correction

Particularly when using "Optimize dark frames," *PixInsight*'s calibration process is not good at getting rid of hot pixels (those that always read the maximum value). For that, cosmetic correction (automatic hot pixel detection) is needed. You can combine cosmetic correction with BatchPreprocessing or BatchPreprocessingFD, but the way it's done is slightly surprising.

Crucially, you can't open or adjust the cosmetic correction tool while the BatchPreprocessing script is running; scripts prevent you from touching anything outside themselves. What you do is choose Process, ImageCalibration, CosmeticCorrection and open up the cosmetic correction tool beforehand (Figure 12.21). Within it, don't choose any files, but check "CFA," then open up the hidden "Use auto detect" panel and check the boxes for detecting hot and cold pixels (or maybe just hot pixels). The default sigma values are generally good.

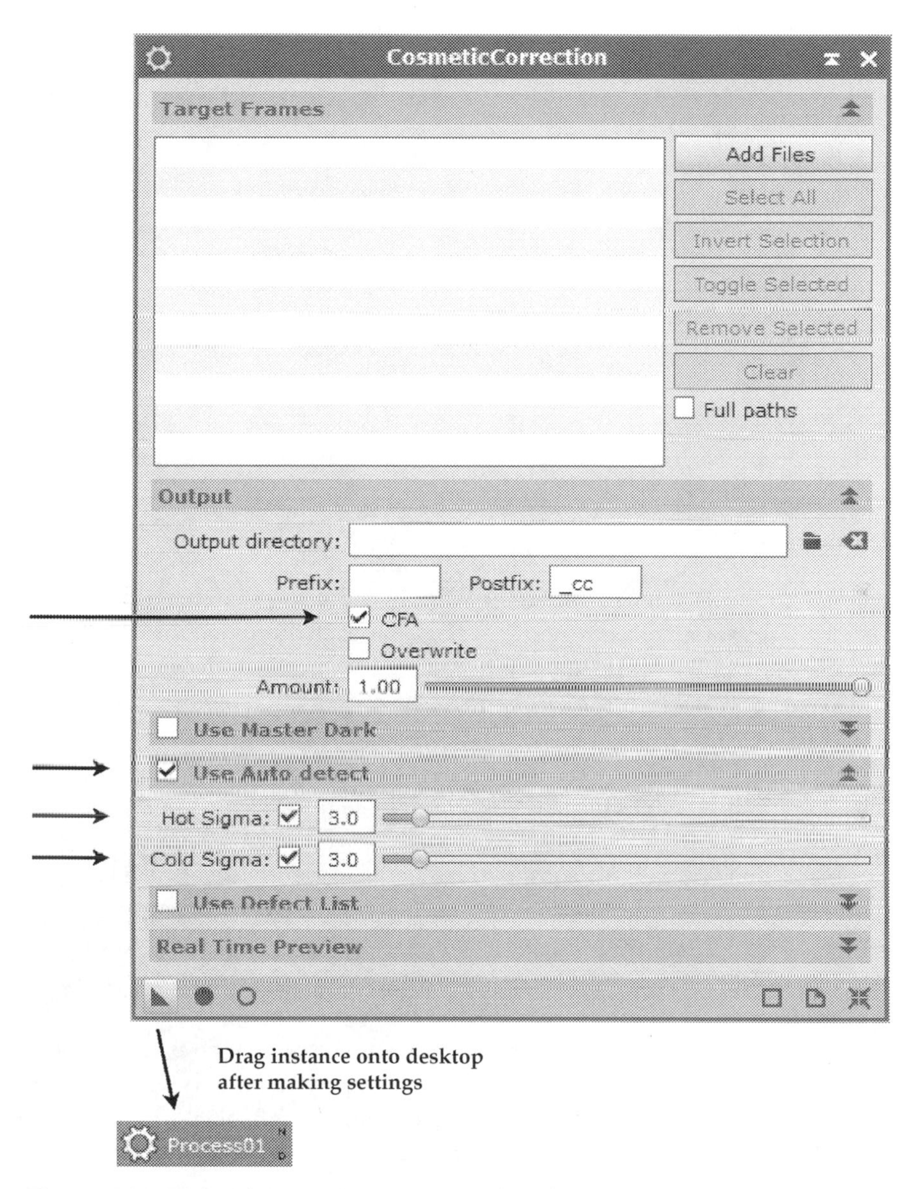

Figure 12.21. *PixInsight* cosmetic correction tool can be used as input to the BatchPreprocessing script. No need to select files when doing so; you are just making settings to be communicated to the script.

After doing so, drag the instance triangle onto the desktop, creating an instance, and take note of its name, or rename it if you want.

Then launch BatchPreprocessing or BatchPreprocessingFD and check "Apply" in the Cosmetic Correction panel. Under "Template icon" choose the instance that you just created on the desktop. That instance will be applied to your images.

Now there will be a fifth output subfolder containing cosmetically corrected image files, created after calibration but before registration. You need not keep it after the stacking has been done.

In BatchPreprocessingFD (the script you can download from this book's web site), you have an alternative: simply check "Apply" and choose "<default>" instead of the name of an instance. You will get simple detection of hot and cold pixels with default settings like those in Figure 12.21. You don't have to create an instance yourself.

12.5.6 Stacking (Integration) as a Separate Step

The stacked ("master light") image produced by BatchPreprocessing or BatchPreprocessingFD is usually satisfactory but not always the best you can achieve. Many *PixInsight* users choose to stack the registered images again, with different settings.

That is done by choosing Process, ImageIntegration, ImageIntegration. The tool window has more tabs than there is room to open at once. Figures 12.22 and 12.23 show the important ones. First choose your files (from the "registered" folder) and choose how to weight them (normally by noise evaluation).

Then close the Input Images and Image Integration tabs so you can open the first two Pixel Rejection tabs. There, choose an algorithm to reject deviant pixels (normally Winsorized Sigma Clipping) and adjust thresholds if you need to (normally the defaults are fine). Finally, run the integration and look at the results.

Note that the default settings are likely to be identical to what you get if you stack within the BatchPreprocessing script. In both cases the images are weighted by evaluating their noise levels and giving priority to the less noisy images. That helps correct for changes of temperature during a long session and other factors that might cause the noise level to change.

The output consists of three images on the screen. One is the stacked astronomical image. The other two are pixel rejection maps, showing where deviant pixels were rejected. They are normally almost all black, but you may see airplane trails in them, or other evidence of substantial changes.

12.5.7 Stretching

The most straightforward way to do stretching in *PixInsight* is by choosing Process, IntensityTransformations, HistogramTransformation (Figure 12.24). This works very much like the familiar *Photoshop* Levels tool (Section 4.3); what you're going to do is move the middle slider a long way to the left. Before doing so, be sure to choose the image to work on, *turn off screen stretch* (which would show you a blank white window!), and activate real-time preview.

Stretching is best done in several steps (at least three; you won't be able to do full gamma correction accurately in one step). After applying each change,

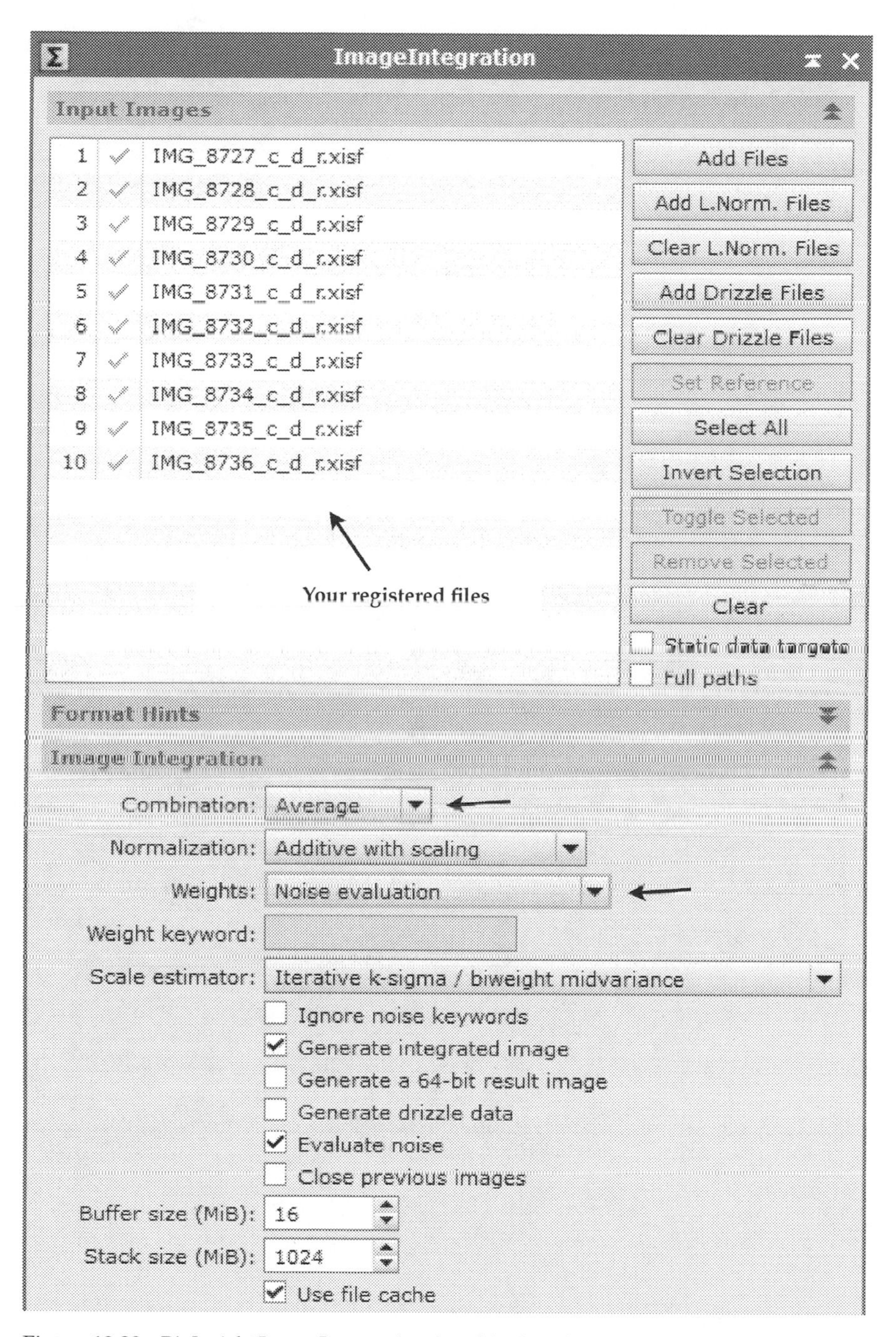

Figure 12.22. *PixInsight* ImageIntegration (stacking) tool (upper part). Images are normally weighted by noise evaluation. Close some tabs in this window to be able to see others.

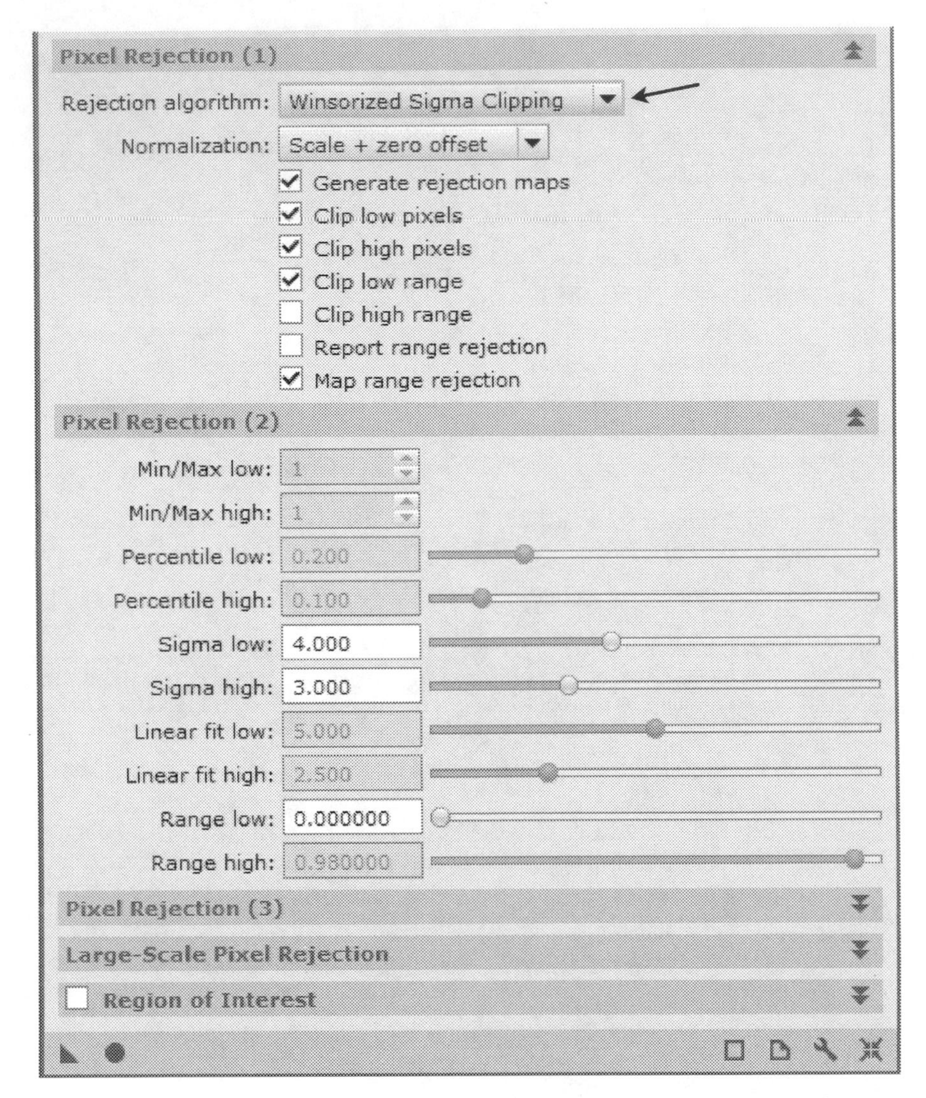

Figure 12.23. *PixInsight* ImageIntegration (stacking) tool (lower part). This is where you specify rejection of deviant pixels, normally by sigma clipping.

immediately hit the reset button at the lower right corner of the tool window so that the next correction will start from scratch. Otherwise real-time preview will assume you're about to apply the same change again, which is not the case.

If the image has a color cast, adjust the red, green, and blue layers separately to bring their three humps into about the same position.

12.5.8 *PixInsight* Workflow Summary

To keep things simple, I have introduced some of the easier steps first, even though they are not done first. The full workflow, to get from a stack of raw DSLR files to a finished image in *PixInsight*, is as follows:

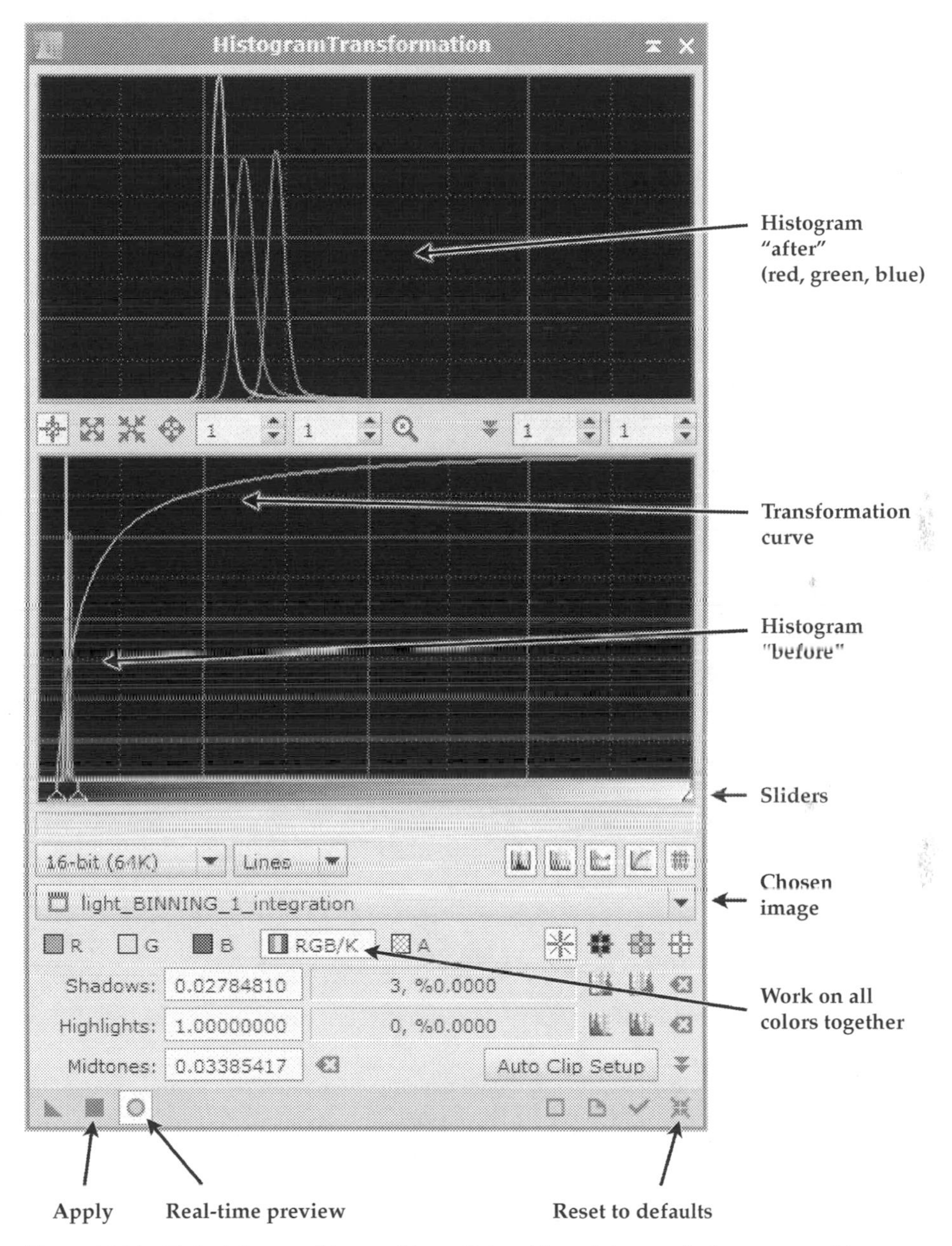

Figure 12.24. *PixInsight* stretching tool is traditional but elaborate; it shows you a histogram before and after the proposed adjustment.

1. Use Process, ImageInspection, Blink, or perhaps Script, Batch Processing, SubframeSelector, to decide which images you are going to stack.
2. If the number of raw files to be converted is large, consider converting them all to FITS (to save memory) using Script, Batch Processing, BatchFormatConversion.

3. Open Process, ImageCalibration, CosmeticCorrection, make settings as in Figure 12.21, and drag an instance onto the desktop.
4. Open Script, Batch Processing, BatchPreprocessingFD (or BatchPreprocessing) and proceed as in Figure 12.19, but:
   - Check "Bias frames are flat darks." (If you don't have BatchPreprocessingFD, and hence don't have this checkbox, proceed as described under "More accurate calibration" above.)
   - Under "Cosmetic correction" check "Apply," and under "Template icon" choose the CosmeticCorrection instance you created on the desktop.
5. Run the script.
6. Optionally, re-stack the images with more adjustments using Process, ImageIntegration, ImageIntegration.
7. Perform stretching (histogram adjustment).
8. Crop and resize your image as desired and save it in a format you can use for other purposes (TIFF or, for the Web, JPEG).

Chapter 13
More Image Processing Techniques

Beyond the basics of image calibration, stacking, and gamma correction, many other techniques are also used to improve astronomical images. This chapter quickly surveys some of the most important. But one important technique, multiscale sharpening, is so closely connected with planetary imaging that I'm saving it for the next chapter.

Several of the techniques in this chapter borrow a principle from artificial intelligence and machine vision: If human beings can see something in an image, it should be possible to program a computer to "see" the same thing. For example, if a human being can see that the bottom of Figure 13.1 is too light, a computer should be able to "see" the same thing, and correct it.

There is a fine line that we must be careful not to cross. I balk at techniques that amount to retouching or painting on the image. I think images should be processed by algorithms. The human artist should supply parameters, but within those parameters, the computer should treat similar pixels and image features alike. It won't do to "enhance" one part of a nebula and suppress another, unless you clearly explain why you did so and your reason is convincing. When you start changing the content of an image, you're doing fantasy space art, not astrophotography. The techniques introduced in this chapter do not run afoul of that guideline.

13.1 Flattening the Background

13.1.1 The Concept

Even after flat-fielding, many astronomical images contain unwanted gradients such as the one in Figure 13.1, or slight uncorrected vignetting. Further, some images need to be flat-fielded, but the corresponding flat-field exposures were not taken.

To see whether your image needs this correction, you may need to use strong screen stretch. It also helps if you reduce the size of the image to considerably

Figure 13.1. (Top:) Even after flat-fielding, this image still is lighter at the bottom and at the upper right than elsewhere because of city lights. (Bottom:) Background flattening removes the gradients.

smaller than the whole screen so that discrepancies between one side and the other are obvious.

Fortunately, there are algorithms to flatten the background of an image by constructing and applying a *synthetic flat* based on analysis of the brightness range and especially the dark areas between the stars. As the figure shows, such algorithms can be very effective.

13.1.2 Subtract or Divide?

You have to choose whether to subtract the synthetic flat or divide by it. Subtraction is appropriate when the gradient comes from external light, such as city lights near the horizon or even amp glow in the sensor (although dark frames should take care of the latter). Division is appropriate for dealing with vignetting, dust specks, or anything else that removes a fraction of the light in the affected areas.

13.1.3 Linear or Gamma-corrected?

Theoretically, background flattening should be done on a linear, non-gamma-corrected image; like flat-fielding itself, both subtraction and division should only work correctly on a linear image.

In practice, although working on linear images is better, background flattening of a nonlinearly stretched image is often successful. Experiment to determine whether to subtract or divide, and if necessary, do more than one iteration.

13.1.4 *Nebulosity*

In *Nebulosity,* background flattening is done by choosing Image, Synthetic Flat Fielder. The instruction manual explains this tool very well. Default settings usually work well.

The number of sample points that you specify is in addition to the minimum of one point; for example, if you select 3 (the default), there will be 4 sample points along each axis, making a 4 × 4 grid, such that there is no sample point in the center. If you had chosen 2 or 4, the deep-sky object in the center of the picture might be mistaken for background.

Nebulosity's synthetic flat fielder also corrects any strong color cast that the background may have. It does this by flat-fielding the three color channels (R, G, and B) separately.

13.1.5 *MaxIm DL*

In *MaxIm DL,* the three menu choices for background flattening use a picture of a flatiron as their icon. They are on the Filter menu.

Flatten Background prompts you to choose where you'd like the background to be sampled, and to make other choices; this is very useful if your picture contains a large nebula or other real detail that could be mistaken for extraneous light.

Auto Flatten Background performs flattening automatically, usually very well. Auto Remove Gradient is similar but confines itself to a one-dimensional gradient and is less likely to mistake image features for unwanted light.

Unlike *Nebulosity* and *PixInsight*, *MaxIm DL*'s background flattener does not remove color casts. There is another tool, Remove Background Color on the Color menu, that converts the darkest parts of the picture to monochrome without affecting brighter areas.

13.1.6 *PixInsight*

PixInsight has two background flatteners, both under Process, Background-Modelization. The AutomaticBackgroundExtractor (Figure 13.2) is, as its name implies, largely automatic. The DynamicBackgroundExtraction tool lets you select where the background is to be analyzed. Both give you a plethora of settings that can almost always be left at the default values. Like *Nebulosity*, *PixInsight* removes color casts when flattening the background.

13.2 Removing Noise

13.2.1 The Concept

Looking at an image with your eyes, you usually have no trouble distinguishing noise (grain) from genuine image detail (Figure 13.3, top). The computer can usually make the same distinction. Noise consists mostly of random variation between individual pixels, while genuine features of the image span more than one pixel in a systematic way.

The simplest way to denoise an image is simply to blur it, but it is much better to recognize the characteristics that separate noise from genuine detail. Algorithms to do this are not infallible, and if denoising is overdone, or if an algorithm simply guesses wrong, you can end up with broad swirls and mottles throughout the image. A good rule of thumb is to do slightly less denoising than seems to be needed.

13.2.2 Luminance vs. Chrominance

On a sensor with a Bayer matrix, adjacent pixels represent different colors, and pixel-to-pixel variation results in speckles of color, not just speckles of varying brightness. Color fluctuation is called *chrominance noise* as distinct from *luminance noise*. Of the two, chrominance noise is usually much more bothersome, and reducing it aggressively does not lose image detail. Accordingly, I usually tell the software to cut chrominance noise more than luminance noise.

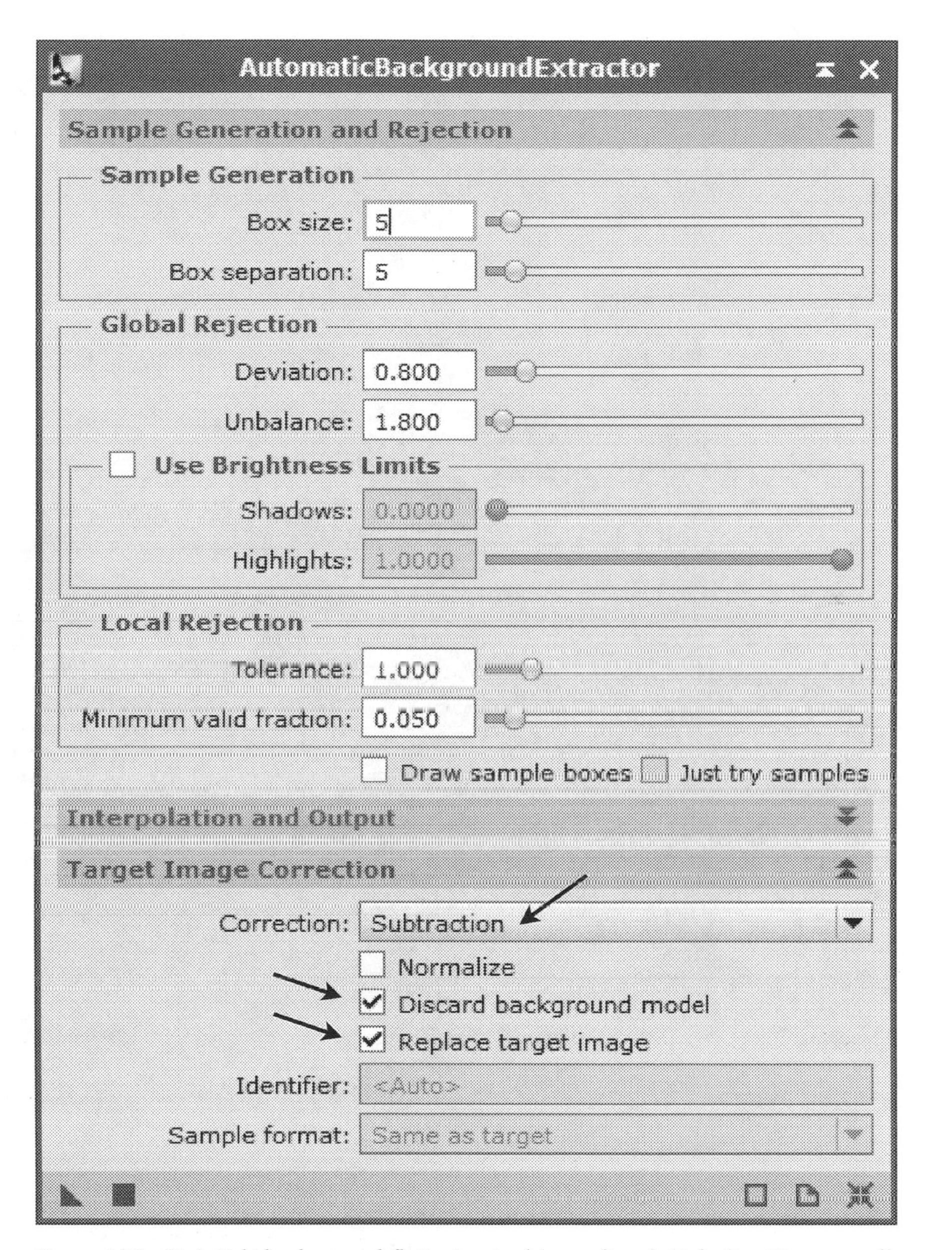

Figure 13.2. *PixInsight* background-flattening tool (one of two). Default settings usually work well; arrows point out settings you must make.

13.2.3 Linear or Gamma-corrected?

Theoretically, noise reduction algorithms should be used on unstretched, linear images, but in practice, I perform noise reduction on the finished, stretched, gamma-corrected image. The reason is that before stretching, in a typical deep-sky image, the pixel values are much lower than the noise reduction

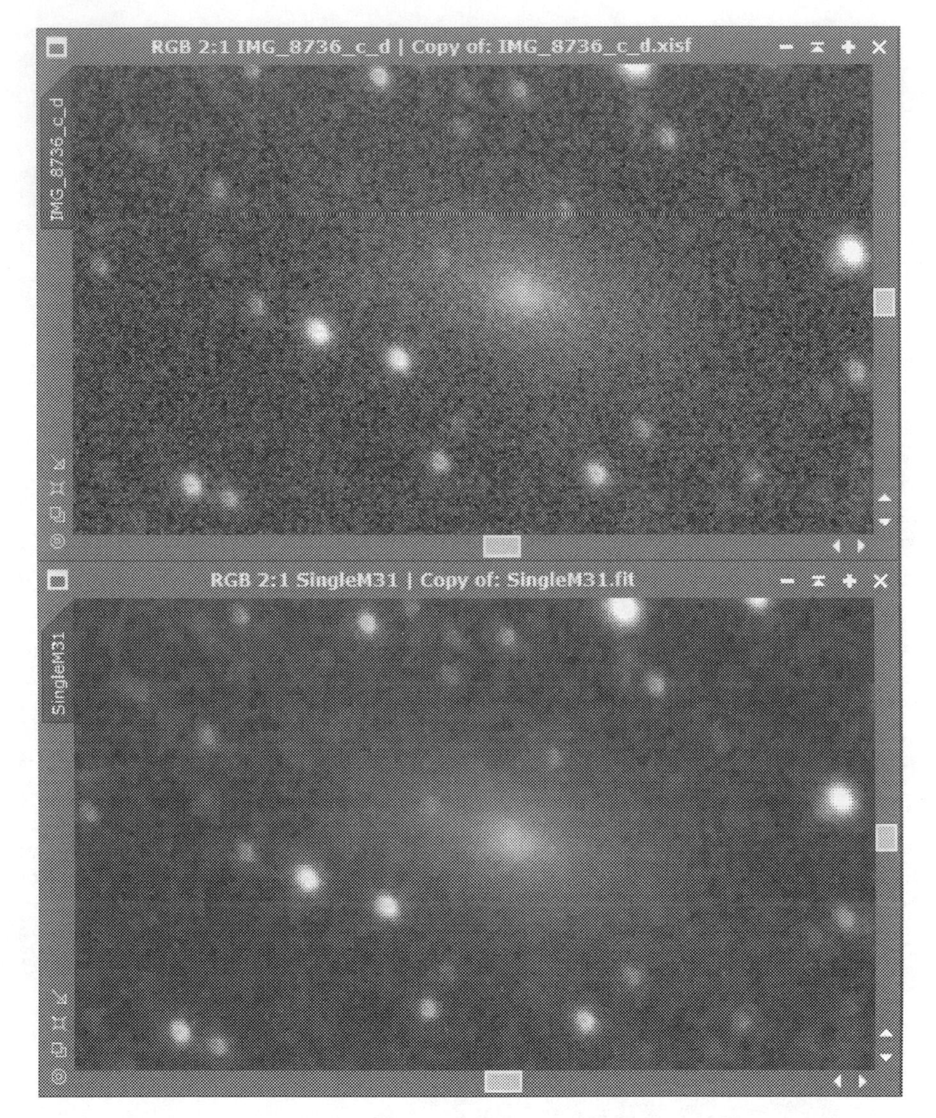

Figure 13.3. Noise reduction removes grain from image without losing image detail.

algorithm expects, and results are erratic. The noise may be theoretically modeled for a linear image, but it is not mathematically well-behaved in the first place, and gamma stretching does not really make it any harder to correct.

13.2.4 *Nebulosity*

Two noise reduction tools are on the Image menu. Adaptive Median Noise Reduction simply blurs the darker areas of the image while leaving the lighter areas unchanged; you choose the threshold. In my experience, on large DSLR images, it often has little effect.

The other tool, GREYCstoration, has many parameters, which you can adjust, but the defaults often work well. It is described in detail in the instruction manual.

13.2.5 *MaxIm DL*

At first sight there seems to be no noise reduction tool in *MaxIm DL*, but actually, noise reduction by multiscale image analysis (Section 2.6.4) is easy to do. Choose Filter, Wavelet Filter, and move the sliders for W1 and maybe W2 (the smallest image features) all the way to the left. This filters out single-pixel variations while preserving everything larger.

13.2.6 *PixInsight*

PixInsight provides six noise reduction tools under Process, NoiseReduction. The one I usually use is TGVDenoise (for Total Generalized Variation), with settings roughly as shown in Figure 13.4. In *L*a*b* mode (which separates

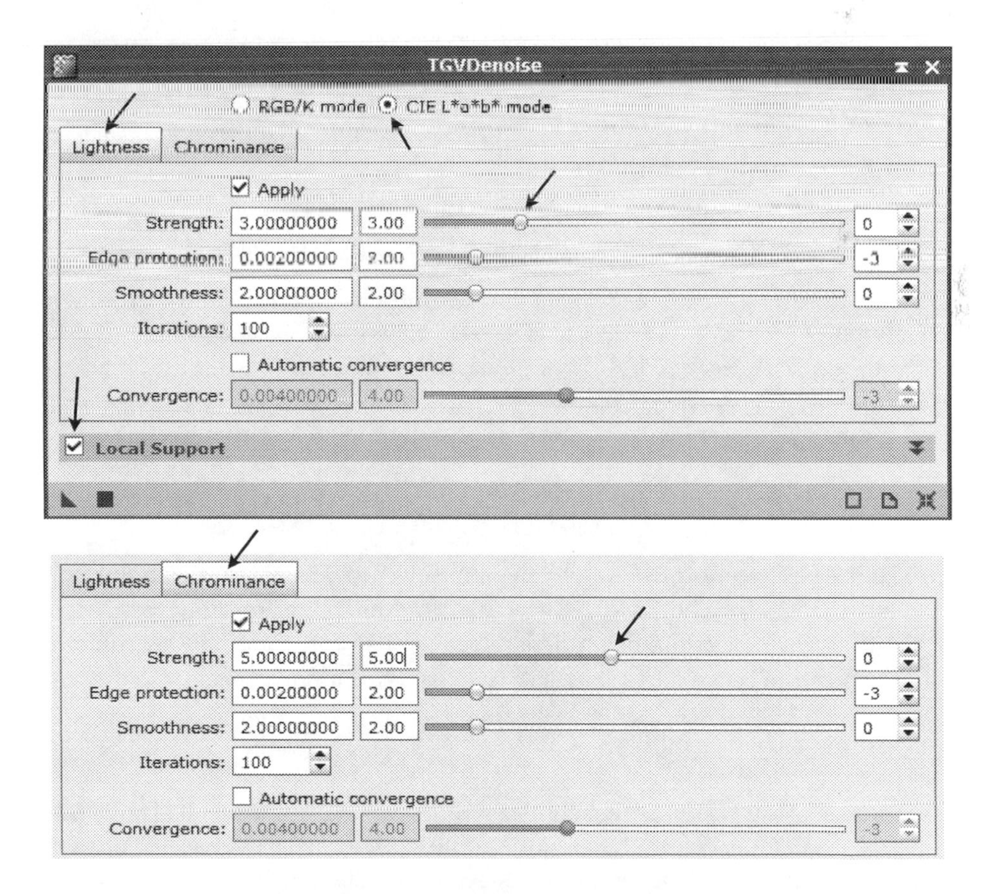

Figure 13.4. These settings for *PixInsight*'s TGVDenoise tool often work well with DSLR images. Note that chrominance noise is to be reduced much more than luminance noise.

luminance from color), you can process chrominance noise separately from luminance noise and reduce it more drastically. Local Support means to remove more noise from darker areas than from lighter ones.

13.3 Color Saturation

13.3.1 The Concept

Color saturation is the amount of difference between colors. To increase the saturation of an image is to make the reds redder, the greens greener, and the blues bluer – that is, to increase the difference between the R, G, and B values of each pixel.

Increasing the color saturation often helps bring out yellowish dust lanes in galaxies and the distinction between different minerals on the moon. Reducing the color saturation is one way to reduce color casts and chrominance noise; taken all the way, it turns a color image into black-and-white.

13.3.2 Linear or Gamma-corrected?

Adjustment of color saturation is always done on the final, stretched image, since stretching itself changes color saturation. Unstretched images are often disconcertingly colorless.

13.3.3 *Nebulosity*

Under Image, choose Adjust Hue/Saturation. That gives you a tool with three sliders for hue, saturation, and lightness (HSL). Move the S slider to the right to increase saturation. The L slider changes the overall brightness of the image, and the H slider can sometimes be useful if you need to correct a strong color cast.

13.3.4 *MaxIm DL*

Choose Color, Adjust Saturation, and select the desired saturation numerically (e.g., 150% for a boost, 75% for a cut). A hidden panel lets you restrict the change to pixels in a certain brightness range, so that, for example, you can reduce the saturation of the darkest areas to combat chrominance noise.

13.3.5 *PixInsight*

The *PixInsight* color saturation tool is under Process, IntensityTransformations, and is complex enough to require some explaining (Figure 13.5). You can adjust the saturation of different colors separately, which is almost never desirable. (Discrepancies between colors should be attacked through the histogram adjustments for R, G, and B.) To adjust the saturation of all colors together, you must

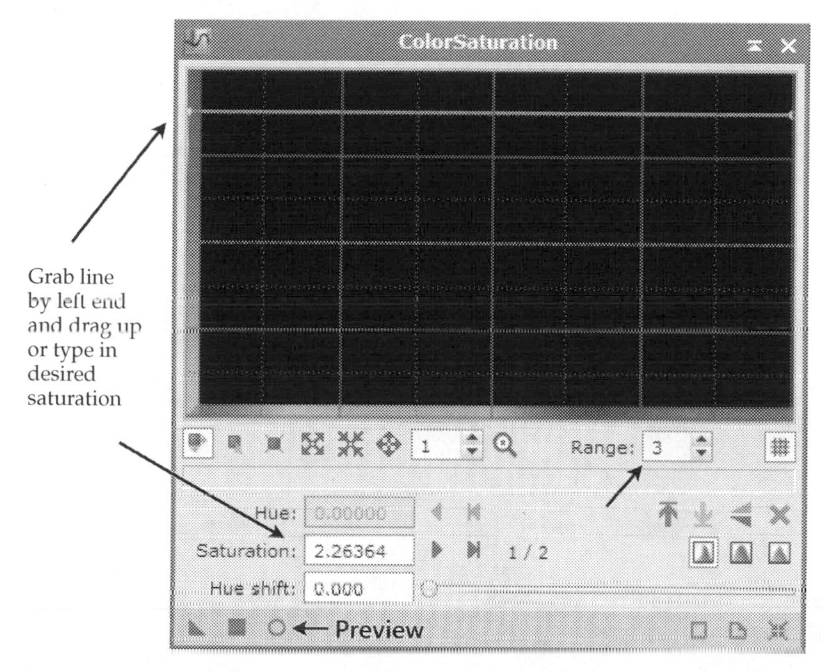

Figure 13.5. Color saturation in *PixInsight* is adjusted by moving a horizontal line up and down; "Range" setting determines how far you can move it. An alternative is to type in the desired saturation (relative to 0 for no change, positive to boost, negative to cut).

use the mouse to grab the end of the horizontal line and pull it upward or downward, keeping it horizontal; grabbing it in the middle will bend it. Alternatively, you can type in the desired saturation (initially 0; positive to boost, negative to cut).

13.4 Masks

Often you may want to process the brighter parts of a picture differently than the darker parts. An example is increasing the color saturation in the bright parts of a galaxy, but not in the dark background, where it would only bring out chrominance noise.

Masks are a *PixInsight* feature that make this easy (not provided in *Nebulosity* or *MaxIm DL*). A mask is an image, the same size as a picture, that blocks processing of part of the picture. In the mask, white means yes and black means no; that is, the white parts of the mask allow the processing to proceed fully, and the black parts block processing.

Here is an example of how to use a mask to confine a color saturation adjustment to the bright parts of an image:

1. Open the image you want to process.
2. Make a copy of it. To do this, right-click on the image and choose Duplicate. Now you have another copy of the image, open in a different window.

3. Convert the new image to monochrome (Image, Color Spaces, Convert to Grayscale).
4. Histogram-stretch the new image so that it is just black and white, with little or no gray in between, white in the areas where you want to raise the color saturation and black elsewhere (Figure 13.6).

Figure 13.6. High-contrast copy will serve as mask to confine processing to light areas of original image (*PixInsight*).

5. Now select the original image again. Choose Mask, Select Mask, and choose the stretched monochrome image (the mask) to be the mask for the original image.
6. At this point the masked parts of the image, where processing is blocked, turn bright red. You can turn off the red coloring by choosing Mask, Show Mask (which toggles on and off).
7. Still with the original image selected, perform the color saturation adjustment. It takes effect only where the mask permits it.
8. Choose Mask, Remove Mask.

This particular adjustment would normally be done to an image that is already gamma-corrected, but masks can be used at any stage of the processing.

13.5 Who Moved? The Difference between Two Pictures

13.5.1 The Concept

The difference between two images is often of interest. Figure 13.7 shows the difference between two images in one picture, making the motion of the asteroid Iris obvious.

You might guess that the difference is computed by subtraction, and you'd be right, except that pure subtraction doesn't do the job – there are no negative numbers in pixel arithmetic. Further, subtraction would make the matching parts of the images disappear, but in fact, we want them to remain faintly visible so that we can see where things are.

Accordingly, what is wanted is an image that is about 60% of the first image minus 40% of the second image. To avoid negative numbers, after subtracting the second image, we will add its maximum value back in (40% of maximum white). Specifically, we want each pixel value in the final image to be:

$$\text{New image} = 0.6 \times \text{First image} - 0.4 \times \text{Second image} + 0.4 \times \text{Maximum pixel value.}$$

Pixel values may range from 0 to 1, 0 to 255, 0 to 65 535, or something else depending on how the arithmetic is done.

13.5.2 Preparing the Images

Before being combined in this way, the images will normally already be fully gamma-corrected and stretched so that they span a normal brightness range.

They must of course be aligned. Images taken during the same session and guided on the same star may already be aligned well enough. If there are only two, you can align them in *Photoshop* by pasting a copy of the second image on a copy of the first, making the new layer partly transparent, nudging it pixel by pixel until it is positioned perfectly, and then making it completely opaque. Both *MaxIm DL* and *PixInsight* can output aligned (registered) star field images without, or in addition to, stacking them.

Figure 13.7. Top and middle: Two images of the asteroid Iris taken 90 minutes apart. Bottom: Difference image showing its two positions, one white and one black.

13.5.3 *PixInsight*

In this case I will tackle the software packages in reverse order because *PixInsight*'s implementation of pixel arithmetic is, perhaps surprisingly, the easiest to understand. Under Process, PixelMath, PixelMath is the tool shown

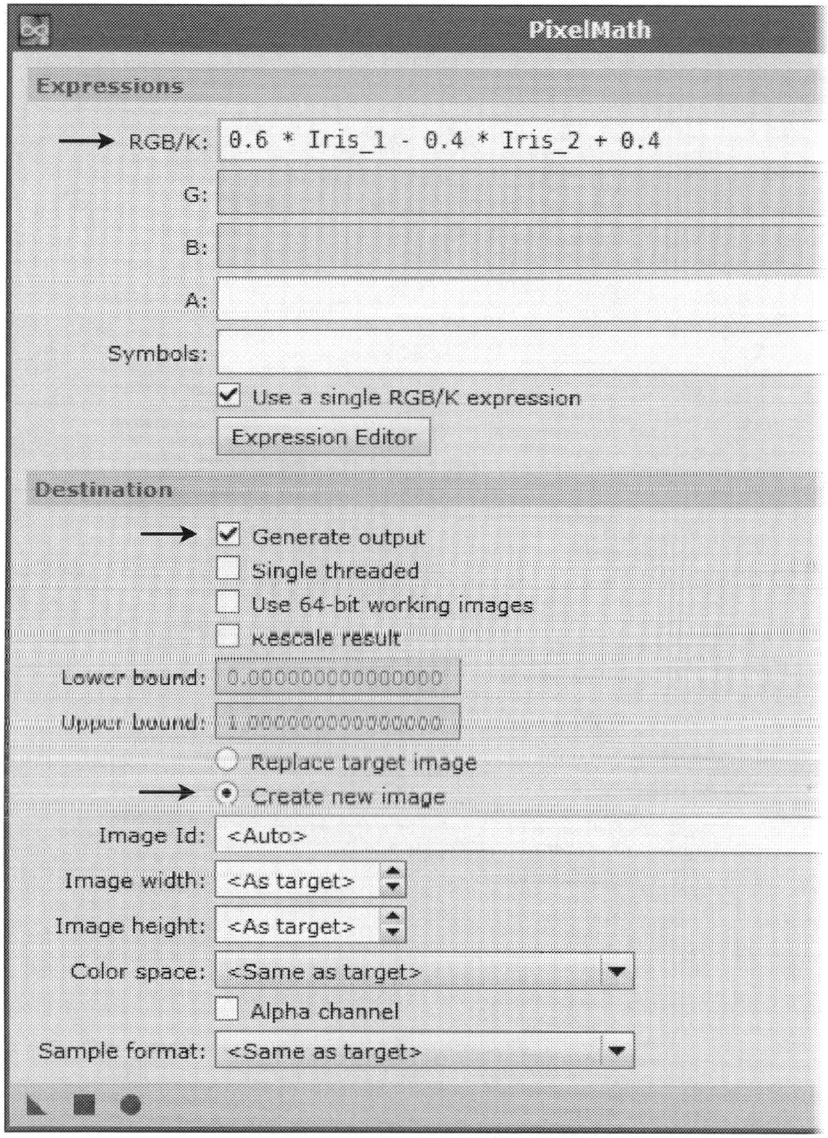

Figure 13.8. *PixInsight* PixelMath tool lets you type expressions for computations to be performed. Pixel values are treated as ranging from 0 to 1 regardless of bit depth. Note settings to create a new image rather than modify an existing one.

in Figure 13.8. With both images open, simply type in the expression for what to compute. The expression editor, available at the click of a button, will help you confirm the names of the images and the available operators, but you can type or edit the formula without using it.

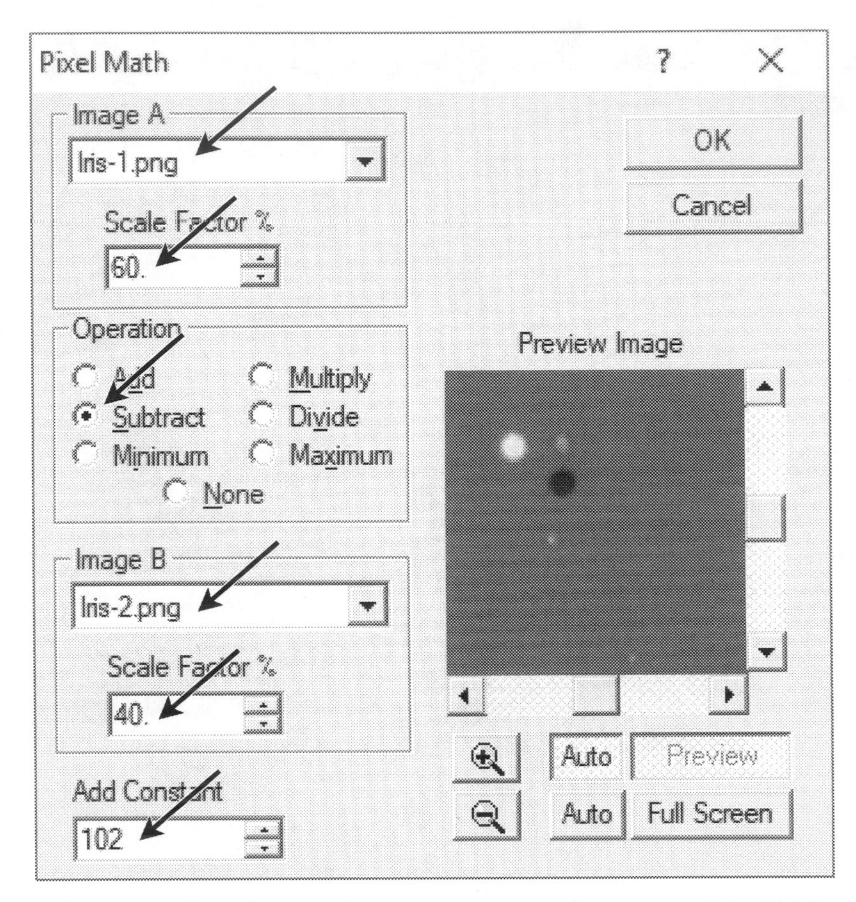

Figure 13.9. Pixel arithmetic in *MaxIm DL* is expressed as menu settings. This is equivalent to Figure 13.8; the number 102 is 0.4×255.

13.5.4 *MaxIm DL*

MaxIm DL performs the same computation, but instead of a formula, the computation is expressed as a set of settings in a tool (Figure 13.9). The first image is scaled 60%, the second image is scaled 40% and subtracted, and the constant 102 is added (which is 40% of 255, the maximum pixel value in these 8-bit images). The output replaces the first image.

13.5.5 *Nebulosity*

Pixel arithmetic in *Nebulosity* is broken up into even more steps, but the calculation can still be done. This is how:

1. Open the first image.
2. Choose Image, Scale Intensity, Multiply by X, 0.6.

3. Save the first image, which has been scaled.
4. Open the second image.
5. Choose Image, Scale Intensity, Invert.
6. Choose Image, Scale Intensity, Multiply by X, 0.4.
7. Save the second image.
8. Choose Batch, Align and Combine Images. Make the settings: Save stack, Alignment method: None, Stacking function: Average, and uncheck Fine-tune star location.
9. Choose the two files you saved. Save the resulting stacked image.
10. Stretch the stacked image because it was averaged rather than summed.

To avoid the stretching step, you might want to multiply by 1.0 and 0.8 respectively (not 1.2 and 0.8, to avoid going out of range).

13.5.6 *Photoshop*

Finally, this kind of image combining happens to be easy to do in *Photoshop*. Invert the second image, copy it, and paste it over the first image with 40% opacity. You can even fine-tune the alignment at the same time, nudging it so that the stars that show in both images are perfectly aligned.

13.6 High Dynamic Range (HDR)

Many digital images cover a greater brightness range than a computer screen or printer can properly display. The usual way to make them displayable is to reduce the really large-scale differences in brightness – the large differences between parts of the picture that are far apart – while preserving the small-scale differences. Artists' paintings work that way; the local contrast is normal but the large-scale contrast is much less than in the real scene.

A high-dynamic-range image may be a single picture from a camera with good dynamic range, or it may be a stack of exposures of different lengths. In terrestrial HDR photography, the latter is normal practice.

To make a high-dynamic-range image viewable, perform multiscale processing (Section 2.6.4) with enough layers that some of them cover quite large scales, and then reduce the residue component, as in Figure 2.12 (right). What's left is a viewable HDR image. Using M42, which is astronomers' favorite HDR target, Figure 13.10 shows what HDR processing achieves.

Any software that offers multiscale image processing can do this, but *PixInsight*'s HDRMultiscaleTransform tool (under Process, MultiscaleProcessing) is particularly effective and easy to use; it is almost self-explanatory.

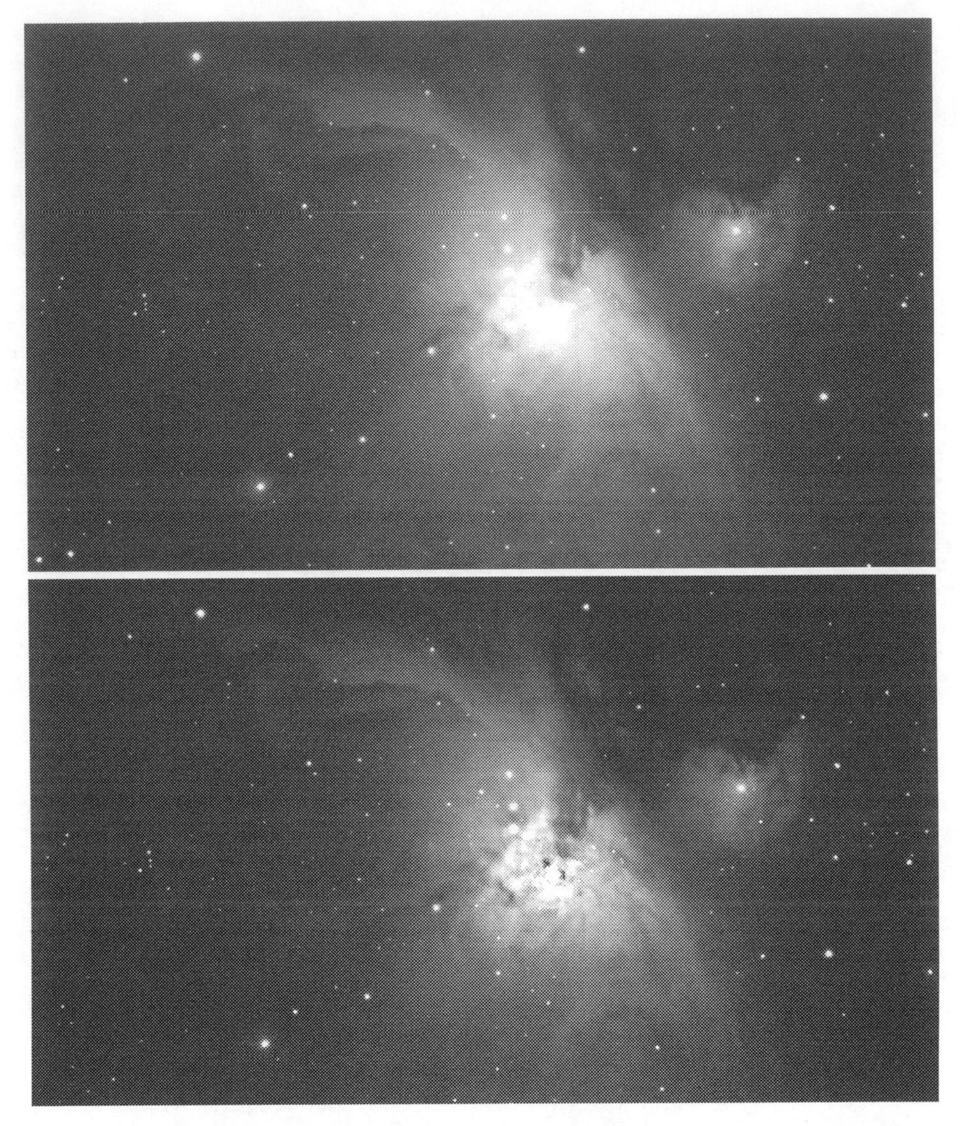

Figure 13.10. High-dynamic-range (HDR) image relies on multiscale processing to bring out detail in light areas that would otherwise merge with maximum white.

The image must, of course, actually contain the detail you want to bring out. That means it must not have been stretched too much. Nonetheless, I get the best results if the image has been stretched enough that it looks good without screen stretch. The effect of HDRMultiscaleTransform on a dark, unstretched image is likely to be too extreme.

Chapter 14
Sun, Moon, Eclipses, and Planets

Digital technology has revolutionized solar, lunar, and planetary imaging even more than deep-sky work. Amateurs with portable telescopes don't just rival the best work done by observatories 50 years ago, they surpass it. Thanks to video imaging, we get planetary images better than anything that had been taken from earth before 1985 or so.

Accordingly, this chapter addresses – at last – the sun, moon, and planets. The applicable techniques are such a vital part of astrophotography that they deserve coverage even if a DSLR is not always the ideal camera for practicing them.

To be precise, video astrocameras (such as your autoguiding camera) are ideal for high-resolution lunar and planetary work; DSLRs vary in suitability for this, although some come close. For full-face sun and moon shots, DSLRs are ideal only if they have electronic first-curtain shutter. Other DSLRs are at something of a disadvantage.

Nonetheless, let's take a look at what DSLRs can accomplish. Because the art of high-resolution imaging could fill another book the size of this one, I shall have to be brief.

14.1 Full-face Lunar and Solar Images

Full-face images of the sun and moon, including eclipses, are more like daytime photography than any other kind of astronomical imaging. The camera is not pushed to its limits; the JPEG file from the camera requires little or no processing; and the challenge is mainly in choosing a suitable optical system and taking pictures without vibration from the shutter.

14.1.1 Optics and Field of View

The sun and moon happen to have the same apparent diameter, about 0.5°. More precisely, the moon is always less than 0.57° (0.01 radian), depending on its position in its orbit, and the sun is always less than 0.55°. That means the diameter of the image of the sun or moon on the sensor is slightly less than 1/100 of the focal length of the lens or telescope.

It follows that the full face of the sun or moon will fit on a full-frame sensor when the focal length is less than 2400 mm, and on an APS-C sensor when it is less than about 1500 mm. To avoid an excessively tight fit, in practice I recommend maximum focal lengths of 2000 and 1250 mm respectively. Focal lengths as short as 300 mm give pleasing images.

The optical system, then, can be anything from a long telephoto lens (Figure 14.1) to a medium-sized telescope (Figure 14.2). Field flatness is important, and field-flattening focal reducers are beneficial. With the typical field curvature of an $f/10$ Schmidt–Cassegrain, it is best to focus on a point about two thirds of the way from the center of the sun or moon to the limb.

It can also help to stop the telescope down. You do not need the full aperture of an 8-inch (20-cm) telescope to get a sharp full-face picture of the sun or moon. Stopping it down to 2 inches (5 cm) off-axis will improve field flatness and also reduce the effect of unsteady air. Telephoto lenses, also, are often sharpest when not wide open.

For a solar eclipse with the corona, of course, you will want a much wider field, and that implies a focal length of 300 mm or even less.

14.1.2 Exposure

Exposures of the sun and moon are easy to evaluate by trial and error, but for a table of suggested exposures, see Appendix B.

The main challenge is overcoming shutter and mirror vibration. All the tricks in Sections 3.5.2–3.5.5 are applicable. If all else fails, a really short exposure (1/500 or faster) may give satisfactory results by "stopping" the vibration.

14.1.3 Tracking

Most solar and lunar exposure times are so short that the telescope need not track the sky, but with total solar and lunar eclipses, tracking may be necessary.

In detail: The rotation of the earth is one arc-second per 1/15 second of time, and one arc-second is close to the practical limit of resolution. It follows that even the highest-resolution imaging does not require tracking if the exposure is 1/15 second or less. With relatively modest telescopes and telephoto lenses, you can go appreciably longer; typically, a 300-mm telephoto lens will not show any motion blur in a 1/4-second exposure on a fixed tripod.

Photographing a total solar eclipse, you may need to expose a few seconds, and photographing a total lunar eclipse, as long as a minute or more. In those situations, tracking becomes important. The sun is of course best tracked at solar rate, an option on most computerized mounts, but sidereal rate is close enough

Figure 14.1. 1980s-era Olympus 600-mm lens gives pleasing image of the moon at $f/11$. Single 1/250-second exposure at ISO 500, Canon 60Da, taken from camera as JPEG file, unsharp masked and cropped in *Photoshop*. A newer-technology lens would be sharper.

to make no difference. Lunar eclipse exposures longer than a couple of seconds require lunar-rate tracking, but even that isn't perfect because the moon's orbital motion is not uniform and there is a declination component. Now you know why really sharp pictures of the totally eclipsed moon are rare.

Figure 14.2. Celestron 8 EdgeHD with flat-field $f/7$ focal reducer gives critically sharp image of the full face of the moon, aperture 8 inches (20 cm), focal length 1400 mm. Single 1/320-second exposure at ISO 400, Canon 60Da, processed like Figure 14.1.

14.1.4 Stacking

Full-face lunar and solar images can be stacked to reduce grain. *AutoStakkert* (Section 14.2.4) does a good job with JPEG files straight from the camera or TIFFs converted from raw images.

14.1.5 The Moon

The moon's surface has two peculiarities worth noting. First, because of its porous structure, it tends to reflect light in the direction from which it came, like a highly reflective road sign. As a result, the surface brightness of the quarter moon is appreciably less than that of the full moon. The exposure table in Appendix B.2 takes this into account.

Second, there are colors, though they are subtle. Increasing the color saturation of a lunar image (Section 13.3) often reveals that Mare Tranquillitatis is bluish and the other maria are brownish.

Images of the moon lend themselves very well to unsharp masking. The underlying detail is very crisp, and the magnification of a typical lunar photograph is low enough that processing artifacts are not a problem.

14.1.6 The Sun

Safe imaging of the sun requires a filter of suitable density *in front of the telescope* so that direct sunlight never enters the optical system. Densities are measured logarithmically, so that D = 5.0 means the light is cut by a factor of $10^5 = 100\,000$, and D = 6.0 means 1 000 000. It is not safe to look at the sun through filters not designed for solar work; in particular, Wratten 96 neutral density filters, even when stacked to give a high density, transmit too much infrared to be safe. The same is true of photographic dye filters and crossed polarizers.

The filter material in eclipse glasses is fine for photographic use; you can mount it in a larger piece of cardboard and attach it to the front of your telephoto lens. (The opening need not be round, nor as large as the lens itself.) Sun filters for telescopes are made by Thousand Oaks (www.thousandoaksoptical.com), Baader (www.baader-planetarium.com), and others.

When photographing the sun through a filter, you may need to cover the eyepiece of your camera to keep unwanted light from entering (Section 3.4).

Thousand Oaks solar filters give an orange-colored image, and when photographing through them, it is important to check the histograms of all three colors (red, green, blue), since there is a risk of overexposing the red layer of the image, causing loss of detail, while getting a picture that looks normal. Err on the side of underexposure.

Figure 14.3 is a sharp image of the sun with sunspots, taken with a 5-inch (12.5-cm) telescope and Thousand Oaks filter. The surface of the sun is granulated, and solar granulation is sometimes hard to distinguish from grain in the image. The granulation you see in the picture is genuine (see Figure 14.4).

14.1.7 Eclipses, Solar and Lunar

Partial eclipses of the sun and moon are photographed the same way as if there were no eclipse going on, except that during an eclipse of the moon, the moon dims appreciably while in the penumbra, before we see the earth's shadow falling on it, so the exposure needs to be longer.

Total eclipses are more of a challenge. Total lunar eclipses are unpredictably dark; sometimes the moon is easy to see, and sometimes it is so dark that it

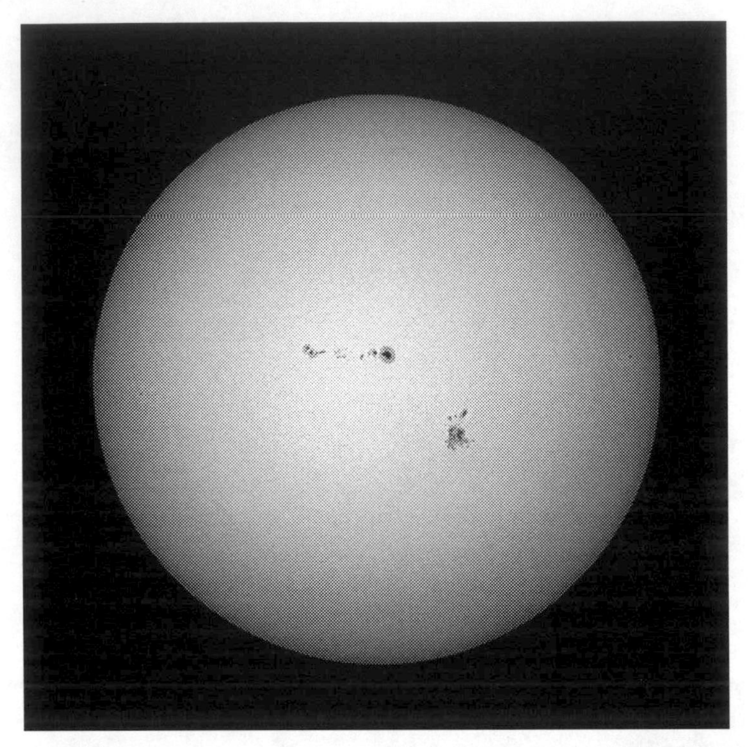

Figure 14.3. Sunspots, granulation, and faculae. 5-inch (12.5-cm) $f/10$ Schmidt–Cassegrain telescope with Thousand Oaks RG-6000 filter (D = 5.0, orange). Six 1/1000-second exposures at ISO 100 were taken from the camera as JPEG files, stacked with *AutoStakkert*, and unsharp masked and cropped with *Photoshop*.

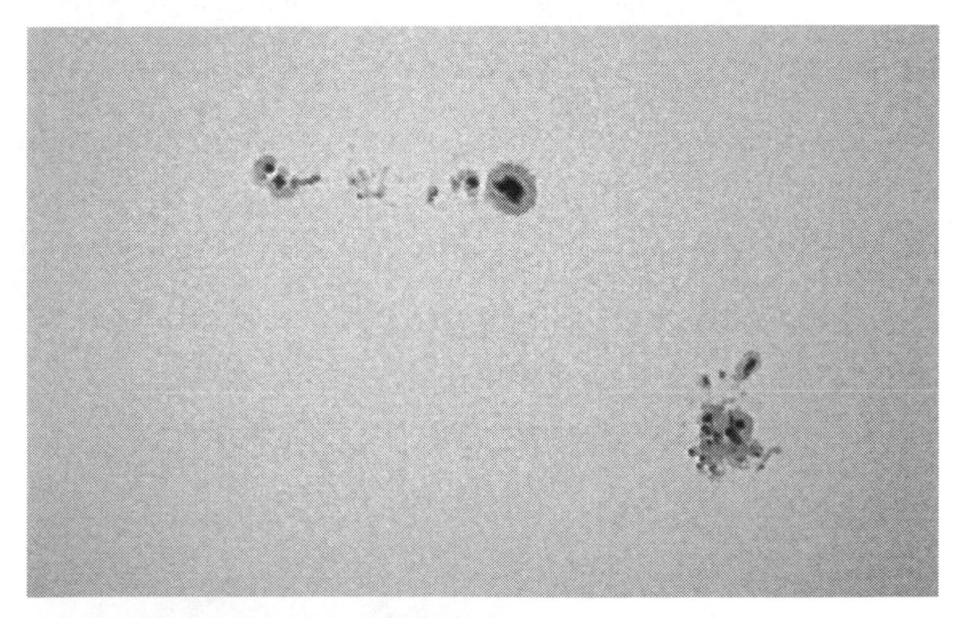

Figure 14.4. Center of Figure 14.3 enlarged. Solar granulation is clearly visible.

Figure 14.5. Total lunar eclipse of 2008 February 20 with moon very unevenly illuminated. See text for details.

cannot be seen even with a telescope, depending on how much light is reaching it edgewise through the earth's atmosphere. Often the illumination across the face of the totally eclipsed moon is uneven. Accordingly, the exposures in Appendix B.2 are approximate. You may have to expose as long as for a star field. In fact, a total lunar eclipse is your only opportunity to photograph the moon surrounded by stars.

Figure 14.5 is an example. During this total lunar eclipse, the moon was very unevenly illuminated. I piggybacked the camera and lens on a telescope for tracking. The setup consisted of a Canon 300-mm $f/4$ telephoto lens, $\times 1.4$ converter, and Canon 40D camera, set for an effective aperture of $f/6.3$ and ISO 200. I took 1-, 2-, and 4-second "hat trick" exposures and stacked the JPG files with *Registax* to get greater dynamic range, then postprocessed the resulting image with *Photoshop* to adjust brightness and contrast and perform unsharp masking.

Total solar eclipses are always high-dynamic-range subjects; the vast brightness range of the solar corona is as challenging as any nebula. To photograph the eclipse of 2017 August 21 (Figure 14.6), I used the same Canon 300-mm $f/4$ lens, this time without a converter, with a Canon 60Da camera and Celestron AVX mount tracking at solar rate (and none too well polar-aligned; doing the drift method on the sun in the daytime was tedious). Figure 14.7 shows a single

Figure 14.6. Solar eclipse of 2017 August 21, partial phase before totality. Nikon D810, Nikon 300-mm $f/4$ lens at $f/5.6$, 1/200 second at ISO 250 with Thousand Oaks sun filter; picture cropped and postprocessed in *Adobe Lightroom.* (Wellford Tiller)

Figure 14.7. Solar eclipse of 2017 August 21 in Hiawassee, Georgia, as the picture came out of the camera. Canon 300-mm $f/4$ lens at $f/10$, Canon 60Da, 1/80 second at ISO 800.

Figure 14.8. Stack of three exposures like Figure 14.7, opened as raw files, stacked with *PixInsight*, processed with multiscale median transform, and finally gamma-corrected and stretched.

JPEG file as it came out of the camera with minimal stretching. Figure 14.8 is the result of stacking three identical exposures (to reduce grain) and then using the Multiscale Median Transform tool in *PixInsight* to weaken the residue layer (Sections 2.6.4 and 13.6), bringing out dramatic detail in the corona.

14.2 High-resolution Video: How it's Done

14.2.1 Overview of the Process

High-resolution video imaging consists of taking thousands of images successively (as frames of a movie), then aligning, stacking, and sharpening. The results are spectacularly better than film photography or single digital images.

Figure 14.9 is an example. This is a stack of the best 75% of 5430 video images taken with a Canon 60Da in Movie Crop mode (using the full resolution of a small part of the sensor) with an 8-inch (20-cm) Celestron EdgeHD telescope and a $\times 2$ extender, giving a focal length of about 4000 mm. The exposure was 1/160 second at ISO 1600, and video was recorded for 90 seconds at 60 frames per second. The images were enlarged $\times 1.5$ and stacked using *AutoStakkert*, then sharpened with *RegiStax*.

The image on the left, without sharpening, would have been considered a good image during the film era; the one on the right shows the result of the digital magic. In what follows, I'll first tell you quickly how all of this is done, and then come back to some technical issues.

Figure 14.9. Jupiter through an 8-inch (20-cm) telescope; see text for details. Left: Stack of video images. Right: Same, after multiscale sharpening.

14.2.2 Acquiring the Images

Planetary imaging generally works best with a focal length between 2000 and 6000 mm, a point to which we will return. That generally means using ×2 or ×3 negative projection with the typical portable telescope. That, in turn, is easy to do – just put the camera in the eyepiece tube, with a Barlow lens or telecentric extender ahead of it. You can even leave the diagonal in place if you make sure to flip the image later to correct the mirror-image effect.

There are four ways to record the movie:

(A) Use a video astrocamera (such as your autoguiding camera) with software such as *FireCapture* (www.firecapture.de).

(B) Use your DSLR in its normal movie mode. This gives you downsampled video, using less than the full resolution of the sensor, because standard video images are no larger than 1920 × 1080 pixels ("Full HD,"), and the camera resamples the images down to this size while recording.

(C) Use your DSLR in a digital zoom mode that gives higher resolution while using only the central part of the sensor. "Movie Crop" on the Canon 60D is ideal; it uses the central 640 × 480 pixels at full resolution and seems to have been designed for scientific work. On other cameras, if a digital zoom mode is available, experimentation may be needed to find the effective pixel size.

(D) Use a computer to record the Live View image of the planet at maximum magnification. This can be done with *BackyardNIKON* and other software. The camera doesn't know it's taking a picture; the computer is recording the preview. A variation on this method uses the HDMI video interface rather than the USB interface to receive the image from the camera.

Method A is generally the best because it gives you uncompressed video and full control of the camera, but it does not involve a DSLR. The camera can be color or monochrome. Monochrome cameras with deep red or IR-pass filters often give surprisingly sharp images because the air is steadier in red and infrared than over the rest of the visible spectrum.

The big advantage of methods B and C is that you can use a DSLR without even bringing a computer along. Another advantage of the DSLR is that you can use its large sensor and wide field to locate and center the planet, even if you are going to switch into digital zoom mode to do the recording. That is much easier than centering a planet on a tiny sensor on which it barely fits. It is why I have ended up doing most of my planetary work with a Canon 60Da and Method C rather than a video astrocamera.

When the DSLR does the recording, the video files undergo lossy compression, but I have had no trouble making up for the loss by simply recording more frames. Note that your exposure is limited by the frame rate; if you record 60 frames per second and your exposure is 1/40 second, you will get duplicated frames – which is better than no video at all, but far from ideal. Newer DSLRs also allow you to select 24 frames per second.

Using native video mode, Method B, makes the camera work as if the pixels were much larger than they actually are. By experimenting, I found that the Nikon D5300 downsamples its images by a factor of 3.1; this is typical, and on the D5300 the effect is that instead of their real size, 3.9 μm pixels, the pixels have an effective size of 12 μm, large enough to require a long focal length (maybe 12 000 mm) for high-resolution images of planets.

One advantage of Method B is that it can cover the full face of the moon or sun, giving you a way to overcome vibration if you can't do so in a still picture. It is also very good for wide fields of the moon (Figure 14.10). Good planet images are more of a challenge, but with care, worthwhile results can be obtained (Figure 14.11).

Method C uses the pixels at, or near, their real size (Figures 14.9, 14.12). Notice that you must use a digital zoom mode, one that expands the middle of the field to fill the picture. Merely selecting a smaller number of pixels for the full field doesn't do what is needed; it just downsamples the original image further.

Method D seems to be the worst of both worlds, but actually, it can work quite well. It may or may not yield uncompressed video, depending on the camera. How well you can use Method D depends on whether your camera lets you control the effective exposure while in Live View. Some cameras always try to auto-expose in Live View, regardless of the manual exposure settings. That is generally fine for lunar work but frustrating with small bright planets on a large black background.

14.2.3 How Long to Expose

Suggested exposures for the moon and planets are in Appendix B. If, instead of a DSLR, you are using a video astrocamera that doesn't have ISO settings, start with the exposures suggested for ISO 1600.

Figure 14.10. Region of the lunar crater Tycho. 8-inch (20-cm) $f/10$ telescope with $\times 3$ extender, giving a focal length of 6000 mm at $f/30$. Nikon D5300 in 1920 $\times$ 1080 video mode, best 75% of 1800 frames, each exposed 1/80 second at ISO 400, converted with *PIPP*, stacked with *AutoStakkert*, and sharpened with *Registax*.

Figure 14.11. Left to right, Io, Europa, and Jupiter. Same optics as Figure 14.10, same camera and video mode, best 75% of 1800 frames, each exposed 1/80 second at ISO 1600, converted with *PIPP*, stacked with *AutoStakkert*, and sharpened with *Registax*.

With a DSLR, you must, of course, record video with manual exposure settings; with some Nikons, you will need to enable "manual movie settings" in a menu, and then disable it again when you go back to conventional deep-sky imaging (see Section 3.2.1).

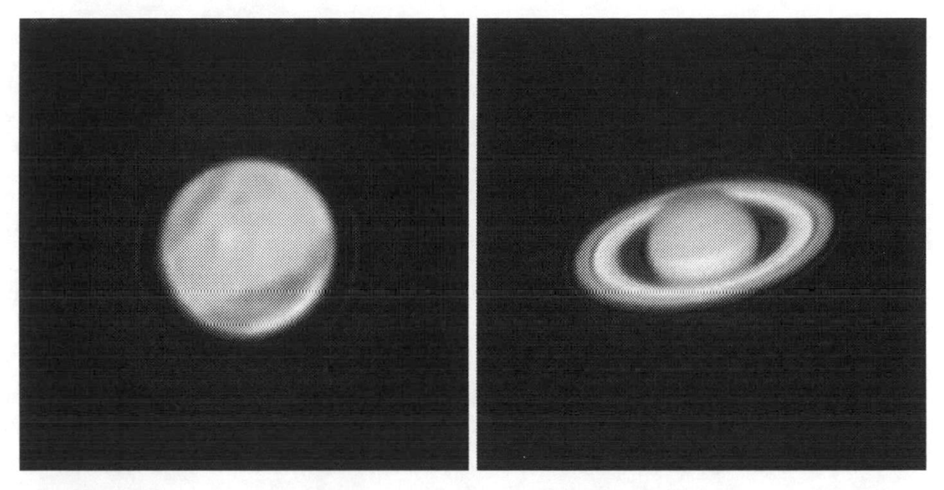

Figure 14.12. Mars and Saturn with an 8-inch (20-cm) $f/10$ telescope, ×2 extender (giving a focal length of 4000 mm at $f/20$), and Canon 60Da in Movie Crop mode. Mars: best 75% of 5534 frames exposed 1/400 second at ISO 1600. Saturn: best 75% of 7149 frames exposed 1/60 second at ISO 1600. Converted with *PIPP*, stacked with *AutoStakkert*, and sharpened with *Registax*.

Your telescope should track the sky, but perfect tracking is not necessary; periodic error and slight drift can be helpful because they move the planet image around on the sensor and reduce the effect of dust specks and defective pixels.

The rotation of the planet limits how long you can record video. Jupiter rotates around its axis in slightly less than 10 hours. Obviously, if you recorded video frames of Jupiter for an hour, and then stacked them, your image of the surface of Jupiter would be very smeared by the rotation. Even ten minutes would be too long. To calculate how long you can record without smearing, use the formula:

$$\text{Max. recording time (minutes)} = \text{Acceptable smear ('')} \times \frac{60}{\pi} \times \frac{\text{Rotation period of planet (hours)}}{\text{Apparent diameter of planet ('')}}.$$

For example, if you decide you can tolerate 0.5″ of smearing (which is near the limit of earth-based resolution), and Jupiter is near minimum distance from the earth, with an apparent diameter of 50″, then you can record for up to 1.9 minutes. That is not a strict limit, so 2 minutes is a reasonable limit for Jupiter and Saturn. With Mars, you can record for at least 9 minutes (usually longer), and with the moon, as long as you want.

In the formula, the factor of 60 converts hours into minutes, and π accounts for the fact that the planet rotates around its whole circumference in one rotation, and hence the speed at the center of the disk seems to be one circumference per the given number of hours. The apparent diameter and acceptable smear, though measured in arc-seconds, are angular sizes as seen from the earth, not related to the planet's rotation.

14.2.4 Preparation and Stacking

In what follows, I shall concentrate on widely used free software, though related functions are also available in commercial software packages. Presently, free software leads the wave of innovation, and commercial software tries to keep up.

Most astronomical software does not accept the MOV and MP4 video files that DSLRs produce; the files need to be converted to AVI. Several video editors can do the conversion, but the handiest is *PIPP* (Planetary Imaging PreProcessor, freeware from https://sites.google.com/site/astropipp). *PIPP* not only changes the file format but can also stabilize the image (to reduce movement) and, most importantly, crop the picture to make smaller video files that do not require as much memory for subsequent processing. The output should be saved as uncompressed AVI for maximum compatibility with other software. This does not remove any losses caused by earlier lossy compression, but it does prevent further loss. The free command-line utility *FFmpeg* (www.ffmpeg.org) can also do the format conversion, but not cropping.

Once the files are converted, they must be stacked. That can be done in *Registax* (to which we'll return shortly), but a more full-featured stacking package is Emil Kraaikamp's *AutoStakkert* (freeware from www.autostakkert.com, donations accepted). *AutoStakkert* comes with detailed instructions which space precludes repeating here. Suffice it to say that in order to stack images, you have to specify alignment points, which you can pick individually or have *Autostakkert* choose automatically.

14.2.5 Multiscale Sharpening

Now for multiscale sharpening. For the theory, see Section 2.6.4. What we're going to do is emphasize features of sizes that contain the most detail, but not noise; usually, this means most of the emphasis will be at a filter radius of about 1 or 2 pixels.

The software package that introduced multiscale sharpening to the world – along with planetary video stacking – is *Registax* (Figure 14.13), freeware from www.astronomie.be/registax, still quite usable, although as I write this it has undergone no major development for several years. Similar functionality is offered in *PixInsight* and *MaxIm DL*, among others; the corresponding *PixInsight* tool is shown in Figure 14.14.

To start sharpening an image, you have to choose the scale of the wavelets (the filter radii): dyadic (1, 2, 4, 8, 16), linear (1, 2, 3, 4, 5), or something else (even 1, 1, 1, 1, 1 is offered in *Registax* and has its uses). I suggest starting with the traditional dyadic spacing.

Then you emphasize the individual layers by trial and error; in *Registax* this is a simple matter of moving sliders and looking at what happens. Usually, one of two things happens. Either layer 1 needs a lot of emphasis, or else layer 1 contains all the noise and layer 2 needs a lot of emphasis. For planetary imaging,

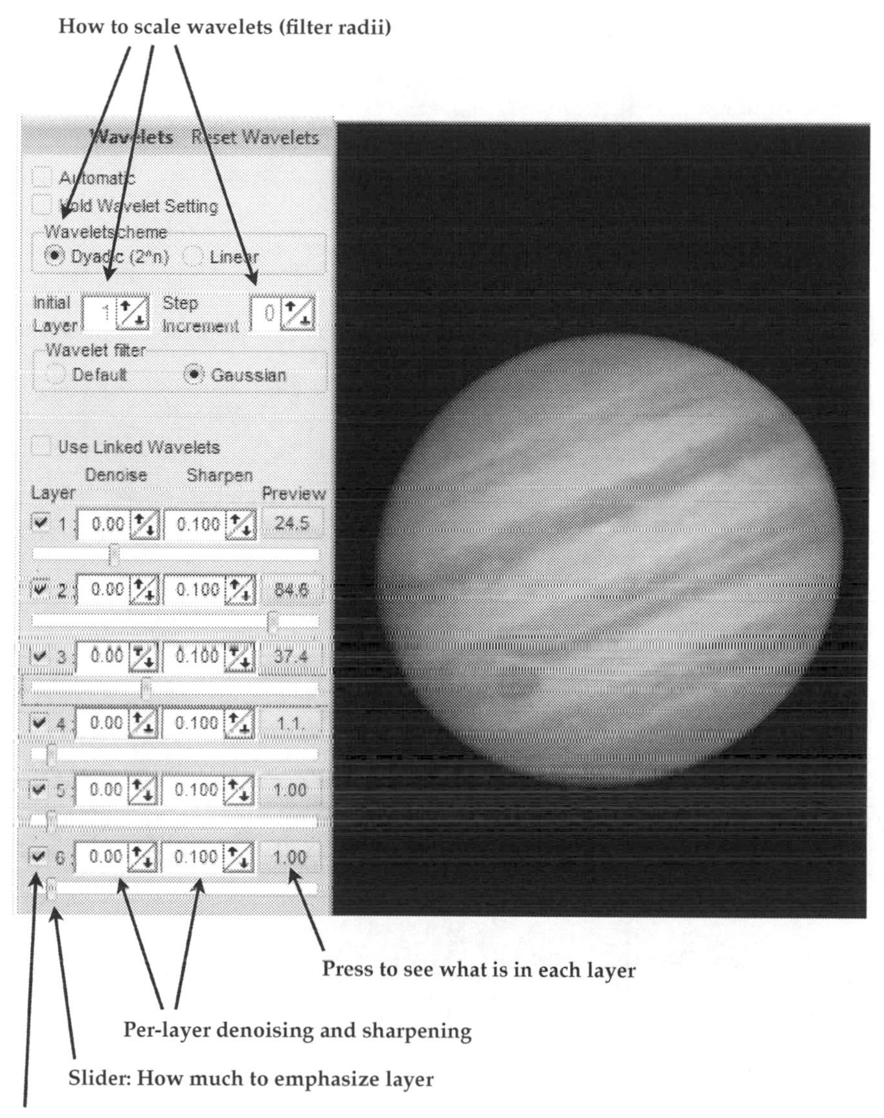

Figure 14.13. Multiscale sharpening adjustments in *Registax,* the classic software package for this purpose.

there is rarely any need to emphasize any of the layers with larger filter radii. However, multiple layers with a radius of 1 can be useful; they are not redundant, because each one works on the residue from the previous one, and quite often, the first one traps almost all the noise. That is why *Registax* allows you to specify a starting radius of 1 and an increment of 0.

Along the way, you can denoise and/or sharpen individual layers. In *Registax,* be sure to press Do All after making the settings, so they will be applied all over the image.

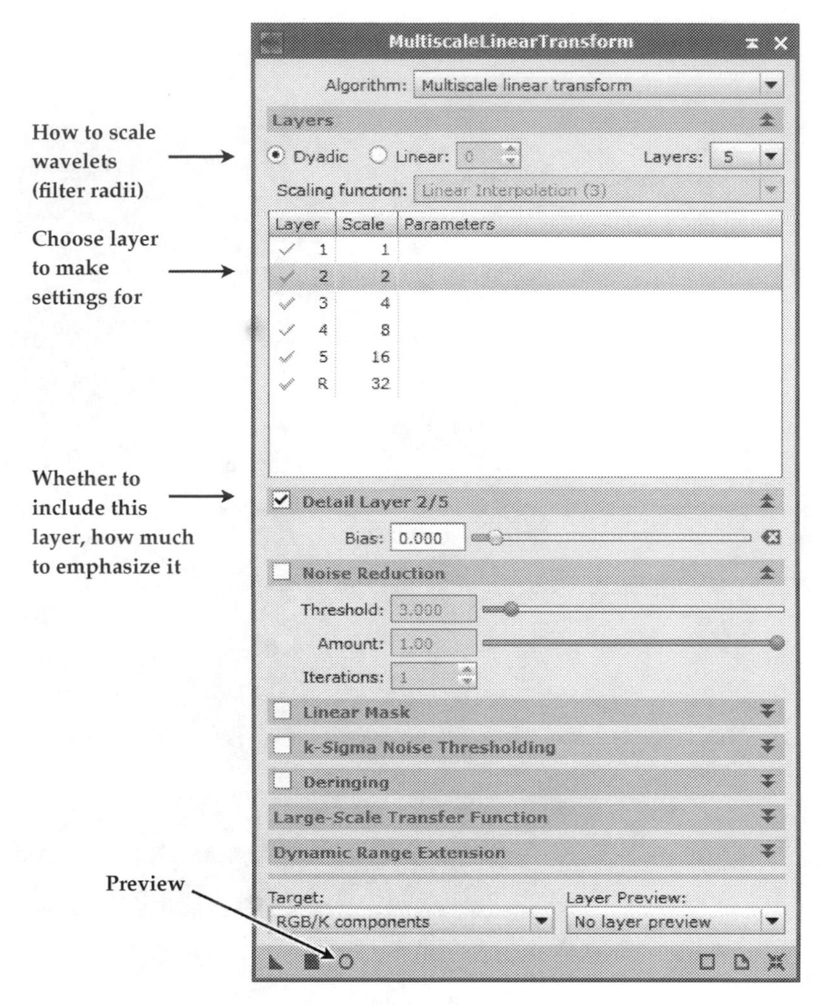

Figure 14.14. *PixInsight*'s multiscale sharpening tool is simple and straightforward.

14.2.6 RGB Alignment

Sharpening an image often reveals blue and orange fringes on opposite sides, or occasionally other colors. These are caused by atmospheric refraction, which spreads light out into a spectrum like a prism, and also by asymmetries in deBayering and image processing algorithms. You can correct the fringes and sharpen the whole image by moving the red, green, and blue layers of the completed image relative to each other. This is handled by the RGB Align button in *Registax,* which can estimate the needed correction for you, after which you can continue to nudge the color layers in any direction until they're perfect.

A similar function is provided in *MaxIm DL* as Color, Realign Planes. At time of writing, there is nothing directly analogous in *PixInsight.* You can also do RGB alignment during stacking with *AutoStakkert.*

14.3 High-resolution Video: Technical Matters

14.3.1 Matching Focal Length to Pixel Size

How big does the image on the sensor need to be in order for the camera to pick up all the planetary detail that the telescope can resolve? The bigger the better? No – remember that enlarging the image makes it dimmer, so you want to make it big enough, but not bigger. The optimum depends on the pixel size of your camera.

In very steady air, amateur-size telescopes (anywhere from 8 to 24 inches [20 to 60 cm] in aperture) can record detail somewhat smaller than 1″ across. This limit is set partly by diffraction and partly by the atmosphere.

Accordingly, the pixels in the camera need to be appreciably smaller than 1″. On theoretical grounds many of us initially adopted a pixel size of 0.5″, which is still fine if you need a relatively bright image (e.g., for Saturn), but experienced planetary photographers have found advantages to going smaller, down to 0.2″ or even 0.1″.

It happens that if you adopt 0.2″ as the desired pixel size, the formula for calculating focal length is very easy to remember:

$$\text{Recommended focal length} = 1000\ \text{mm} \times \text{Pixel size}\ (\mu\text{m}).$$

For example, a camera with 4-μm pixels needs a focal length around 4000 mm for optimum planetary work. This is not a value you have to hit exactly, just a rough guide. Anything from half to double the computed focal length will work. Go for the low end of the range when the object being photographed is not bright (e.g., Saturn), the telescope is small, or the air is not extremely steady. Conversely, go toward the higher end if your camera outputs compressed video (losing a bit of resolution), atmospheric conditions are especially good, or the telescope is large.

The focal length of any but the largest telescope will be shorter than ideal, so what you do is use negative or positive projection to enlarge the image. A common 8-inch (20-cm) $f/10$ Schmidt–Cassegrain telescope has a focal length of 2000 mm, a bit short for typical cameras, but adding a ×2 extender changes it to 4000 mm, which is ample. Conversely, with a relatively large telescope (such as a 24-inch [60-cm] $f/15$ in an observatory), you may have more focal length than the formula calls for; it does no harm, but your images may look better if you reduce their size after processing them.

14.3.2 Why High-resolution Video Works

The way video imaging works its magic is largely misunderstood. It is not "lucky imaging" – selecting the few sharpest frames from a long recording. That is how visual planet observers operate; they stare at the planet, waiting for what Percival Lowell called "revelation peeps," moments of unusually steady air in which the blur momentarily clears up.

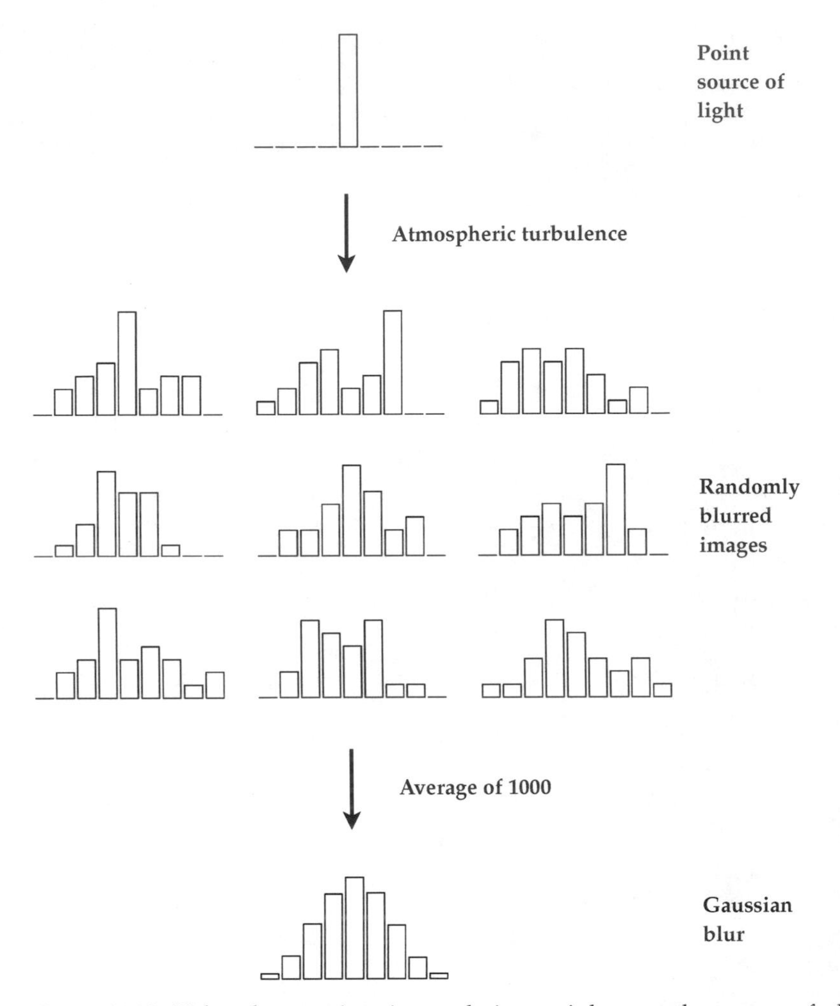

Figure 14.15. Video planetary imaging works its magic because the average of a large number of random blurs is approximately a Gaussian blur, which can be sharpened. Graphs show how many photons fall on each of a simulated row of nine pixels.

But if that were how video imaging worked, best results would come from stacking only a few of the best images and throwing away the rest. Software such as *Registax* does rate the images for sharpness, and I agree that egregiously bad ones should be thrown out, but in fact, stacking the best 75% of the images in a recording is almost always better than the best 10%. That shows that something is going on other than "lucky imaging."

The way the magic works is shown in Figure 14.15. The sum of a large number of random blurs is approximately a Gaussian blur, and Gaussian blurs can be undone. The figure shows a computer simulation of the way the light from a point source (a star, or a single point on a planet) would fall on a row of nine pixels. All of the individual images are blurred randomly, but when a thousand of them are averaged, the intensities are distributed along a bell curve. In effect, we've turned an unknown blur into a known blur of a type on which multiscale enhancement works very well. A surprising amount of the image detail can be brought back by simply strengthening the wavelet sizes that were weakened by the Gaussian blur.

Part IV
Advanced Topics

Chapter 15
Sensor Performance

The image sensor is the heart of the camera. The two things we most want to know about sensors are whether one camera is better than another, and what ISO setting works best with a given camera. The quick answers are that you can judge cameras by published test results, and that the best ISO for deep-sky work is usually 200 to 400 for newer (ISO-invariant or "ISOless") DSLR sensors and 800 to 1600 for older ones.

This chapter and the next will explore sensor performance in detail. Let me start, though, with a cautionary note. You don't have to have the best possible sensor to get good pictures. Both second-generation and third-generation DSLRs take excellent astrophotos. If a camera is serving you well, it will not become obsolete when a slightly better sensor becomes available.[1]

15.1 Generations of DSLRs

Roughly, we can divide DSLRs into three generations. First-generation DSLRs have relatively noisy sensors and include the Canon EOS 10D, 20D, 20Da, and 300D (Digital Rebel), Nikon D50, and Nikon D80. They were in their heyday when the first edition of this book was written, and we got surprisingly good astrophotos with them.

Second-generation DSLRs have appreciably less noise and usually have a more mature set of features, including live focusing; these include the Canon EOS 40D, 60D, 60Da, and Nikon D300. Their noise level depends on the ISO setting; in deep-sky work, these cameras usually work best at ISO 800 to 1600 to overcome noise.

Third-generation DSLRs have even lower noise, and the noise level is *ISO-invariant* or (to use terms less precisely) "ISOless." That means the noise level, measured in electrons, is *relatively* (not perfectly) independent of ISO setting.

---

[1] In this chapter and the next, I am particularly indebted to William J. Claff, www.PhotonsToPhotos.net, for helpful suggestions and information.

That, in turn, means you can underexpose a photograph, then brighten it up with digital postprocessing, and get the same result as if you had set the ISO higher in the camera, except that you're less likely to overexpose the highlights. Daytime photographers notice this because it allows them to adjust exposure after taking the picture, rather than before. In astrophotography, it means you can take deep-sky images at ISO 400 or even 200 for greater dynamic range.

Initially, these ISO-invariant sensors were a Sony breakthrough. Cameras of this type include the Sony A7R (mirrorless) and the Nikon D5300 and D810(A), all of which use Sony sensors. (Because of its unusual amplifiers, the Sony A7S does not actually qualify as ISO-invariant, though it performs better than many that do.) Canon, which makes its own sensors, entered the third generation with the 80D, SL2 (200D), 5D Mark IV, and later products (not, however, the 6D Mark II, an excellent camera that is the last of the old breed). The Pentax Kx, with its Samsung sensor, also belongs to this class.

The division into generations is not sharp, of course. You should evaluate each camera on its merits rather than just what generation it belongs to. And do not assume that all of a manufacturer's products enter a new generation at the same time. It is quite common for different models to have different sensors, even with similar megapixel counts.

15.2 How Sensors Work

15.2.1 Photoelectrons

Electronic image sensors work because light can displace electrons in silicon. Every captured photon causes a valence electron to jump into the conduction band; when that happens, we call it a *photoelectron*. In that state, the electron is free to move around, and the image sensor traps it in a capacitive cell.

The number of electrons in the cell, and hence the voltage on it, is an accurate indication of how many photons arrived during the exposure. That is, the response to light is almost perfectly linear, which makes subtraction (e.g., of darks from lights) possible and accurate.

Photographic film works the same way, except that the photoelectrons are collected in silver iodide crystals, where they cause a latent chemical change that makes the crystals react with developer later. The process is rather unreliable, and many of the photoelectrons leak away. That is why film "forgets" many of the photons in a long exposure, a phenomenon known as *reciprocity failure*, and a 5-minute exposure may only record twice as much light as a 1-minute exposure. Also, film's response to light levels is exquisitely nonlinear – every combination of film and developer has a different nonlinear response curve, giving it a distinctive look.

The best digital image sensors achieve a *quantum efficiency* well above 50%, which means they capture an electron for nearly every photon that hits them. The highest quantum efficiency achieved with film was about 10%.

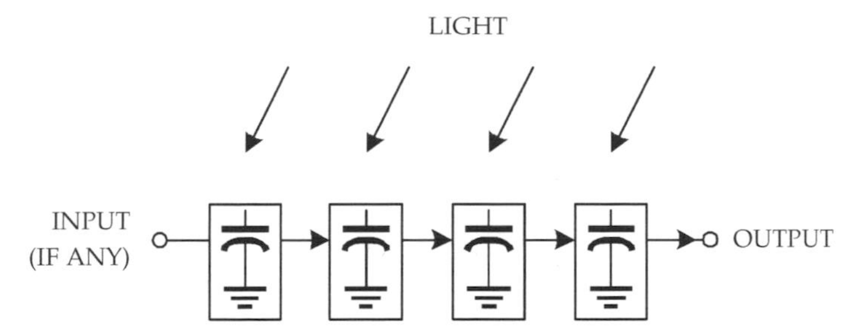

Figure 15.1. Charge-coupled device (CCD) consists of cells (pixels) in which electrons can be stored, then shifted from cell to cell and retrieved. Light makes electrons enter the cells and slowly raise the voltage during an exposure. (From *Astrophotography for the Amateur.*)

15.2.2 CCD and CMOS Sensors

Today, there is no systematic difference in performance between CCD and CMOS sensors, and DSLR makers are steadily switching to the latter. CCD technology was developed earlier, but CMOS has caught up. Canon has devoted a lot of effort to making and perfecting its own CMOS sensors; many other camera makers use CCD and CMOS sensors made by Sony, Toshiba, and Samsung.

The difference between CCD and CMOS sensors has to do with how the electrons are read out. *CCD* stands for *charge-coupled device*, a circuit in which the electrons are shifted from cell to cell one by one until they arrive at the output (Figures 15.1, 15.2); then the voltage is amplified, digitized, and sent to the computer. The digital readout is not the electron count, of course, but is exactly proportional to it.

CMOS sensors do not shift the electrons from cell to cell. Instead, each cell has its own small amplifier, along with row-by-column connections so that each cell can be read out individually.There is of course a main amplifier along with other control circuitry at the output.[2]

15.2.3 What We Don't Know

DSLR manufacturers do not release detailed specifications of their sensors. Almost everything we think we know about them reflects a certain amount of guesswork.

What's more, the "raw" image recorded by a DSLR is not truly raw. Some image processing is always performed inside the camera – but manufacturers are extremely tight-lipped about what is done. For example, most DSLRs add a positive bias to every digital value, and some early Nikon DSLRs (D50, D70,

[2] The term *CMOS* means *complementary metal-oxide semiconductor* and describes the way the integrated circuit is made, not the way the image sensor works. Most modern ICs are CMOS, regardless of their function.

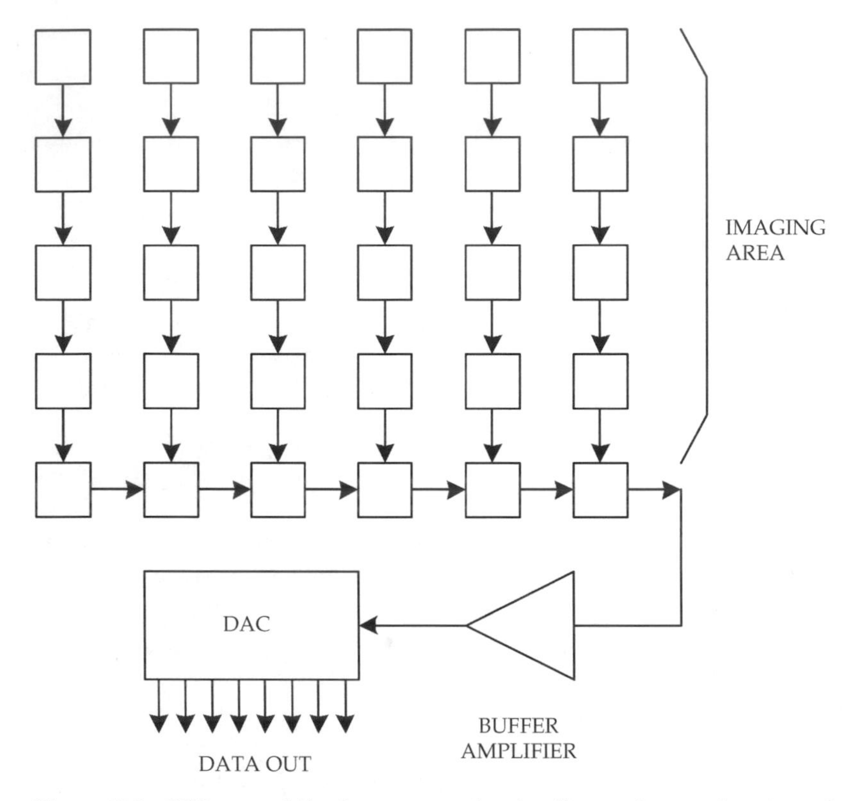

Figure 15.2. CCD array shifts the contents of each cell, one after another, through other cells to the output for readout. CMOS arrays have an amplifier for each pixel, with row and column connections to read each pixel directly. (From *Astrophotography for the Amateur.*)

D80) had particularly heavy-handed preprocessing of raw images, including a notorious "star eater" speckle-removing algorithm.

We can hope that in the future, DSLRs will come with the equivalent of film data sheets, giving the spectral response and various characteristic curves. Until that happens, we have to rely on third-party tests.

For pioneering investigations of astronomical DSLR performance, see the web sites of Christian Buil (www.astrosurf.com/buil) and Roger N. Clark (www.clarkvision.com). Buil focuses on astronomical performance and spectral response, while Clark sets out to measure sensor parameters such as full-well electron capacity and signal-to-noise ratio.[3] On the technical performance of DSLR sensors in general, see especially the thorough set of web pages maintained by Emil Martinec at the University of Chicago (www.PhotonsToPhotos.net/Emil%20Martinec/noise.html).

[3] Clark is also the author of *Visual Astronomy of the Deep Sky* (Cambridge University Press, 1990, now out of print), a groundbreaking study of the factors that determine the visibility of faint deep-sky objects to the human eye.

Technically oriented sensor tests of newer cameras are published regularly by *DxOmark* (www.dxomark.com), and *DxOmark*'s data are regularly reanalyzed from a more technical point of view at www.sensorgen.info and www.PhotonsToPhotos.net, which also presents tests of its own.

Another way to test a DSLR is to ignore the internal details and use well-established measures of picture quality, such as dynamic range and color fidelity. This is the approach taken by *Digital Photography Review* (www.dpreview.com) and other photography websites and magazines.

All of these tests point to one important result: DSLRs are improving steadily and rapidly. The latest models have more dynamic range and less noise than those even two years older, and all current DSLRs perform much better than film.

15.3 Sensor Performance Basics

15.3.1 Pixel Size

The pixels in a DSLR sensor are much larger than those in a compact digital camera or smartphone. That's why DSLRs perform so much better. Typically, a DSLR pixel is 3 to 6 μm square and can accumulate over 40 000 electrons. Besides the desired signal, a few electrons always leak into the cell accidentally at random, but each of them will constitute only 1/40 000 of the total voltage. In the smaller pixels of a compact digital camera, the same electrons would do much more harm.

15.3.2 Quantization and DNs (ADUs)

Each cell has to contain an integral number of electrons. You can have 23 624 electrons or 23 625, but not $23\,624\frac{1}{2}$ because there's no such thing as half an electron. This is why the camera cannot distinguish an infinite number of different brightness levels. But then, neither can light itself; there's no such thing as half a photon, either.

A much coarser kind of quantization takes place at the output of the amplifier, where the cell voltage is turned into a digital signal. Good DSLRs perform 14-bit digitization, meaning that each cell voltage is rendered into a whole number between 0 (black) and $2^{14} - 1 = 16\,383$ (white).

These 14-bit numbers are often called *DNs* (digital numbers) or *ADUs* (analog-to-digital units). They can be stored in a 16-bit TIFF file, where they do not use the whole available range. Such a file, viewed directly, looks very dark until it is "stretched" by multiplying all the values by a constant. Even then, it will still look dark because it needs gamma correction.

Some cameras record only 12-bit digital numbers in their raw files. Those 12-bit numbers range only from 0 to 4095, losing much-needed distinctions between brightness levels in underexposed images. In the future, we will have DSLRs that record 16-bit numbers.

Note that stacking multiple images overcomes quantization. The average of several frames can perfectly well correspond to $23\,624\frac{1}{2}$ electrons if the last electron was present in half of them. The DNs in a stacked image are either floating-point numbers, or at the very least, integers with more significant digits than those delivered by the camera.

15.3.3 Bias (Offset), Dark Clipping, and Compression

The starting value of the DN scale is not 0, for two reasons: you can't get all the electrons out of the sensor before taking an exposure, so the electron counts won't start at 0, and many DSLRs add a constant to the DNs to rule out the possibility of negative numbers. For example, many Canons add 2048. If, due to random noise in the amplifier, a particular pixel shows a negative voltage, it will not be clipped at zero; instead, it will come out slightly less than 2048, and subsequent image processing algorithms will handle it correctly.

Strictly speaking, *bias* is the inability of the sensor to deliver exactly 0 volts, and the number added in postprocessing is an *offset* or *pedestal*. However, in this book I use the terms *bias* and *offset* interchangeably.

The actual situation is even more complicated. Before adding a constant offset, some cameras subtract an estimate of the real bias, which is exposure-dependent and can be estimated by statistical analysis or by measuring special pixels at the edge of the sensor that are not exposed to light. That is why the bias of a dedicated astronomical camera goes up in longer exposures but the artificial bias of a DSLR does not.

Some Nikons compress the raw file in a slightly lossy way. Instead of storing 14-bit numbers from 0 to 16 383, they store more concise codes that represent numbers throughout that range but cannot actually hit every number exactly. Some exact values are unavailable (for instance, there might be a representation for 9584 but not 9585). This is not as disastrous as it sounds because values were already being skipped due to the discreteness of photoelectrons. To understand why, imagine an ideal sensor with no bias or noise, whose output is 3 DN per electron (which is typical). Obviously, such a sensor could output 2100, which is a multiple of 3, but not 2101 and 2102, which are not. In any case, losses of this type are overcome immediately and accurately by stacking multiple images.

The bottom line is that raw files are a linear representation of the image, but they are not the actual numbers that came out of the analog-to-digital converter. A variety of transformations may have been applied to them.

15.3.4 Linearity

Linearity is one thing we don't have to worry about. The digital number that comes out of each pixel is almost exactly proportional to the number of photons that went in, except for bias and other measurable errors. This is one of the best things about digital sensors. It makes calibration and many other kinds of image processing possible. Even uneven sky fog gradients from streetlights can be subtracted out of the image.

To be precise, digital image sensors are linear to within 1% or better, except for a few special sensors designed for enormous brightness ranges. Nonlinearity due to distortion in the amplifiers is something sensor manufacturers have to worry about, but we don't.

Film photographers are familiar with film's nonlinear response to light. Its response is only roughly proportional to the amount of light reaching it, and different films have different "curve shapes" leading to different artistic effects. That is not a concern with digital imaging.

15.3.5 ISO Speed Adjustment

To provide adjustable ISO speed, the camera lets you amplify the voltages ahead of the the digitization process. That is, you don't have to use the full capacity of the cells. Suppose your cells hold 50 000 electrons. You can multiply all the output voltages by 2, and then 25 000 electrons will be rendered as maximum white (and so will anything over 25 000). This means you need only half as much light, but the image will be noisier because every unwanted electron now has twice as much effect.

In some cameras, "whole-stop" ISO settings (100, 200, 400, 800, etc.) are implemented one way, and the "in-between" values (125, 160, 250, 320, 500, 640) are implemented a different way. This is evident because their dynamic range curves bob up and down between the whole-stop settings. An educated guess is that the whole-stop settings are implemented by changing the amplifier gain, and the in-between settings, by digital postprocessing. Accordingly, I usually use only the whole-stop settings. Some published tests of second-generation Canons, on the other hand, show better dynamic range at 640 than at either 800 or 400 (www.PhotonsToPhotos.net). In any case, the difference is very slight.

Further, the very lowest ISO settings may not be genuine. For example, on the Canon EOS 5D Mark IV, tests show that ISO 50 behaves like ISO 100 with 1 stop of adjustment in postprocessing. Such ISO settings are called "extended." Bearing in mind that other cameras may try the same trick, you probably should not judge the performance of a camera at settings below ISO 200. The very highest ISO settings are often "extended" too.

15.3.6 Gain

The number of electrons corresponding to each digital number (DN) can be measured (see Section 15.5.3). This measurement is traditionally given in one of two ways, either:

$$\text{Digital numbers per electron} = \text{DN}/\text{e}^- = \text{ADU}/\text{e}^-,$$

or else:

$$\text{Electrons per digital number} = \text{e}^-/\text{DN} = \text{e}^-/\text{ADU}.$$

Given the first way, gain is less than 1 at the lower ISO settings and increases with ISO. Given the second way, it is greater than 1 at the lower ISO settings and decreases with ISO. You will see both in published reviews, but this book will always use the first one.[4]

At moderately high ISOs, the gain in DN/e^- is greater than 1, which means every single photoelectron shows up in the resulting DN. It used to be thought that "unity gain" or "unity ISO" (the ISO setting that gives one digital number per electron) was ideal because it guarantees you won't miss any electrons, but in fact some cameras benefit from a higher ISO setting to overcome read noise. The choice of ISO setting should be based on measured camera performance, not gain by itself.

15.3.7 Color Balance (White Balance)

The raw image file records your color balance settings but does not apply them. Color balance is applied when the raw file is converted into a JPEG image (inside the camera or later) or when the raw file is deBayered by the software that will process it. Some software pays attention to the settings made in the camera, but other software, particularly of a scientific nature, ignores them.

The red, green, and blue pixels have to be treated differently because the light sensitivity of the sensor is inherently uneven, stronger in red than in green or blue, and this imbalance is only partly corrected by the infrared-blocking filter in the camera. In some software packages, such as *MaxIm DL*, you get to specify multiplication factors for red, green, and blue yourself.

If camera color balance settings have an effect on your processed image – and they may or may not, depending on your software – then the obviously correct choices are either daylight balance, to get the same color rendition all the time, or automatic white balance, to try to avoid a strong overall color cast, particularly with a filter-modified camera.

15.3.8 The Anti-aliasing Filter

A big difference between film and digital photography is that light spreads sideways in film but not on digital sensors. If light falls on one point on a piece of film, it will always diffuse somewhat into the surrounding region. With a digital sensor, however, you can have light on one pixel and complete darkness on the next one.

The lack of spreading causes two problems in digital photography. First, the Bayer matrix assumes that the brightnesses and colors of adjacent pixels are

[4] My rationale is that, throughout electronics, gain is always output divided by input. Here the DN is obviously the output. In parts of the electro-optics profession, though, the other definition of gain is solidly entrenched.

similar, to make interpolation possible. Second, when you photograph a striped object, such as a distant zebra, a digital sensor can produce moiré effects (illusory stripes and bands) as the stripes of the zebra interact with the vertical and horizontal rows of pixels. Moiré effects in art are instances of what is called *aliasing* in signal processing.

To prevent these problems, most DSLRs include an anti-aliasing filter that blurs the image slightly, right in front of the sensor. Your first impression might be to want the filter taken out, but I do not recommend removing it; it's there for good reasons. Nonetheless, some newer DSLRs, such as the Nikon D5300 and D810(A), do not use an anti-aliasing filter; they rely on the fact that sensor pixels are now so small that no lens is sharp enough to hit them one at a time. In effect, the lens provides the anti-aliasing.

15.4 Image Flaws

15.4.1 Bad Pixels

Even a good sensor can have a few flaws. *Hot pixels* are pixels that always read out as an excessively high number due to excessive electrical leakage (dark current) or other flaws; *dead pixels* always read out as zero.

Of the two, hot pixels are more of a problem. You can see them by making a 5-minute exposure with the lens cap on. Most of the hot pixels will look bright red, green, or blue because they hit only one spot in the Bayer color matrix. Indeed, vividly colored stars in a DSLR photograph should be viewed with suspicion; they may not be stars at all.

In general, the hot pixels in a sensor are reproducible; they will be the same if you take another exposure soon afterward. That's why dark-frame subtraction is effective at eliminating them. They can also be deleted simply by recognizing extreme values in individual pixels. This is done internally in some cameras, and also in image processing software. It is called *cosmetic correction.*

Hot pixels and dead pixels are much less common in today's cameras than they were a decade ago. This is partly because sensors have improved and partly because of cosmetic correction taking place in the camera before the raw file is written out.

15.4.2 Pixel Inequality

Even among the pixels that are not hot or dead, there is inequality. No two pixels have exactly the same sensitivity to light, the same leakage, or the same bias. These three effects together constitute *fixed-pattern noise* in the image. The term PRNU (pixel response non-uniformity) is also sometimes used. Fixed-pattern noise is mostly random grain, but in some cameras, it may have a striped or tartan-like pattern due to inequalities between whole rows and columns (Figure 15.3).

Figure 15.3. Row and column unevenness in a Canon 300D bias frame, made visible by extreme contrast stretching. Sometimes this kind of "banding" intrudes into a highly stretched picture, especially if flat darks were not included in the calibration.

Whether striped or random, fixed-pattern noise is, to a great extent, correctable by calibration with dark frames, flat fields, and flat darks, especially the latter two. To deal specifically with banding, some newer sensors have additional pixels at the edges that are never exposed to light so that the camera can measure and correct the bias of each row or column. Banding can also be reduced by postprocessing; *PixInsight* includes a "Canon banding reduction" script.

Banding should be visible only on images whose contrast was greatly enhanced, and then it should be more or less randomly scattered all over the picture. If a sensor has a few prominent streaks, those are defects, not normal banding.

Dust motes and other contamination on the sensor, and also vignetting by the lens, produce pixel inequalities just like those just mentioned, except that they aren't the fault of the sensor.

15.4.3 Blooming

Blooming (Figure 15.4) is a phenomenon that makes an overexposed star image stretch out into a long streak (mainly in just one direction, not two or four matched spikes due to diffraction). It is uncommon with DSLRs though common with older astronomical CCDs. Blooming occurs when electrons spill over from one cell into its neighbors or when the circuitry for a row or column is overwhelmed by a strong signal. It is usually positive (expanding the star into

Figure 15.4. Negative blooming darkens the columns that contain highly overexposed stars in this image taken with a Canon 1100D. The four spikes around each star are caused by diffraction, not blooming. (Gianluca Belgrado)

a streak) but sometimes negative (darkening a row or column as the camera overcompensates).

15.4.4 Amplifier Glow (Electroluminescence)

The main amplifier for the CCD or CMOS sensor is located at one edge of the chip, and, like any other working semiconductor, it emits some infrared light. Some sensors pick up a substantial amount of "amp glow" in a long exposure (Figure 15.5). Dark-frame subtraction removes it.

Pesky glows similar to this can sometimes come from a display in the camera that is constantly illuminated (perhaps even inside the eyepiece) or light leakage into the eyepiece (e.g., while taking dark frames in a bright room). These problems are infrequent but worth knowing about.

15.4.5 Cosmic Rays

Even if the sensor is perfect, pixels will occasionally be hit by ionizing particles from outer space. These often come two or three at a time, byproducts of collisions of the original particle with atoms in the air (Figure 15.6). Like hot

Figure 15.5. Amplifier glow (*arrow*) mars this image of the Rosette Nebula. Single 10-minute exposure at ISO 800, unmodified Nikon D50, 14-cm (5.5-inch) *f*/7 TEC apochromatic refractor. Dark-frame subtraction was later used to remove the amp glow. (William J. Shaheen)

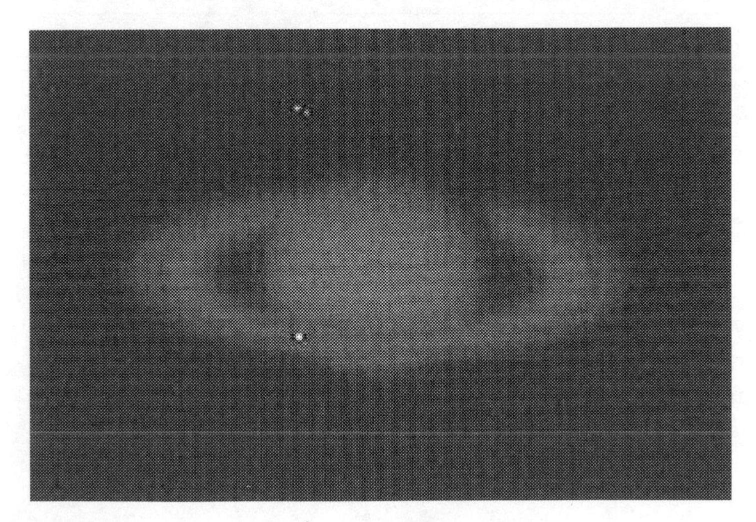

Figure 15.6. Cosmic ray impacts in a single frame of a webcam video recording of Saturn. All three particles arrived during the same 1/30-second exposure, indicating their origin in a single cosmic ray. With DSLRs, the effect of cosmic rays is usually much less noticeable than this.

pixels, cosmic ray hits are likely to be vividly colored because of the Bayer matrix.

Cosmic rays are a source of non-reproducible hot pixels. They are also the reason you should not believe just *one* digital image if it seems to show a nova or supernova. When conducting a search for new stars, take every picture at least twice.

15.4.6 Degradation with Age

It is widely reported that CCD and CMOS sensors slowly degrade with age, suffering an increase in dark current and pixel inequality, but noticeable degradation normally takes years. Some of the postulated aging mechanisms are proportional to exposure to light or electricity, and some, simply to age.

The effects of aging are small, and current thinking is that it may take decades for a camera to degrade noticeably. Don't keep using calibration frames that you took two or three years ago, but don't turn down a secondhand 5-year-old camera that seems to be working well just because you think it may have aged. You can use the sensor tests described later in this chapter to confirm that it is working as specified.

15.5 Noise, in Detail

15.5.1 What Noise Is

If the same amount of light reaches all parts of the sensor, every pixel ought to register the same digital number. For various reasons, this does not happen; the image looks grainy, and the error is called *noise,* further classified as *fixed-pattern noise* if it recurs in the same pixels each time and *random noise* if it doesn't. This crosscuts several other classifications of noise based on its causes.

15.5.2 Signal-to-noise Ratio (SNR)

Given an image of a perfectly smooth, uniform object, in which every pixel should give the same digital number, noise can be measured by comparing the variation in the digital numbers with their average value:

$$\begin{aligned}\text{Signal-to-noise ratio (SNR)} &= \frac{\text{Signal level}}{\text{RMS noise level}} \\ &= \frac{\text{Average pixel value}}{\text{Standard deviation of pixel values}}.\end{aligned}$$

This is sometimes expressed in *decibels* (*dB*), which are logarithmic units:

$$\text{Signal-to-noise ratio (SNR) (dB)} = 20 \log_{10} \frac{\text{Signal level}}{\text{RMS noise level}}$$

$$= 20 \log_{10} \frac{\text{Average pixel value}}{\text{Standard deviation of pixel values}}.$$

Here 6 dB is equivalent to doubling the ratio, and 20 dB is equivalent to multiplying it by 10. Here RMS means standard deviation, just as in the measurement of other varying signals (Section 9.3.1).

Caution: Not everyone agrees about the factor of 20 in these formulae. According to the definition of the decibel, it should be 20 if the signal represents a quantity like voltage, but 10 if the signal represents power (wattage). The DNs of a digital camera do represent voltage – which represents luminous intensity, which is power. You can see why there is no consensus! In this book I use 20, but it is probably better to avoid the use of decibels altogether because of the confusion.

Signal-to-noise ratio can be measured with any software that supports pixel statistics. With a DSLR, be sure to isolate one of the Bayer color channels (preferably green) because the red, green, and blue pixels are obviously going to be different, and the difference between colors should not be mistaken for noise in the image. More about this in Section 16.6.

What all this means in practice is that an image with SNR > 100 (that is, 40 dB) looks fine; at SNR 10 (20 dB), some grain is visible, and at SNR 4 (12 dB), the grain is overwhelming, though the image is still easy to see. Note that SNR can go down to 1.0 (which is 0 dB) and even lower; at SNR 1.0 the signal equals the noise. Even an image with SNR 0.1 (which is -20 dB) can be recognizable under the noise.

15.5.3 Shot Noise

The most fundamental cause of noise in underexposed images is that there is no such thing as part of a photon. You can only capture whole numbers of photons. Likewise, there is no such thing as part of a photoelectron.

Noise arising from these facts is called *shot noise.* Imagine having an army of riflemen pelt the side of a barn with bullets. Because the riflemen shoot independently, the rate of impacts is variable, and instead of an exactly fixed number of bullets per second, the barn receives a somewhat different number of impacts in each 1-second interval. Shot noise is the same phenomenon, but with photons or photoelectrons.

What this means to us is that low light levels *are* noisy, no matter what we do about them, and no matter how good the sensor. Grainless imaging at low light levels will never happen. Naturally, it is better to pick up more of the photons; a more efficient sensor is less harmed by shot noise because, from its point of view, the light level is not quite so low.

This is also why the darker (less exposed) parts of a picture are noisier than the brighter parts – and that, in turn, is why numerous processing techniques rely on lightness masking or on algorithms whose effect is different at different brightness levels.

The one good thing about shot noise is that it makes it possible to count photoelectrons. By nature, shot noise has a Poisson distribution, which means that the square of its standard deviation is the same as its mean. The digital numbers that come out of the sensor are not the same as the electron counts, but they are proportional to them. Knowing that their distribution is a Poisson distribution multiplied by a constant, you can find the constant, which is the gain of the amplifier. Specifically:

$$\text{Gain (DN per electron)} = \frac{(\text{Shot noise in DN})^2}{\text{Signal level in DN}},$$

where the shot noise is expressed as a standard deviation.

This assumes you can distinguish shot noise from all other kinds of noise. We will return to this point in Section 16.7.3.

15.5.4 Read Noise

Read noise is randomness in the signal that is not due to photon shot noise. It is called read noise because it arises in the process of reading the values of the sensor pixels and delivering them as digital numbers. Some writers define read noise more broadly to include fixed-pattern noise (Section 15.4.2) and dark current (see next section), but I do not.

Read noise originates in the sensor, in the pixel amplifiers of the CMOS chip, in the subsequent main amplifier, and in the analog-to-digital converter (ADC).

The relation between read noise and ISO setting reveals a lot about the sensor. *Upstream* read noise originates before the amplifier at which the ISO setting is applied, and *downstream* noise originates after it. Turning up the ISO setting amplifies upstream noise (along with the signal) but overcomes downstream noise.

Downstream noise can originate in circuitry outside the sensor itself. This is apparently the case with the Canon 750D and 800D, for example; the two models apparently have the same sensor, but the latter has less downstream noise due to improvements elsewhere.

Third-generation sensors are ISO-invariant because they have little downstream noise. Their noise originates when the photons are being sensed, not when the signal is being amplified. That's why you don't have to use a very high ISO setting to overcome noise.

Figuring all this out from published measurements of read noise can be confusing. *Measured in DN*, upstream noise goes up with higher ISO settings but downstream noise is constant. However, some testers report read noise *measured in electrons* (e^-). Measured that way, upstream noise is unaffected by ISO, but downstream noise goes *up* at lower ISO settings because the same level of noise is equivalent to more unamplified electrons. For examples see Table 16.1.

15.5.5 Dark Current (Thermal Noise)

Leakage of electrons into the pixels is called *dark current* and is largely a fixed-pattern phenomenon – some pixels are affected more than others – though it also has a random component. If nothing else, the flow of discrete electrons has its own shot noise, introducing some randomness.

Dark current is affected by temperature; that's why dedicated astronomical cameras have thermoelectric coolers or at least large heat sinks. It's also why dark frames should always be taken at the same temperature as the images from which they will be subtracted. Even ordinary DSLRs are noticeably less noisy in the winter than in the summer. Note that sensors warm up during use, since they are dissipating electric power.

Theoretically, the dark current of a silicon CCD sensor doubles for every 8° C (15° F) rise in temperature.[5] This relationship is affected by the way the sensor is fabricated, and I have not investigated whether it is accurate for the latest DSLR sensors. What is definite is that there's less noise in the winter than in the summer.

15.5.6 Chrominance Noise

Any kind of random or irregular noise, combined with a Bayer matrix, produces random or irregular colors. This is called *chrominance noise* and is not a distinct kind of noise, but merely the interaction of any noise with the Bayer matrix. It is distinguished from *luminance noise* (variation in brightness) by noise removal algorithms that operate after deBayering, such as those in *PixInsight*.

15.5.7 Effect of Stacking, Binning, and Downsampling

Recall that when you stack N images, the random noise goes down by a factor of $\sqrt{N}$. That means the SNR improves by this factor; it doubles (that is, it gains 6 dB) every time you quadruple the number of images stacked.

Thus a stack of 64 barely-usable images with SNR 2.0 (= 6 dB) makes a good stacked image with SNR 16.0 (= 24 dB) which can be denoised further by image processing to remove grain.

---

[5] Extrapolating from the curve on p. 47 of Steve B. Howell, *Handbook of CCD Astronomy*, 2nd ed. (Cambridge University Press, 2006), to normal DSLR camera temperatures. At lower temperatures the change per degree is greater. See also the formula on p. 202.

This is the same principle that makes longer exposures have less shot noise than shorter exposures. Noise rises less rapidly than signal level. Recall that shot noise *squared* is proportional to signal level. That means shot noise itself is proportional to the square root of the signal level. Quadruple the signal level, either by stacking four images or by quadrupling the exposure, and you have 4 times as much signal but only 2 times as much noise. The same is true of the random component of read noise. When stacking images, we usually average them rather than add them; thus we see the noise going down rather than the signal getting stronger, but the principle is the same.

Binning (combining adjacent pixels) is equivalent to stacking; when you group every N pixels into one, you improve the SNR by a factor of $\sqrt{N}$. In cameras that include a Bayer matrix, binning can introduce unexpected color fringes on objects that hit the edge of a "bin" and might activate, say, a red pixel but not the adjacent greens and blues. It is better to reduce such images by downsampling, which is also equivalent to stacking. Noise is reduced the same way, and the effect can be dramatic. For example, a 24-megapixel image, reduced to 1 megapixel for display on a computer screen, gains SNR by a factor of $\sqrt{24} = 4.9 = 13.8$ dB.

Note that the bigger your sensor, the more you will bin or downsample. Because of the $\sqrt{N}$ function, a 24-megapixel sensor can play the role of a 6-megapixel sensor with half as much noise and one stop greater dynamic range. More megapixels buy you more performance.

Chapter 16
Testing Sensors

This chapter describes how to interpret published tests of DSLR sensors and how to test your own. We start by focusing on published tests. By the time you read this, the cameras tested here will not be the latest on the market, but they are good examples.

At least outside the manufacturers' own labs, the art of DSLR sensor testing is in its infancy. We are still learning what methods give valid results and what assumptions can and cannot be made. Because raw image files are not truly raw, not every laboratory procedure for testing sensors is valid when the sensor is inside a DSLR. Accordingly, you will see substantially different published test results from the same kind of camera tested by slightly different methods, even when the parameters being measured are the same. You may even see published test results that are demonstrably wrong.

We also don't know how much variation there is between individual cameras of the same make and model, especially if they were made some time apart. Not only do sensors vary, but manufacturers can even introduce improvements without announcing them.

Accordingly, the only tests you can really compare are tests of different cameras done by the same people using the same methods. Tests done separately by different people are not comparable.

It is also wise to remember that photography is a logarithmically scaled enterprise, and a factor-of-1.5 or even factor-of-2 change in most parameters has little photographic effect. Back in the film era, the ISO speed of different batches of Kodak Ektachrome 200 Professional film ranged from 160 to 240. Presumably so did the amateur version of the same film, but no one noticed.

16.1 ISO Invariance

A quick and simple test that distinguishes second- from third-generation sensors is to take identically exposed pictures of a daytime subject at different

ISO settings (all with the same shutter speed and f-stop), postprocess them digitally so they are all equally bright, then compare noise levels. Third-generation (ISO-invariant) sensors produce approximately matching pictures; with second- and first-generation sensors, the underexposed pictures are much grainier.

This test is routinely done by *Digital Photography Review* (www.dpreview.com), and you can also do it yourself. Be sure to work with raw image files, not JPEGs.

16.2 True ISO Speed

Tests of true ISO speed are published by *DxOmark* (www.dxomark.com) and other reviewers. This test requires a calibrated light source and is not a do-it-yourself option. *DxOmark* defines the ISO speed of a sensor as $78/H$ where H is the exposure in lux-seconds that produces an image just short of saturation (maximum white).

True ISO speed measurements are useful to confirm that a camera's full range of ISO settings works as advertised. But bear in mind that "just short of saturation" is not a precisely defined concept. If you apply the $78/H$ formula to the actual saturation level, but the manufacturer applied it to images half a stop short of saturation, then your tests will show that all the true ISOs are half a stop low. That is not far from what we actually see in published tests.

More to the point, true ISO speed is not a measure of camera quality or even sensitivity to light. Suppose two cameras, both set to ISO 400, have true ISO speeds of 300 and 500 respectively. All this shows is that the effect of the ISO setting upon the amplifier gain is slightly different. The cameras could even have identical sensors with different amplifiers. They just need to be set differently to produce matching results.

16.3 Dynamic Range

The *dynamic range* of a sensor is the range of brightness levels that it can distinguish in a raw image file. JPEG files output by the camera have had their dynamic range reduced after gamma correction and other automatic adjustments; only the raw file has the full dynamic range.

Dynamic range is measured in *stops,* where (as in all photography) "N stops" means a ratio of 2^N to 1. Stops are also called EV (exposure values). Dynamic range in stops can be no greater than the bit depth of the raw file, which is why I expect that we will soon have 16-bit DSLRs. Current DSLRs have nearly 14 stops of dynamic range, which means that their 14-bit file format will soon bump into a limit.

However, the real limit on dynamic range comes from noise rather than bit depth. Underexposed pictures become too grainy before they become

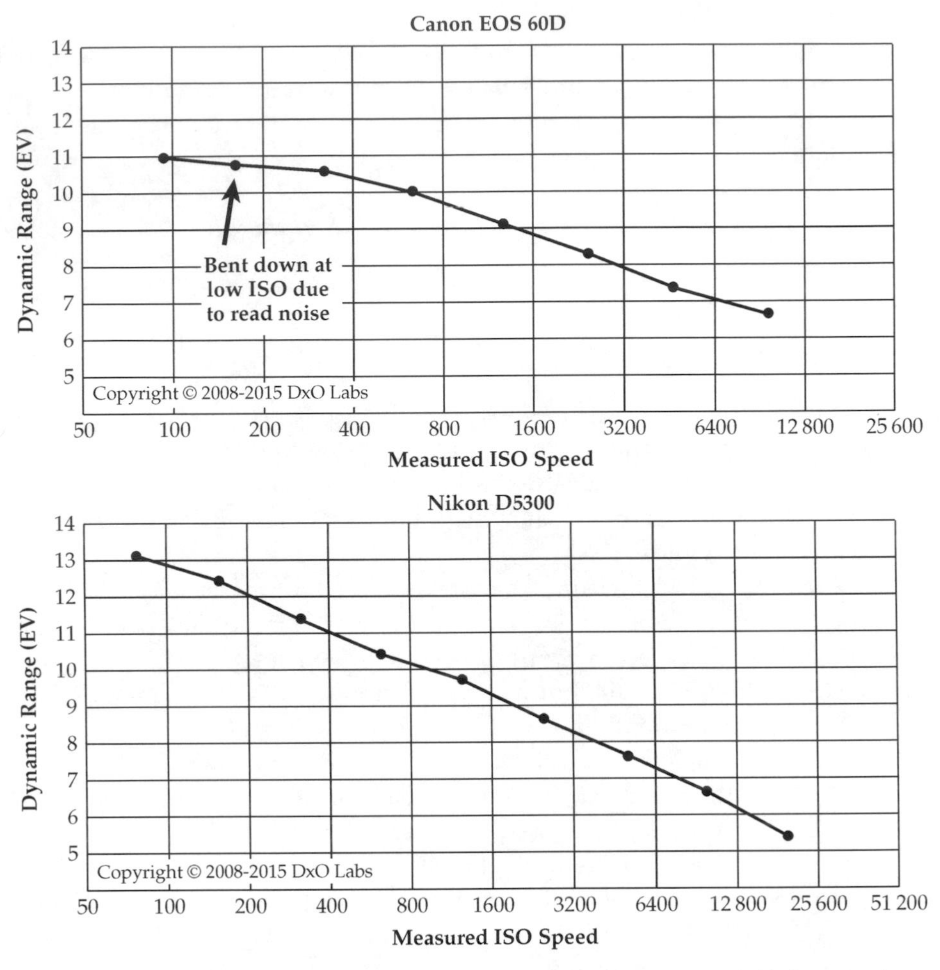

Figure 16.1. Dynamic range versus ISO setting of second-generation and third-generation DSLRs. (Canon's third-generation 200D is similar to the Nikon tested here.) Plots copyright 2015 DxO, Inc., and reproduced by permission (redrawn); arrow and explanatory note added.

extremely, irrecoverably dark. Dynamic range is the range between maximum white and the "noise floor." Accordingly, any test of dynamic range requires the tester to decide how much noise in an image is acceptable, and different testers may make different assumptions.

Three facts follow from the fact that dynamic range is limited by noise:

- Dynamic range is reduced by dark current in long exposures.
- Dynamic range is increased, sometimes dramatically, by stacking and calibrating images, since doing so lowers the noise floor.
- Measured dynamic range depends on how much noise you consider tolerable.

The third fact is very important when you compare tests of cameras with different sensor sizes, even if they are done under identical conditions. Commonly, reviewers downsample all pictures to the same size when they compare noise (grain). This means a picture taken with a full-size sensor will be downsampled more than an APS-C sensor and will show better dynamic range, even if the sensor performance is identical. Sometimes you will see the same camera tested both ways for better comparability. The test procedure in Section 16.7.1 avoids downsampling.

Figure 16.1 shows a graph of dynamic range versus ISO setting as published by *DxOmark*. (On *DxOmark*'s site, be sure to choose "Screen" rather than "Print" data so you'll see the camera's native dynamic range, not a somewhat controversial adjustment intended to account for the noise reduction produced by downsampling.) Published camera tests from other sources often include a similar graph.

If the sensor is third-generation (ISO-invariant), the dynamic range is highest at the lowest ISO setting; from there, it goes down one stop for every stop of ISO. With a second-generation or earlier sensor, however, the dynamic range is the same at several of the lowest ISO settings, where downstream read noise is dominant.

In Figure 16.1, the upper graph shows a second-generation sensor. The dynamic range slopes down in the expected fashion on the right, but on the left, it levels off. The best ISO setting for deep-sky work is the lowest one at which the graph is sloping downward at the full rate – in this case, either 800 or 1600.

The lower graph shows a third-generation ISO-invariant sensor. Its dynamic range slopes downward across the whole graph. With that camera, you can do deep-sky work at ISO 400 or even 200 (perhaps not 100, where the slope seems a bit reduced).

16.4 Noise Analysis

DxOmark publishes a full analysis of signal-to-noise ratio that tells you even more about the camera (Figure 16.2). On this chart, a perfect sensor (with nothing but shot noise) would produce a set of equally spaced diagonal lines, all sloping upward 10 dB per factor-of-10 increase in brightness.

The actual graph consists of curves that are bent down slightly at the right by fixed-pattern noise and at the left by read noise. In first- and second-generation sensors, they are not only bent but also compressed together at the lower left by downstream read noise. If the sensor is ISO-invariant, the curves will remain well separated at the lower left, even though bent. This does not necessarily apply to the very lowest and highest ISO settings, which may not be produced the same way as the others.

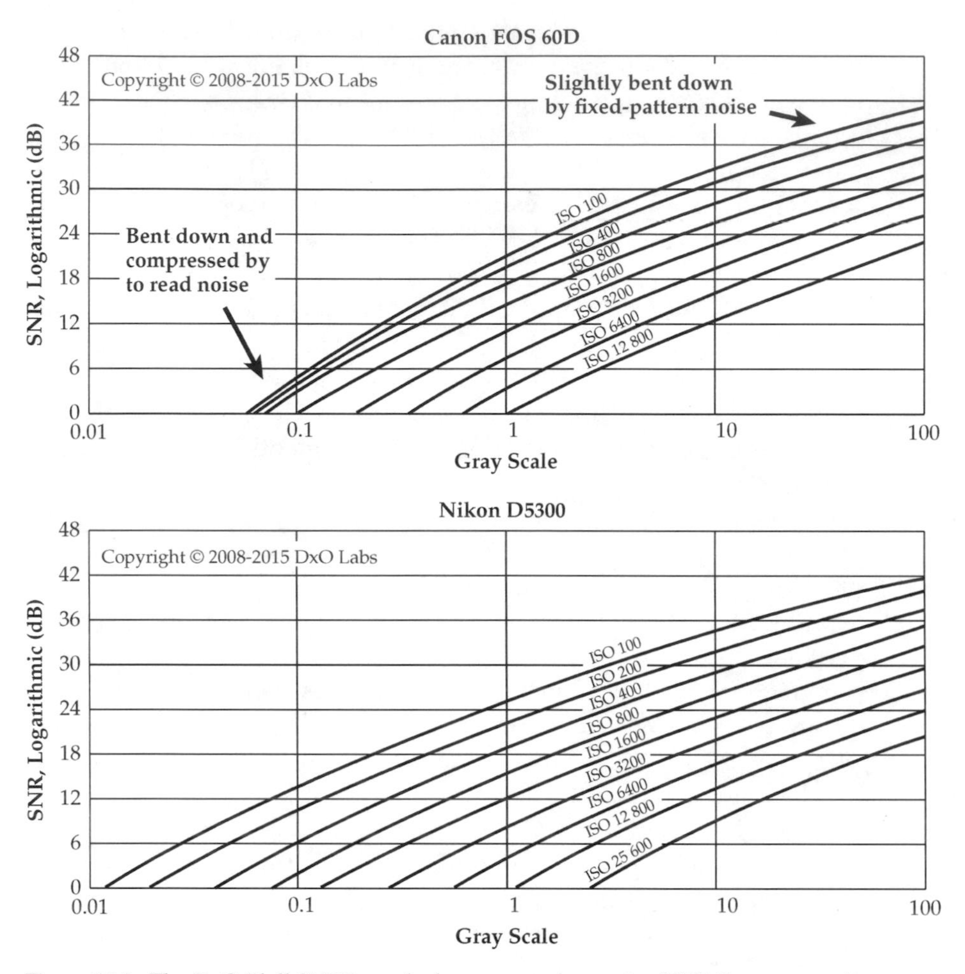

Figure 16.2. The DxO "full SNR" graph shows several aspects of DSLR sensor performance at all ISO settings. (Canon's third-generation 200D is similar to the Nikon tested here.) Graphs for most popular cameras are available at www.dxomark.com. Plots copyright 2015 DxO, Inc., and reproduced by permission (redrawn); arrows and explanations added.

16.5 Quantum Efficiency and Other Parameters

The data gathered by *DxOmark* are reanalyzed by an anonymous group of experts at www.sensorgen.info and also by William J. Claff at www.PhotonsToPhotos.net. The latter is more thorough and up-to-date and also includes some of the author's own measurements. Mr. Claff is always glad to hear from owners of cameras he has not yet tested who would like to collaborate on testing them.

By quantifying the shape and position of each *DxOmark* curve, or comparable measurements, it is possible to deduce the relation between DNs and electron counts, the electron capacity of each pixel, the read noise measured in electrons

(e^-), and, most importantly, the quantum efficiency (the fraction of photons that produce photoelectrons).

Quantum efficiency (QE) is of great interest to astrophotographers; obviously, the higher the better. For two reasons, though, it is not as important a measure of sensor quality as you might think.

First, the highest possible quantum efficiency, 100%, is less than one stop higher than the values we are already getting with good sensors. There is not much room for growth, and the difference between 50% and 100% is not dramatic.

Second, we don't know how accurately QE is being measured, especially since it is an unofficial quantity not published by *DxOmark*. Occasionally the two sites that reanalyze *DxOmark* data report widely differing conclusions. We also don't know how much difference there would be between two cameras of the same make and model, or even the same camera tested twice. Accordingly, don't judge cameras on small differences in their measured QE.

16.6 Obtaining Data from Your Own Sensor

16.6.1 Overview

A do-it-yourself test for ISO invariance has already been described (Section 16.1). Apart from that, sensor tests are usually analyses of flat fields and/or flat darks. Useful tests can be done with images, flats, and flat darks that you already have on hand from astrophotography.

Once you have the images, you will need software that can read out linear pixel values numerically. *Photoshop* is not suitable because it does not process un-deBayered, non-gamma-corrected raw image data. My examples will use *PixInsight* and *MaxIm DL*.

Depending on which test you are performing, the process of "taking the numbers" will include roughly these steps:

1. Open the raw image file without deBayering it.
2. View the image with strong screen stretch. If it's a flat field or dark frame, it will look *awful* because insignificant variations are being amplified and made visible; don't panic.
3. Crop it or select an area that is uniformly illuminated and free of dust specks. A 400-pixel-square region in the middle is big enough.
4. For tests that involve exposure to light, you must extract just one of the four components of the Bayer matrix so you're not comparing pixels of different colors. Green is recommended because it is usually the brightest.
5. Some tests require you to subtract one image from another. When this is done, *you must add a constant* (such as 3000) because negative numbers are not available and would be clipped at zero, giving very incorrect results. Adding the constant does not change the standard deviation that you are going to measure next.

6. Depending on the test you're doing, measure the average (or, preferably median) and the standard deviation of the pixel values in the area of interest.
7. Find out whether your software scaled the pixel values up (e.g., from 14 to 16 bits), because if so, the results of the measurement must be divided by the appropriate factor.

16.6.2 *PixInsight*

To split a camera raw image into the four components of the Bayer matrix (red, green, the other green, and blue), use Process, Preprocessing, SplitCFA. Select the file and specify where the output should go. The results will be four XISF image files. With most DSLRs, the second and third ones (CFA1 and CFA2, the brightest) are green; use either one for further analysis.[1]

Open the file you want to analyze. Turn on screen stretch in the normal manner so you can see what you have.

Crop the image to about 500 × 500 or 1000 × 1000 pixels. The easiest way to crop the image is with Process, Geometry, DynamicCrop. An alternative is to select the image, press Alt-N, and define a portion of it as a "preview," then select that preview as the view to be analyzed.

To subtract images that are open on the screen, use Process, PixelMath, PixelMath. Using the names of the images, construct an expression such as IMG2394 + 3000 − IMG2396 and specify where to put the output (usually Create new image). (Here 3000 is the constant added to avoid negative numbers.) The Expression editor button enables you to construct the expression interactively, picking names and symbols from menus.

Finally, use Process, ImageInspection, Statistics to read the numbers (Figure 16.3). They pertain to the whole image. Note that you will not see the standard deviation (stdDev) until you have clicked on the wrench icon and enabled it. Other measures of dispersion (MAD and avgDev) are not the same.

My experience is that, with the settings in Figure 16.3, *PixInsight* does not scale the raw digital numbers. To confirm this, open an image that contains bright stars, which are maximum white, and confirm that (for a 14-bit raw file) the maximum value is 16 383 or slightly less. Purists will want to do this test on the green layer using SplitCFA as above, but since you're looking for the overall maximum, a shortcut is to simply open the image un-deBayered and examine its statistics. To set *PixInsight* to open raw files without deBayering them, choose View, Explorer Windows, Format Explorer, then double-click on DSLR_RAW and click Pure Raw.

[1] Currently, a bug in SplitCFA requires the image height and width to be even numbers. If yours are odd numbers, open the image un-deBayered (View, Explorer Windows, Format Explorer, double-click DSLR_RAW, and click Pure Raw), crop it slightly (Process, Geometry, Crop), save it as XISF, and then use SplitCFA on it.

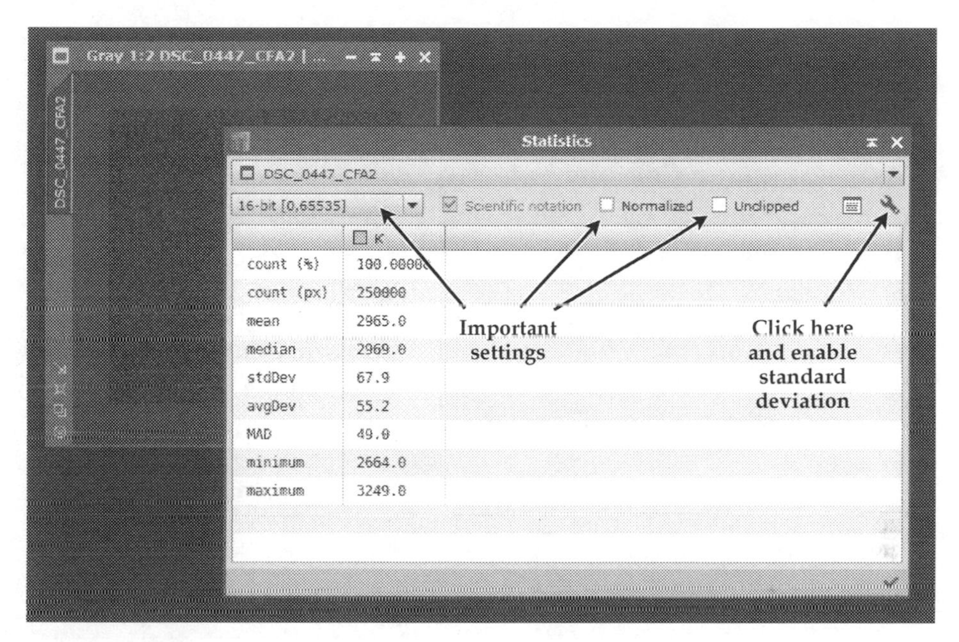

Figure 16.3 Measuring image statistics in *PixInsight*. See text.

16.6.3 *MaxIm DL*

Taking numbers in *MaxIm DL* is appreciably easier. By default, raw files open un-deBayered, with screen stretch on, and the digital numbers are not stretched.

To pick out the green pixels, choose Color, Split Bayer Plane, and (for most cameras) choose plane 2. This makes the image half as big in each dimension. (Be sure to undo this before trying a different plane, or else the software will take your already-reduced image and reduce it further, leading to incorrect results.)

To subtract images that are open on the screen, choose Process, Pixel Math. Choose one image as Image A, the other as Image B, and Subtract as the operation, with scale factor 100% and an added constant of 3000 (to avoid negative numbers). The result replaces Image A.

To read the numbers, use the Information window (Figure 18.8) but with the mode set to Area rather than Aperture. The calibration part of the window is not used. Drag the mouse to select a uniformly illuminated area about 500 to 1000 pixels square near the middle of the image, and its statistics will be displayed.

16.7 Specific Tests

16.7.1 Dynamic Range from One Light Frame and One Flat Dark

You can measure the dynamic range of your camera with nothing more than a raw image of an astrophoto that contains saturated (maximum-white) stars and a single flat dark or bias frame taken at the same ISO setting. In place of the astrophoto, you can use an overexposed daytime photo that reaches maximum white.

A flat dark or bias frame is used rather than a long-exposure dark frame in order to avoid dark current. (On a DSLR, a bias frame and a flat dark are essentially the same thing; any dark frame with an exposure less than 1/10 second will do.) The light frame is used only to determine the value of maximum white (which may not be quite as high as the bit depth allows), and its exposure time is irrelevant.

Strictly speaking, you do not have to isolate one of the Bayer colors for this test, since maximum white is the same everywhere on the sensor, and the dark frame has not been exposed to light (of any color). For strict comparability to other tests, however, you may wish to do so.

Recall that dynamic range is limited by the range of available digital numbers (maximum digital number minus bias) and by total camera noise (not just read noise). Following *DxOmark*'s lead, I take the limit to be the level where signal equals noise. Then:

$$\text{Dynamic range (stops)} = \log_2 \frac{\text{Maximum DN} - \text{Bias}}{\text{RMS noise}}.$$

If your calculator does not provide $\log_2$, use the relation $\log_2 x \approx 3.3 \log_{10} x$. The numbers that you need are the following:

$$\begin{aligned} \text{Maximum DN} &= \text{the highest DN in the astrophoto that includes stars,} \\ \text{Bias} &= \text{the average (or preferably median) of the dark frame,} \\ \text{RMS noise} &\approx \text{the standard deviation of the flat dark.} \end{aligned}$$

This calculation leaves out the part of the fixed-pattern noise that is due to pixel response non-uniformity (PRNU), but the resulting error is small.

Testing both a Canon 60Da and a Nikon D5300 this way, I got dynamic range measurements within about 0.5 stop of *DxOmark*'s, which is reasonably good agreement given that the testing conditions are not the same. You should, of course, do this test with just the green pixels, as described above, and repeat it for every ISO setting that you are interested in.

16.7.2 Read Noise in DN from Two Flat Darks or Bias Frames

A single flat dark (or bias frame) contains all the sensor's read noise and bias, including the portion of the fixed-pattern noise that resides in the bias.

The difference between two flat darks contains none of the bias, but still has more noise than a single frame, by a factor of $\sqrt{2}$, because the random variations in two frames have been combined. (Subtraction works just like addition in this respect, when the quantities being added or subtracted are random and uncorrelated.)

Again, there is no need to isolate one of the Bayer colors because neither of the images was exposed to light, but for strict comparability to other tests, you may wish to do so.

Accordingly, read noise can be measured by subtracting a matched pair of flat darks. Open two such images, subtract (remembering to add a constant), and read the standard deviation. Then:

$$\text{Read noise (DN)} = 0.7 \times \text{Standard deviation of difference of 2 flat darks},$$

where $0.7 \approx 1/\sqrt{2}$.

16.7.3 Gain in DN/e$^-$ from a Pair of Generously Exposed Flats

As already mentioned, gain in DN per electron can be determined by comparing shot noise with signal level. The problem is how to isolate shot noise from other kinds of noise.

A solution is to use the sensor under conditions where shot noise is much greater than noise from other sources. Fortunately, that is exactly what we do when we take a normally exposed flat field at a medium ISO setting (around 400 to 1600), especially if the flat is generously exposed (perhaps three stops above mid-gray). The random noise is then almost entirely shot noise.

To separate the random noise from the fixed-pattern noise, we can use the same method that was used to isolate read noise in the previous test: take the difference of two flat fields that were exposed in immediate succession with the same settings. Then:

$$\text{Shot noise (DN)} \approx 0.7 \times \text{Standard deviation of difference of 2 flats},$$

$$\text{Gain (DN/e}^-) = \frac{(\text{Shot noise in DN})^2}{\text{Signal level in DN} - \text{Bias}}.$$

Bear in mind that we are subtracting a pair of normally or generously exposed flats, not flat darks. The signal level in DN is the average or median of one of the flat fields, measured near the center of the field. The noise is measured on the result of subtracting one image from the other (remembering to add a constant). The bias is the average or median of a flat dark, as in the previous test.

For this test, it is important to isolate one of the Bayer colors so that you are not comparing pixels exposed differently.

Do you need to test this at more than one ISO setting? No, because, by definition, gain in DN/e$^-$ is proportional to ISO. At ISO 400, my Canon 60Da has a gain of 1.68, and my Nikon D5300, 1.85. It's something of a coincidence that the gains are so similar; the Nikon has smaller pixels but also higher quantum efficiency, and the two factors balance out. I also measured the gain of both cameras at ISO 1600, and in each case it was exactly 4 times as much as at ISO 400.

During these tests I discovered that *on Canons, the "Highlight Tone Priority" menu setting cuts the gain in half.* I had always assumed it only affected the

Table 16.1 *Results from the author's tests of his cameras. Gain was measured at ISO 400 and ISO 1600 and interpolated linearly; read noise was measured at all the ISO settings shown.*

Camera	ISO	Read noise (DN)	Gain (DN/e^-)	Read noise (e^-)
Canon EOS 60Da	200	6.5	0.84	7.7
	400	7.7	1.68	4.6
	800	10.3	3.36	3.1
	1600	16.4	6.72	2.4
	3200	29.1	13.44	2.2
Nikon D5300	200	2.3	0.93	2.5
	400	4.0	1.85	2.2
	800	6.8	3.70	1.8
	1600	11.1	7.40	1.5
	3200	19.6	14.80	1.3

exposure meter and/or the creation of JPG files. Not so! Reportedly, Nikon's "Active D-Lighting" menu option also affects the gain under some circumstances, but I have not found any difference between Active D-Lighting "Off" and "Auto" when taking pictures in manual (M) mode.

16.7.4 Read Noise Measured in Electrons

Now that you know the read noise in DN and gain in DN/e^-, you can calculate what you really wanted – the read noise in electrons, which tells you in the most straightforward way possible whether your sensor is ISO-invariant. The conversion is of course:

$$\text{Read noise in } e^- = \frac{\text{Read noise in DN}}{\text{Gain in DN}/e^-}.$$

Table 16.1 shows the results of my tests on my own cameras. You can see that the Nikon is ISO-invariant and the Canon is not, but both perform very well in astrophotography.

16.8 Going Further

To really put a sensor through its paces, you can measure and plot your own version of a *DxOmark* noise study (Figure 16.2). All you need to do is expose a wide variety of flat fields at a variety of ISO settings and brightnesses, ranging from minimum black to maximum white. You don't need to control the exposures, just have a wide variety of them. On the horizontal axis, plot the average pixel value on a logarithmic scale; on the vertical axis, the SNR of each individual image, in decibels (differential pairs are not involved).

A closely related approach is the Photon Transfer Curve (PTC) described by James R. Janesick in *Photon Transfer: DN→ λ* (Bellingham, Wash.: SPIE Press, 2007). A PTC is a plot of average pixel value on the horizontal axis and standard deviation on the vertical axis, both scales logarithmic. From it, you can evaluate gain, read noise, and fixed-pattern noise. It requires a wide range of exposures, but not a calibrated light source or any other form of exposure control, since the average pixel value itself is used as the measure of light intensity.

For copious information about how sensors actually work and factors that affect their performance, see *Image Sensors and Signal Processing for Digital Still Cameras,* a collection of state-of-the-art articles edited by Junichi Nakamura (Boca Raton, Florida: Taylor & Francis, 2006).

Chapter 17
Spectral Response and Filter Modification

Wavelength of light concerns astrophotographers for several reasons. First, wavelength explains the colors of celestial objects in photographs – blue, pink, or red emission nebulae, bluish or yellowish reflection nebulae, and greenish comets.

Second, nebulae emit light at specific wavelengths, and so do streetlights. Appropriate filters can favor one and reject the other.

Third, many astrophotographers use DSLRs that have been modified to extend their response to the deep red end of the spectrum. That makes it easier to photograph thin emission nebulae and, with the addition of other filters, to penetrate light pollution from city lights.

17.1 DSLR Spectral Response

Figure 17.1 sums up a lot of information about the wavelengths of light to which cameras respond. Reading from the top, the first thing you'll notice is that color film has a gap in its response between green and blue, and DSLRs don't. This is important because the strong hydrogen-beta and oxygen-III lines from emission nebulae fall in that gap. That's one reason nebulae that look red on film often come out purplish with a DSLR. Because of the same gap, comets, which have a strong emission at 516 nm from diatomic carbon, look green in DSLR images but neutral on film.

DSLR makers don't publish spectral response curves, so the curves at the top of Figure 17.1 are estimated from CCD data sheets plus a number of published tests and my own experiments. The DSLR normally has an infrared-blocking filter in front of the sensor. We know that with this filter removed, the sensor has strong response out to 700 nm or further.

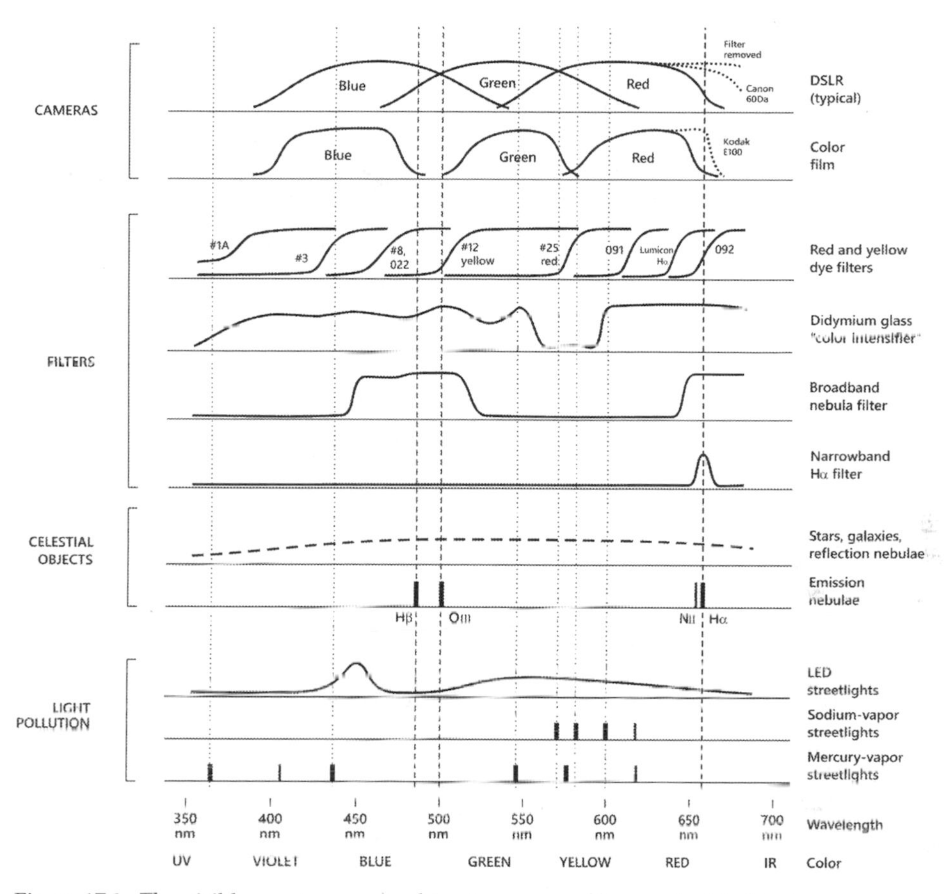

Figure 17.1. The visible spectrum as it relates to cameras, filters, celestial objects, and sources of light pollution. Data are approximate; consult actual specification sheets when possible.

17.2 Filter Modification

17.2.1 What Filter Modification Achieves

Many astrophotographers use DSLRs that have been modified to extend their response to the deep red end of the spectrum. This serves two purposes. One is that it enables the camera to pick up the strong hydrogen-alpha (Hα) emission from nebulae at 656.3 nm. The other is that working in deep red light, with filters blocking out the rest of the spectrum, is a good way to overcome skyglow from city lights.

Canon's EOS 20Da and EOS 60Da and Nikon's D810A, as well as some Fuji DSLRs, have been manufactured with extended red response. Usually, though, astrophotographers rely on third parties to modify their cameras. Reputable purveyors of this service include Hutech (www.hutech.com), Hap Griffin (www.hapg.org), and LifePixel (www.lifepixel.com). For those who have confidence in their ability to work on delicate precision instruments, there are also modification instructions published on the Internet and filters available from various suppliers.

The simplest modification is just to remove the filter. This has three drawbacks. One is that the red response increases so much that the camera no longer produces realistic color images; red always predominates. The second drawback is that the camera no longer focuses in the same plane; neither the autofocus mechanism nor the SLR viewfinder gives a correct indication of focus. The camera can only be focused using Live View or software. The third drawback is that, with such a camera, camera lenses may no longer reach infinity focus at all.

It's better to replace the filter with a piece of glass matched in optical thickness (not physical thickness) to the filter that was taken out. Then the focusing process works the same as before. Better yet, replace the filter with a different filter. Firms that modify cameras offer a selection of filters with different cutoff wavelengths. My preference is to cut off at about 700 nm so that the image quality is not affected by infrared light that is not accurately focused by lenses achromatized for visible light.

How much more Hα response do you get? An increase of 2.5$\times$ to 5$\times$, depending on how much Hα light was blocked by the original filter and whether the new filter is fully transparent at the wavelength. The original IR-blocking filter always has nearly full transmission at 600 nm and nearly zero transmission at 700 nm.

Any filter replacement, even if done by a professional, introduces some risk of trapping dust under the new filter, where it cannot be cleaned off. Such dust normally has no serious photographic effect, since flat-fielding compensates for it.

17.2.2 Is Filter Modification Necessary?

I consider extended red response helpful but not essential. The modification makes cameras less suitable for daytime photography, can bring out chromatic aberration in some lenses, and is little or no help when photographing stars, star clusters, galaxies, or reflection nebulae.

With emission nebulae, it does help, but, often, not spectacularly. Figure 17.2 shows an example. In this case, the modified camera is a Canon EOS 60Da, whose sensitivity at Hα is about 4 times that of the unmodified version.

The picture shows an area in Orion where both emission and reflection nebulae are present. In the picture, the nebula extending downward across the center, with the horsehead-shaped notch in it, is an emission nebula; the brighter nebula at the upper left is a reflection nebula.

What is striking is how well the *un*modified camera works. Yes, the modification helps – but it's not indispensable. The unmodified camera shows the nebula reasonably well.

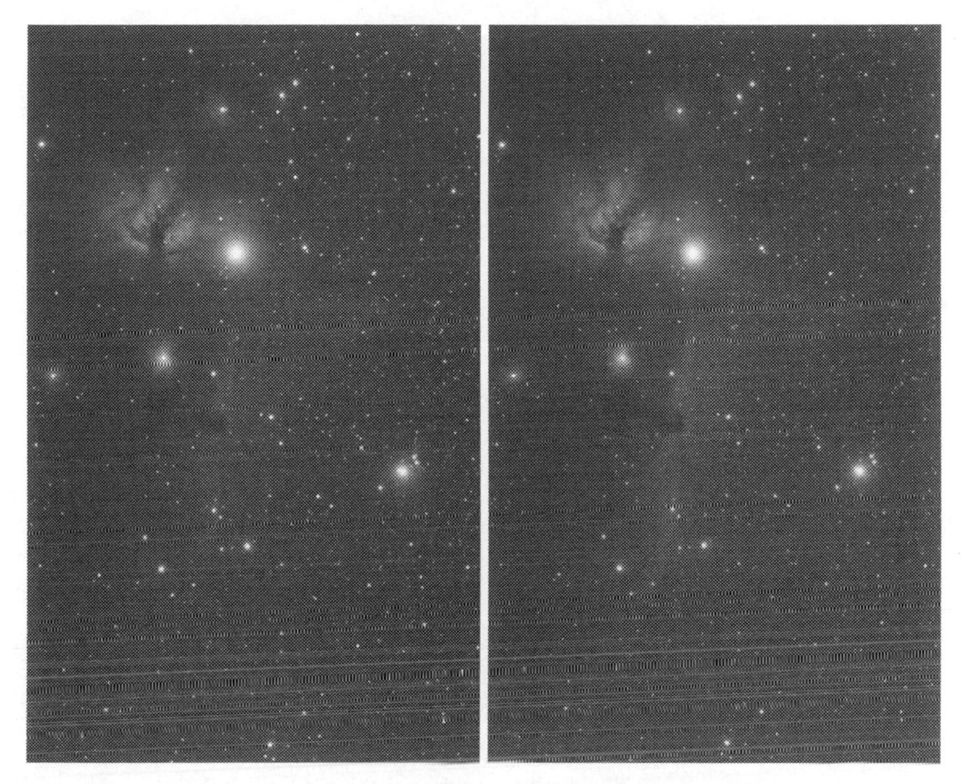

Figure 17.2. Filter modification is helpful but not dramatic. *Left:* unmodified Canon 40D. *Right:* Canon 60Da with extended hydrogen-alpha sensitivity. Each is a stack of 19 or 20 1-minute exposures with a 6.5-cm $f/6.5$ apochromatic refractor (Astronomics AT65EDQ), processed with *Pixinsight* to give similar color balance and overall contrast. See the back cover of this book for the same pictures in color.

To understand why this is so, remember that Hα is not the only wavelength at which nebulae emit light. There are also strong Hβ and O III emissions, which color film did not pick up, leading the previous generation of astrophotographers to think only about Hα. Depending on the type of nebula, there may be more than that. As Table 17.1 shows, the Veil Nebula is actually brightest in blue and near-ultraviolet wavelengths, not hydrogen-alpha.

For that matter, a $4\times$ difference in Hα sensitivity is not gigantic. In photographic terms, it is a difference of two stops, no more than a quarter of the usable dynamic range of the image. The camera would pick up some Hα without modification.

Most importantly, deep red sensitivity matters only when you are photographing emission nebulae or using a deep red filter to overcome light pollution. Even for nebulae, a modified DSLR is not strictly necessary (Figure 17.3). When photographing galaxies, star clusters, or reflection nebulae, modified and unmodified DSLRs will certainly give the same results.

Table 17.1 *Intensities of main spectral lines from emission nebulae (as percentage of Hβ from the same nebula).*

Line	Wavelength (nm)	**M42** (Orion Nebula)	**M16** (Eagle Nebula)	**NGC 6995** (Veil Nebula)	Theoretical thin H II region (2500 K)
O II	372.6 + 372.8	119	157	1488	—
Ne III	386.9	13	2	118	—
Hγ	434.0	41	36	44	44
O III	436.3	—	—	47	—
Hβ	486.1	100	100	100	100
O III	495.9	102	29	258	—
O III	500.7	310	88	831	—
N II	654.8	26	104	124	—
Hα	656.3	363	615	385	342
N II	658.3	77	327	381	—
S II	671.7 + 673.1	9	116	68	—

O II, O III, etc., denote ionization states of elements, each of which emits more than one wavelength. The Hα, Hβ, and Hγ lines of hydrogen are all from H II.

M42 is a dense, strongly photoionized H II region; measurements were taken just north of the Trapezium, very close to the illuminating stars. Data from D. E. Osterbrock, H. D. Tran, and S. Veilleux (1992), *Astrophysical Journal* **389**:305–324.

M16 is a photoionized H II region less dense than M42 and considerably reddened by interstellar dust. Data from J. García-Rojas, C. Esteban, M. Peimbert, *et al.* (2006), *Monthly Notices of the Royal Astronomical Society* **368**:253–279.

NGC 6995 is part of the Veil Nebula or Cygnus Loop, a supernova remnant, and is ionized by the expanding shock wave from the supernova, rather than by light from nearby stars. Each value shown in the table is the mean of the authors' reported fluxes from five positions in the brightest part of the nebula. Data from J. C. Raymond, J. J. Hester, D. Cox, *et al.* (1988), *Astrophysical Journal* **324**:869–892.

Theoretical values for a thin, weakly excited region of pure hydrogen are from D. E. Osterbrock and G. J. Ferland, *Astrophysics of Gaseous Nebulae and Active Galactic Nuclei* (Sausalito, Calif.: University Science Books, 2006), p. 72.

17.3 Filters to Cut Light Pollution

17.3.1 How Light Pollution can be Removed

Now look at the middle part of Figure 17.1, showing the transmission of various filters. Some of them are quite effective at cutting through the glow of city lights. The key idea is to take advantage of differences in the wavelengths emitted by different light sources.

Even without a filter, digital imaging has a huge advantage over film: light pollution can be subtracted out. Simply by adjusting the histogram of the image

Figure 17.3. The Horsehead and other faint nebulae captured under light-polluted skies 10 miles (16 km) from central London with the aid of a broadband nebula filter. Stack of 20 3-minute exposures and 23 2-minute exposures with unmodified Canon EOS 30D camera, 10-cm (4-inch) $f/8$ Takahashi apochromatic refractor, and Astronomik CLS filter. (Malcolm Park, London, UK.)

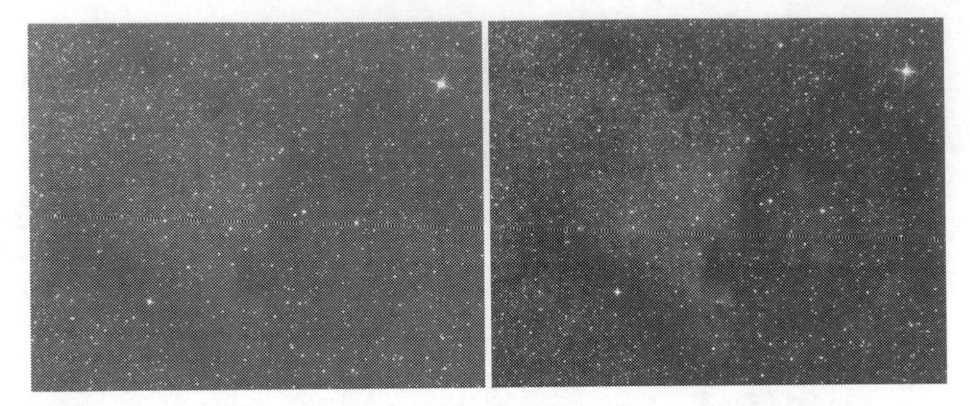

Figure 17.4. Adding a broadband nebula filter darkens the background and brings out nebulosity. North America Nebula (NGC 7000), Canon 60Da, single 30-second exposure at ISO 1600 with Sigma 105-mm *f*/2.8 lens, Astronomik CLS filter added in second picture.

file, we can throw away any constant glow that covers the picture, if it is not too strong. By stacking many short exposures, we can keep gathering light from deep-sky objects without overexposing the image. Even moonlight is much less of an obstacle than it used to be; we can subtract it out.

Even so, filters can still help if they can distinguish one kind of light from another. Incandescent lights, including car headlights, shine at all visible wavelengths, and so do stars, galaxies, and reflection nebulae. Accordingly, filters cannot favor one over the other. But if your light pollution comes from sodium- or mercury-vapor streetlights, which emit only a limited range of wavelengths, you're in luck. Broadband nebula filters or didymium-glass filters can cut the light pollution while not cutting the starlight as much (Figure 17.4).

Alternatively, if you are photographing emission nebulae, which emit only a few wavelengths, but the light pollution covers the whole spectrum, a filter can block some of the light pollution while transmitting all of the light from the nebulae. Broadband nebula filters and didymium glass do this too, as well as the narrowband filters discussed in the next section.

The best situation is, of course, when you are photographing emission nebulae and your streetlights are low-pressure sodium-vapor. In that situation, a filter can easily transmit nearly all the nebula light while blocking nearly all the light from the streetlights.

Unfortunately, that situation is becoming less common as LED streetlights take over. In my location, LED streetlights have been a mixed blessing. They shine downward rather than sideways, so much less of their light goes into the air. My town (Athens, Georgia) has resisted the temptation to make them extra-bright; they are the same brightness as the older streetlights they replace. (This is important because if streetlights are too bright, people driving cars will lose their partial dark adaptation and not be able to see very well when they drive out of the brightly lit area.) Because of the LED conversion, the Milky Way is again visible from my house after a lapse of nearly twenty years.

The problem is that LED streetlights emit a continuous spectrum with a hump around 450 nm (see Figure 17.1). That hump can be reduced with a #8 yellow filter, but the rest of the LED spectrum is not distinguishable from starlight.

All of these filters give the picture a color cast, impair the rendering of the colors of stars, and require the histograms of the three colors (R, G, B) to be adjusted separately so that they end up matching.

17.3.2 Filters to Favor Nebulae

If you can't filter out the light pollution, the alternative is to filter *in* the nebulae – that is, transmit the light from the nebulae but not much else. This only works with emission nebulae, which emit at specific wavelengths, not with stars and galaxies.

Even near cities, there is comparatively little light pollution at wavelengths longer than 620 nm, but nebulae are bright at hydrogen-alpha (Hα, 656 nm). Accordingly, if you use deep red light alone, you can take good pictures of faint nebulae under a bright city sky. Even if your target is not emission nebulae, you may choose to use a narrow band at the red end of the spectrum in order to overcome strong light pollution.

A modified DSLR is helpful but not absolutely necessary. What is important is the right filter, either an interference filter that passes a narrow band around Hα or a dye filter that cuts off wavelengths shorter than 620 nm or so. The common Wratten 25 (A) or Hoya R60 red filter is not red enough; it transmits some of the emissions from sodium-vapor streetlights. Better choices are the B+W 091, Hoya R62, Schott RG630, and the Lumicon Hydrogen-Alpha filter.

Narrowband imaging requires surprisingly long exposures (such as 10 minutes at $f/4$) and is a stiff test of tracking and guiding. Results are generally good if you process the narrowband image like a normal color picture and then convert it to grayscale at the last step. However, with pictures taken in deep red light, the noise level is slightly less if you isolate just the red pixels from the Bayer matrix, as was done with Figures 17.5 and 17.6.

Various software packages provide different ways to isolate the red pixels. In *Nebulosity,* open the raw file and choose Batch, Batch One-Shot Camera, RGB Color, Extract R. In *MaxIm DL,* open the raw file, choose Color, Extract Bayer Plane, and red is usually plane 1. (Unfortunately, you must do this on each raw file separately.) In *PixInsight*, use the SplitCFA procedure described in Section 16.6.2; red is usually plane 0. If you are splitting a heavily red-filtered image, it will be obvious which plane is the right one.

The result will be a set of files containing only the red color plane, which you can then calibrate and stack as black-and-white images. Naturally, a different narrowband filter might use the green or blue pixels rather than red. The idea is to throw away the Bayer pixels that are definitely not used, since they contain only noise. On separating color layers, see also Section 16.6.1.

Figure 17.5. This picture of nebulae in Orion used a modified DSLR and a deep red filter to cut through light pollution and moonlight; the gibbous moon was only 30° away. Stack of fifty 1-minute exposures with a Canon 60Da at ISO 1600, a Sigma 105-mm lens at *f*/2.8, and a B+W 091 filter on an iOptron SkyTracker.

An especially interesting technique is to take three narrowband pictures at different wavelengths, then combine them as if they were red, green, and blue to make a false-color image. The Hubble Space Telescope commonly uses red for sulfur (S II, 672 nm), green for Hα (which is of course red in real life), and blue for O III (the "Hubble palette").

Some narrowband imagers modify their DSLRs by removing not only the IR-blocking filter but also the Bayer matrix. Then all the pixels respond to all colors of light. However, the camera no longer functions as a normal DSLR; it does not take pictures in color, nor focus correctly through the viewfinder; and it might best be described as a dedicated astrocamera made out of a DSLR.

17.3.3 The Middle Ground

Some filters are narrower than the typical broadband nebula filter, but broad enough to give a relatively bright image and take in more than one spectral line. An example is the Astronomik UHC, which transmits wavelengths around 450–500 nm (Hβ and O III) and 650–700 nm (Hα). Using one, Blake Estes took the spectacular nebula picture in Figure 6.4, from central Los Angeles.

Pictures taken with such a filter are processed like normal full-color images, but with some adjustment of the histograms of the three colors to achieve good color balance.

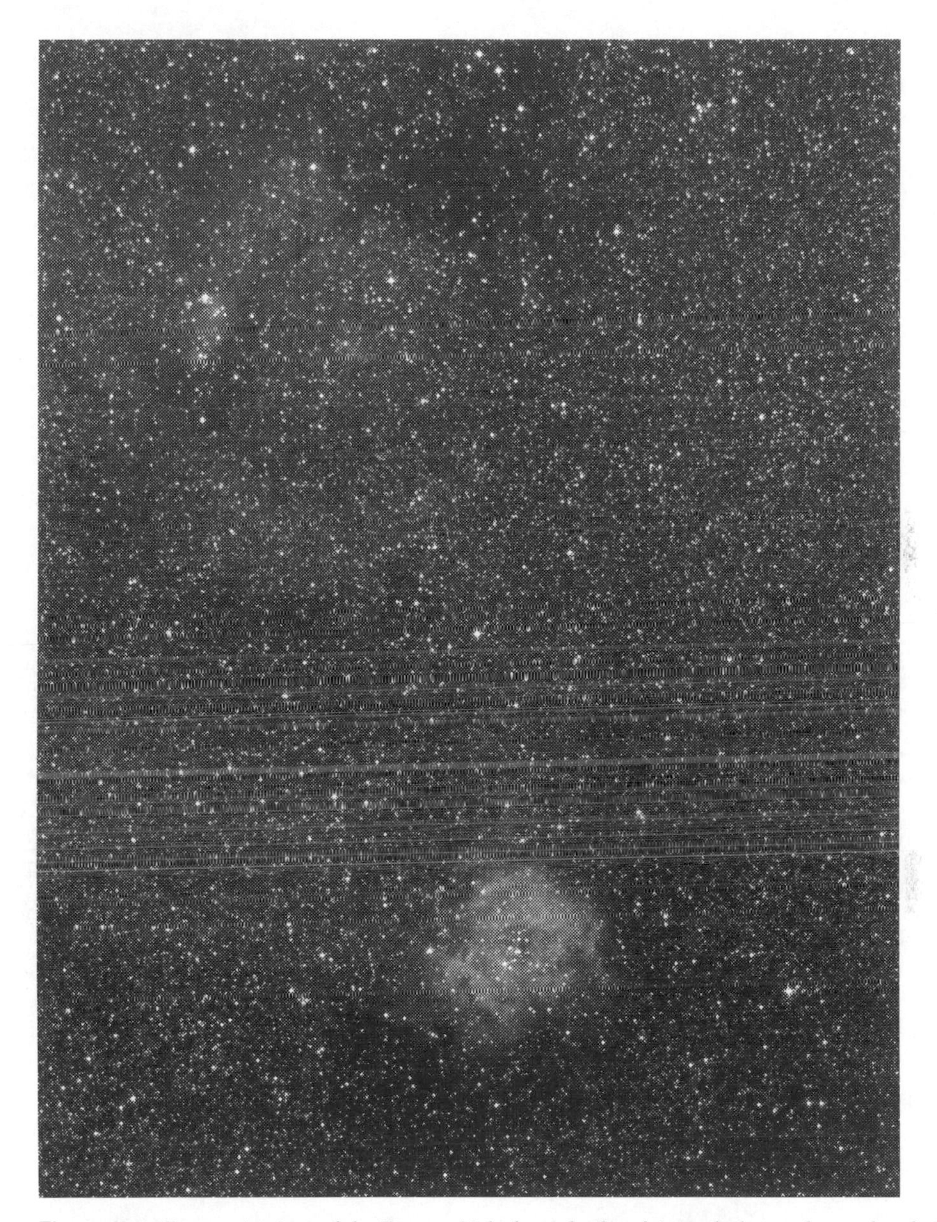

Figure 17.6. Dramatic view of the Rosette Nebula and other faint nebulae and star clouds in Monoceros. Stack of 15 3-minute exposures with same setup as Figure 17.5, at the same site, in town, with substantial light pollution, but without the moon in the sky.

17.4 How Filters Are Made

17.4.1 Dye Filters

Dye filters are made by adding dyes to glass or plastic. As Figure 17.1 shows, the red, yellow, and orange are good at cutting off wavelengths shorter than a threshold. Blue, green, and magenta dye filters, not shown on the graph, are much more uneven in their performance.

There are several systems of dye-filter nomenclature, including traditional Kodak Wratten numbers and letters (#25 or A), German designations starting with 0 (such as 091), Japanese designations starting with letters (L39, Y43, R62, where the digits are the cutoff wavelength in tens of nanometers), and others. For more about filters and their nomenclature, see *Astrophotography for the Amateur.*

17.4.2 Interference Filters

Interference filters are made, not by adding dye or impurities to glass, but by depositing multiple thin, partly reflective coatings on the surface. The thickness and spacing of the coatings resonate with particular wavelengths of light. In this way it is possible to select which parts of the spectrum to transmit or block. "Nebula filters" are normally interference filters. They are easily recognized because they look almost like mirrors, reflecting rather than absorbing the light that they do not transmit.

Two of the most respected vendors of interference filters are Lumicon (www.farpointastro.com) and Astronomik (www.astronomik.com). Both offer a wide variety of filters. The Astronomik CLS broadband nebula filter is especially popular and useful. Other filters pick out specific spectral lines of nebulae.

Interference filters are not always available in sizes to fit in front of camera lenses. Instead, they are sized to fit $1\frac{1}{4}$-inch and 2-inch eyepiece barrels. Those threaded for 2-inch eyepieces will fit 48-mm camera-lens filter rings; the thread pitch is not the same, and only one thread will engage, but that is enough. Astronomik has introduced a line of filters that fit inside some camera bodies, behind the lens (Figure 17.7); besides rejecting light pollution, the filter keeps dust out of the camera.

17.4.3 Didymium Glass

One type of colored glass acts much like a broadband nebula filter but is cheaper and more readily available in large diameters. It is didymium glass, which blocks the brightest emissions of sodium-vapor (orange) streetlights.

Didymium glass looks bluish by daylight and purplish by tungsten light. It contains a mixture of praseodymium and neodymium. Glassblowers look through didymium glass to view glass melting inside a sodium-rich flame.

Today, didymium glass is sold as "color enhancing" or "intensifier" filters. Nature photographers initially used it to brighten colors because it blocks out a region of the spectrum where the sensitivities of the red and green layers of color film overlap.

One widely available didymium filter is the Hoya Red Intensifier (formerly just Intensifier). I recommend that you avoid Hoya's Blue Intensifier

Figure 17.7. Astronomik "Clip-Filters" fit inside some camera bodies, behind the lens. This one, for APS-C Canons, occupies the extra space reserved for EF-S lenses; EF lenses and T-rings fit the lens mount normally, ahead of the filter.

and Green Intensifier, which are less transparent at the desired wavelengths. Hoya's Portrait filter is a weaker kind of didymium glass with less blockage of the unwanted wavelengths.

17.4.4 Precautions

Reflections from filters are more of a problem with DSLRs than with film cameras because the sensor in the DSLR is itself shiny and reflective. The sensor and a filter parallel with it can conspire to produce multiple reflections of bright stars. The reflection risk seems to be much greater when the filter is in front of the lens than when it is in the converging light path between telescope and camera body. Dye filters can and should be multi-coated to reduce reflections. Interference filters are inherently shiny.

Interference filters are affected by the angle at which light enters them. A wide-angle or even medium-telephoto lens usually will not get the full benefit of the interference filter across the entire field, even if the filter is mounted inside the camera body. The symptom is that the corners of the picture are discolored relative to the center. Didymium glass does not have this problem.

One last note: not every filter that works well on an eyepiece will work equally well when placed in front of a long lens. The long lens magnifies optical defects. I have only had problems with the optical quality of a filter once, but it's definitely something to be aware of.

Chapter 18
Tools for Astronomical Research

Digital imaging has narrowed the gap between professional and amateur astronomy. It is nowadays common for amateurs to photograph objects that are not very well known to science. We are no longer confined to just a few hundred prominent targets discovered centuries ago. Accordingly, we need to open up the professional astronomers' kit of tools for identifying and researching celestial objects. The most important data sources are free on the Internet.

18.1 Star Maps

You can't do astronomy without star maps, which nowadays are largely computerized. The time-honored commercial software packages of this type are *TheSky* (www.bisque.com) and *Starry Night* (www.starrynight.com).

Recently, though, a free, open-source competitor has taken the lead, *Stellarium* (www.stellarium.org). Available for Windows, Linux, and macOS, *Stellarium* produces very realistic pictures of the sky (it is even used in planetarium projectors) and, with downloadable catalogs, covers stars all the way down to 18th magnitude, as well as a multitude of deep-sky objects; it even connects automatically to *Simbad* to find objects that are not in its catalogs.

Stellarium is handy for planning photographs because it can calculate and display the field of view given the sensor size and focal length (Figure 18.1).

Also noteworthy is *Cartes du Ciel* (freeware from www.ap-i.net for Windows, Linux, and macOS), which is more oriented toward producing printable maps for astronomers planning observing sessions. *Cartes du Ciel* can connect directly to the *Digitized Sky Survey* (DSS, http://archive.eso.org/dss) and download an observatory image corresponding to what you're seeing on the map (see also *Aladin* in the next section).

Both *Stellarium* and *Cartes du Ciel* (as well as all their main competitors) compute planet, comet, and asteroid positions, downloading orbital data from the Minor Planet Center for the purpose, and can control computerized telescope mounts.

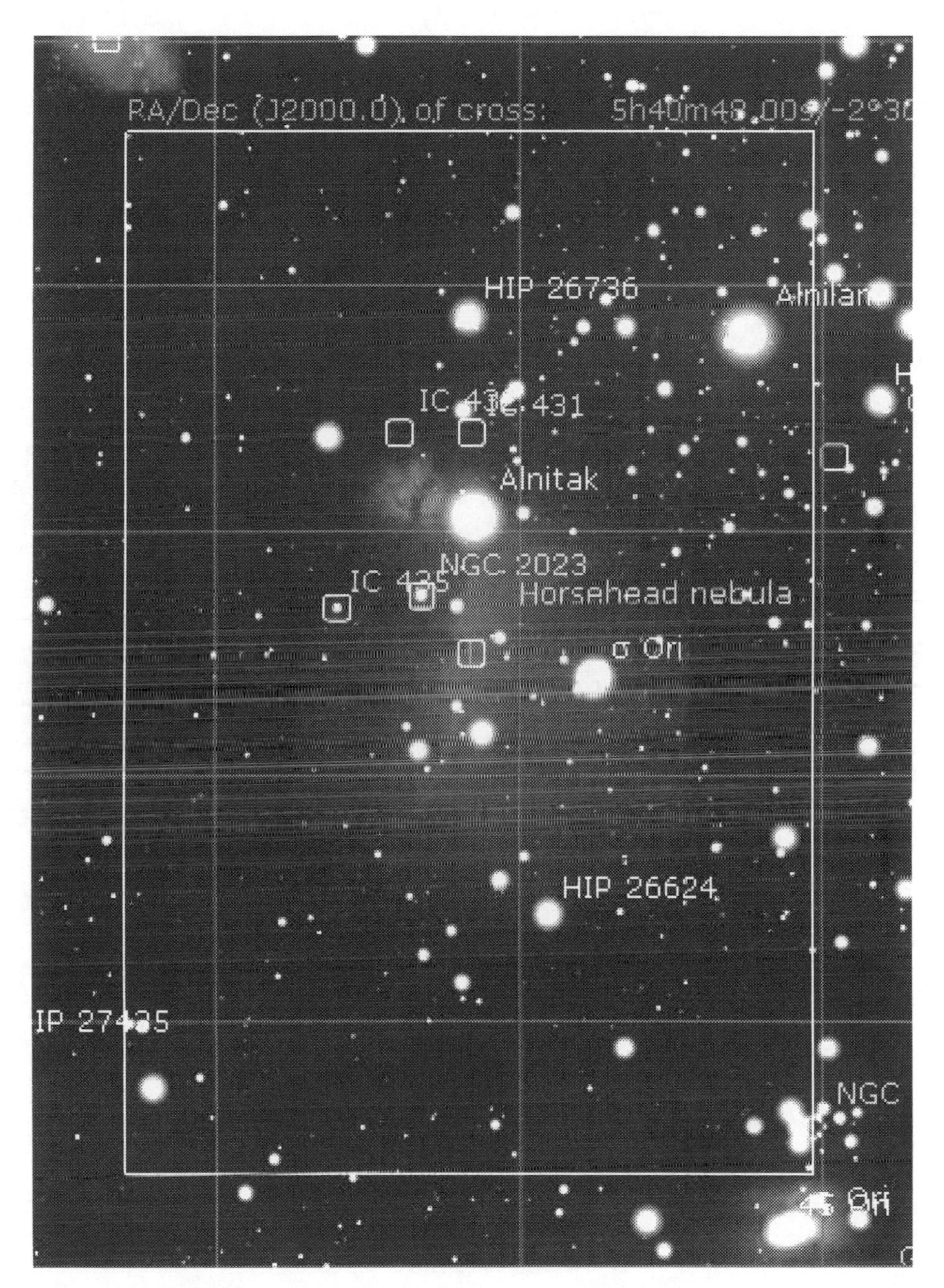

Figure 18.1. Planning a photograph with *Stellarium.* Field of view is automatically calculated from sensor size and lens focal length.

One printed star atlas remains very useful for astrophotographers. It is the *Interstellarum Deep Sky Atlas,* by Ronald Stoyan and Stephan Schurig (Cambridge University Press, 2015; www.deepskyatlas.com; also available in German). Designed for serious telescopic deep-sky observers, this atlas clearly indicates the relative visual prominence of objects. Not only does it identify distant galaxies and faint nebulae, it also tells you which ones are more likely to show up in your pictures.

18.2 *Simbad, Aladin,* and *VizieR*

Simbad, Aladin, and *VizieR* are three components of a huge database for professional astronomers that is maintained at the Centre de données astronomiques de Strasbourg (CDS) in France. The names are from the *Arabian Nights* (with French spelling) and are also acronyms, but obscure ones; the documentation doesn't say why the final *R* of *VizieR* is capitalized.

Simbad is an online lookup service for objects outside the Solar System, ranging from stars and clusters to galaxies and quasars. Access is free of charge through the Web at http://simbad.u-strasbg.fr. Type in the designation or position of an object, and you get a summary of all available information about it, including links to research literature (Figure 18.2). If you know the position only approximately, you can "query around" to get all objects within a specified distance of that position.

Aladin (Figure 18.3) is a free program that you run on your own computer to retrieve images and information through the Web. It is fundamentally an all-sky map made of images from the *Digitized Sky Survey* and other surveys, together with links to *Simbad* for information about the objects.

The third component, *VizieR,* is a combination of catalogs and is easily accessed through *Simbad.* For example, if you're looking at a double star and want to see its entry in the *Washington Double Star Catalog* (WDS), just click on its WDS link on the *Simbad* page, and *VizieR* will show you exactly what the catalog says about it. Not confined just to major catalogs, *VizieR* brings together thousands of lists of objects and observational data published by working astronomers.

18.3 Case Study: An Unnamed Nebula in Monoceros

In the center of Figure 18.4 is a patch of emission nebulosity that caught my eye when I was examining a picture of NGC 2259, the star cluster to the left of center in the picture. The unnamed nebula is not shown in *Stellarium* or the *Interstellarum Deep Sky Atlas* and does not have an NGC or IC number.

Because the nebula turned up in pictures of a known field, I had no trouble determining its position – I just opened up a view in *Stellarium* that matched

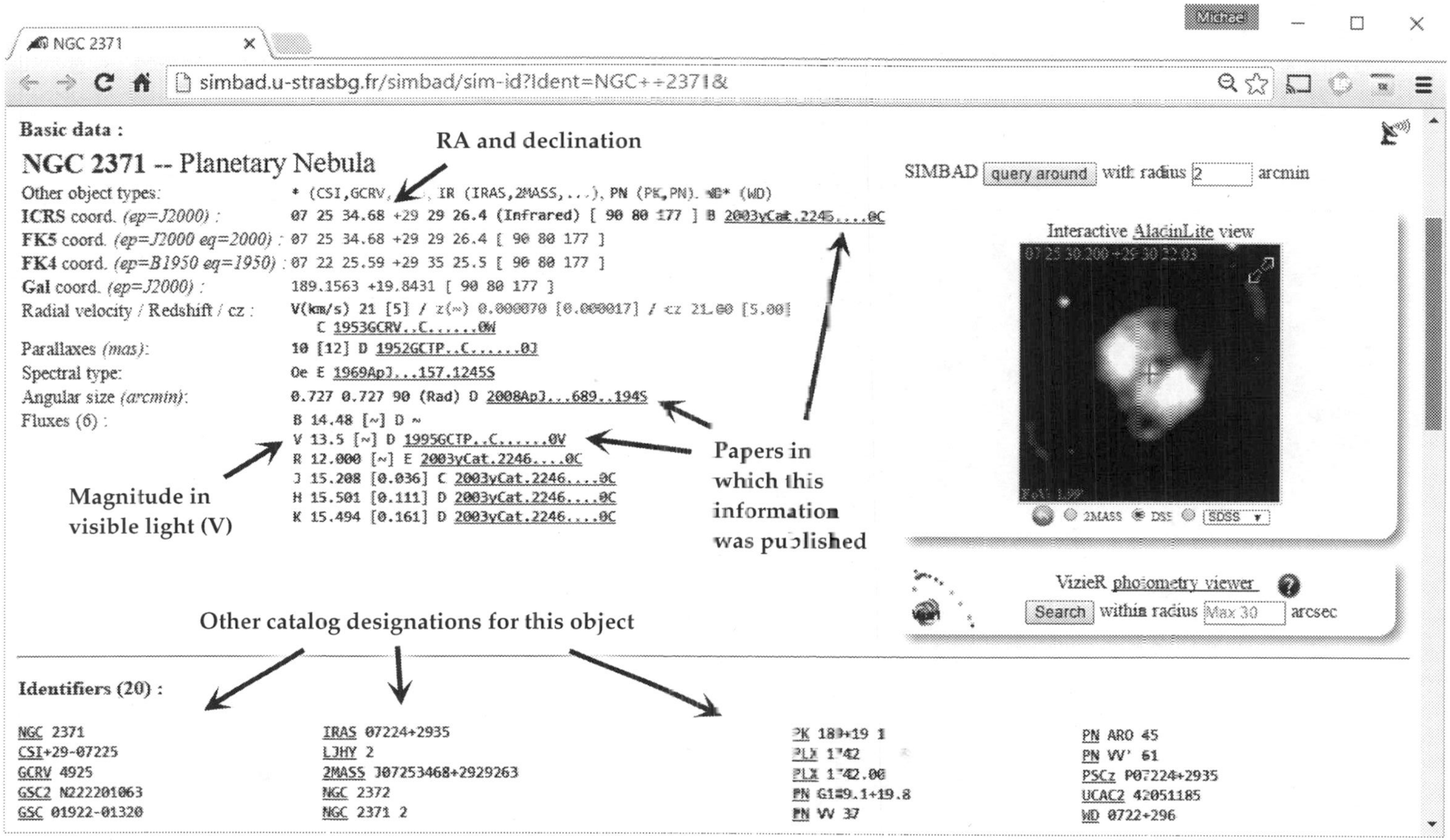

NGC 2371

simbad.u-strasbg.fr/simbad/sim-id?Ident=NGC++2371&

Basic data :

NGC 2371 -- Planetary Nebula

RA and declination

Other object types: * (CSI,GCRV,...), IR (IRAS,2MASS,...), PN (PK,PN), WD* (WD)

ICRS coord. (ep=J2000) : 07 25 34.68 +29 29 26.4 (Infrared) [90 80 177] B 2003yCat.2246....0C

FK5 coord. (ep=J2000 eq=2000) : 07 25 34.68 +29 29 26.4 [90 80 177]

FK4 coord. (ep=B1950 eq=1950) : 07 22 25.59 +29 35 25.5 [90 80 177]

Gal coord. (ep=J2000) : 189.1563 +19.8431 [90 80 177]

Radial velocity / Redshift / cz : V(km/s) 21 [5] / z(~) 0.000070 [0.000017] / cz 21.00 [5.00] C 1953GCRV..C......0W

Parallaxes (mas): 10 [12] D 1952GCTP..C......0J

Spectral type: Oe E 1969ApJ...157.1245S

Angular size (arcmin): 0.727 0.727 90 (Rad) D 2008ApJ...689..194S

Fluxes (6) :
B 14.48 [~] D ~
V 13.5 [~] D 1995GCTP..C......0V
R 12.000 [~] E 2003yCat.2246....0C
J 15.208 [0.036] C 2003yCat.2246....0C
H 15.501 [0.111] D 2003yCat.2246....0C
K 15.494 [0.161] D 2003yCat.2246....0C

Magnitude in visible light (V)

Papers in which this information was published

SIMBAD query around with radius 2 arcmin

Interactive AladinLite view

2MASS DSS SDSS

VizieR photometry viewer

Search within radius Max 30 arcsec

Other catalog designations for this object

Identifiers (20) :

NGC 2371
CSI+29-07225
GCRV 4925
GSC2 N222201063
GSC 01922-01320
IRAS 07224+2935
LJHY 2
2MASS J07253468+2929263
NGC 2372
NGC 2371 2
PK 189+19 1
PLX 1742
PLX 1742.00
PN G189.1+19.8
PN VV 37
PN ARO 45
PN VV' 61
PSCz P07224+2935
UCAC2 42051185
WD 0722+296

Figure 18.2. *Simbad* is the professional astronomers' web database. This is just the top of a long, informative page. Note that RA and declination are for epochs 2000 and 1950, not the current date.

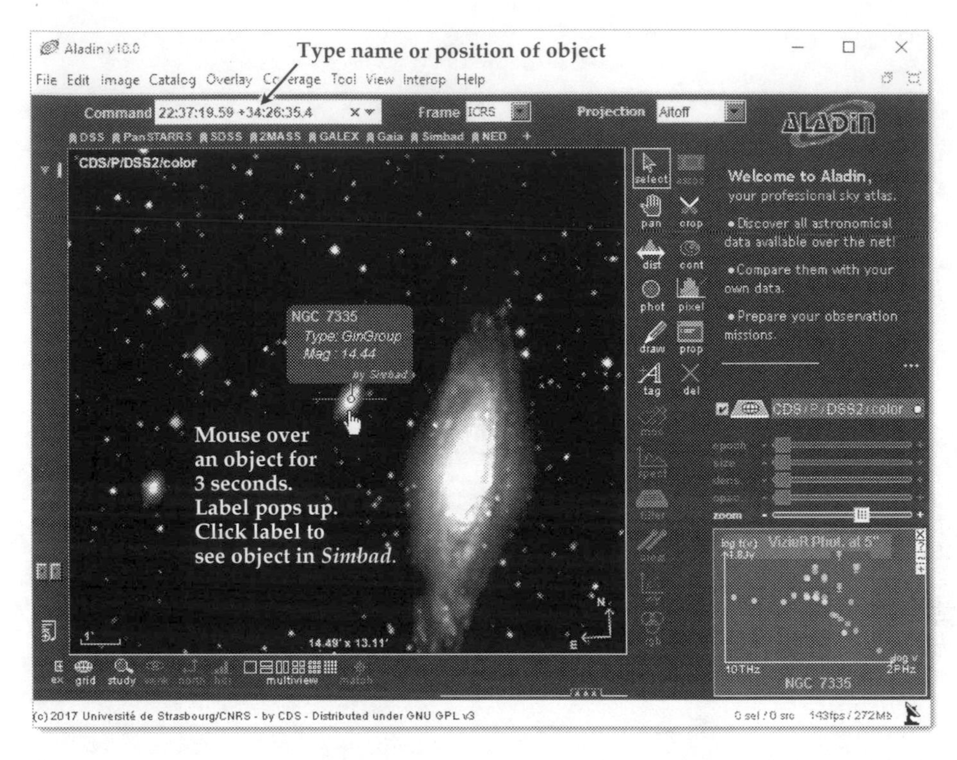

Figure 18.3. *Aladin* runs on your computer, retrieves sky images through the Internet, and gives you instant access to *Simbad*. The pointer with *Simbad* access has to be enabled under "Tool."

the picture, moved the cursor to the position of the nebula, clicked on the sky background, and read off the coordinates.

But what, if anything, is known about the nebula that piqued my interest? To find out, I went to *Aladin* and entered the coordinates I had obtained from *Stellarium*. It displayed a DSS picture showing the nebula. There was no link to an object precisely centered on it, but the best match, toward its westward edge, was an object known to *Simbad* as GAL 201.6+01.6, described as an H II (ionized) region.

That's precisely what an emission nebula is – so I can count my object as tracked down. It has, however, no catalog designation at all; GAL 201.6+01.6 is a positional designation based on galactic coordinates.

The position given by Simbad was determined by radio observation, and it tells us that the thickest part of the nebula is partly obscured by dark nebulosity in the foreground. (Dust doesn't block radio waves.) This nebula does not seem to have been noticed as a visual object at all. Although it is easy to photograph, I have no idea how large a telescope would be needed to view it.

Figure 18.4. In the center of the picture is an unnamed nebula in Monoceros at RA 06:36:36, declination +10°46′ (2000.0). Stack of 15 1-minute exposures, Canon 60Da at ISO 3200, Canon 300-mm $f/4$ lens, picture considerably cropped.

18.4 Plate Solving for Identification and Position

In days of yore, when photographs were taken on glass plates, *plate solving* was the task of figuring out exactly what part of the sky a picture covered, with what scale and orientation. Now that we have computers and digital images, plate solving can be done automatically.

The classic method of plate solving is similar to image stacking. Instead of combining two digital photographs with rotation and stretching, you combine a photograph with a roughly matching chart made by plotting data from a star catalog. The output of the stacking process tells you how the chart and the image line up.

That assumes you already have a pretty good guess of what the image covers. A much more powerful technique is being pioneered at www.astrometry.net with NSF, NASA, and Canadian support. Using methods adapted from machine vision, their plate solver can usually identify pictures of any star field in the sky, regardless of field of view, magnitude limit, or even whether the picture is upside down or mirrored.

The *Astrometry.net* plate solver is free for amateurs as well as professionals to use. All you do is go to the web site, upload your picture, wait a few minutes,

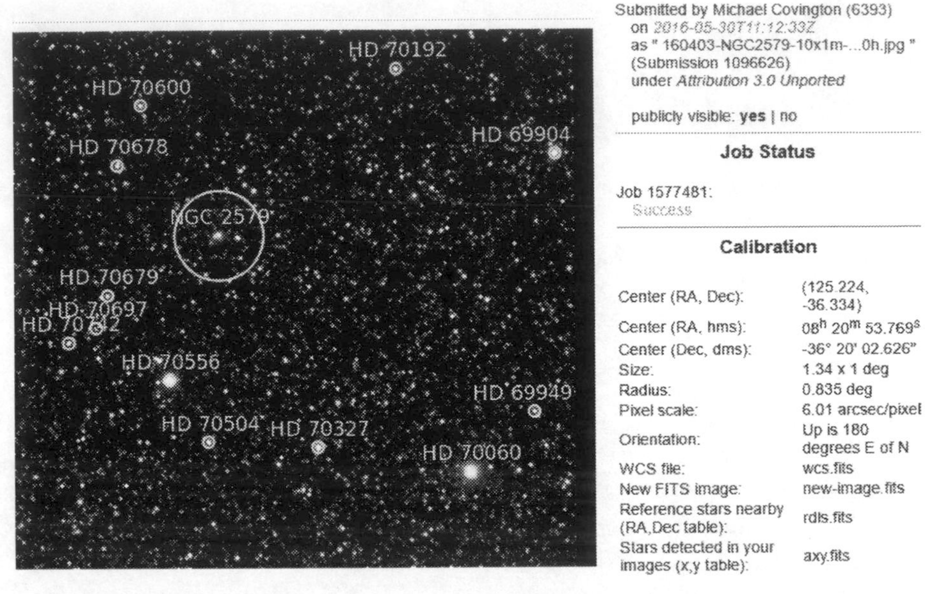

Figure 18.5. *Astrometry.net* identifies stars and deep-sky objects in any picture you upload to it. Note incorrectly reported orientation, a common problem with JPEG files; north is actually up in this picture.

and download the results, which consist not only of an annotated image (Figure 18.5) but also information about the right ascension, declination, field of view, orientation, scale, and even polynomials to account for distortion. You can download a FITS file that contains your uploaded image with that information added to the header.

The *Astrometry.net* software can also be downloaded to run on your own computer – currently only under UNIX, Linux, or macOS.

A limitation worth noting is that *Astrometry.net* does not identify asteroids or other Solar System objects. Nor will it tell you if your picture contains a nova or other unexpected object.

18.5 Case Study: Have I Discovered a Star Cluster?

Figure 18.6 shows two star fields. One of them is the well-known Tau Canis Majoris star cluster (NGC 2362), a sprinkle of stars around a bright star that almost hides them from view in the telescope. The other is a similar-looking group of stars that I found in Virgo while testing an autoguider. Is it a star cluster?

Since I hadn't been paying attention to what I was photographing – just choosing guide stars at random – the second picture was a job for *Astrometry.net*,

Figure 18.6. *Left:* The Tau Canis Majoris star cluster (NGC 2362). Stack of 5 2-second exposures, Celestron 8 EdgeHD, *f*/10, Canon 60Da, ISO 3200. *Right:* Is this also a star cluster? Stars around HIP 61212 in Virgo. Stack of three 2-minute exposures, Celestron 8 EdgeHD, *f*/7 compressor, Canon 60Da, ISO 1600.

which quickly told me that the central star was HIP 61212 in the Hipparcos catalog. I looked up that star in *Stellarium* and found no cluster plotted. Then I went to *Aladin* and called up the same region of the sky. In order to see anything, I had to zoom out quite a bit.

By clicking on them in *Aladin,* I looked up half a dozen of the (apparent) cluster stars in *Simbad.* If the stars were really a cluster, they should be about the same distance from the earth. Unfortunately, *Simbad* did not list distance measurements for these stars (nobody has ever measured them). It did list proper motions for them (thanks to the Tycho-2 catalog), and they were all different. Since the stars in a cluster should have similar proper motions, I think we can declare the case closed – this is not a star cluster.

18.6 Variable-star Photometry

A DSLR is a fine instrument for measuring the brightness of stars. The American Association of Variable Star Observers (AAVSO, www.aavso.org) is intensively developing methods for doing this, and I refer you to *The AAVSO DSLR Observing Manual,* on that web site, for complete and up-to-date instructions. Here, however, I can at least tell you enough to whet your appetite and equip you to take analyzable images if you happen upon a nova or other discovery whose brightness will need to be measured.

18.6.1 Acquiring Images

Photometric DSLR images are different from those needed for taking good pictures. The key requirements are:

- The stars of interest *must not reach maximum white* (or maximum red, green, or blue) in any of their pixels. (A quarter to half of maximum brightness is good.)
- The image should be *slightly out of focus,* so that stars are about ten pixels in diameter (about five pixels FWHM). This spreads the light over multiple pixels, reducing irregularity, and also makes it easier to avoid hitting maximum brightness. You need disks, not doughnuts, so don't defocus too far.
- The image must be *monochrome* and should be *dark-frame corrected* to remove hot pixels. Flat-field correction, if done, improves accuracy.
- The image must be *linear,* with no histogram adjustment or gamma correction, so that the pixel values are proportional to the number of photoelectrons captured.

It is best to use a low ISO setting (400 or less) for greater dynamic range. Exposures are surprisingly short, and tracking need not be perfect; some people are doing good variable-star photometry with 50-mm lenses and fixed tripods.

Spectral response is an issue. The standard *V* filter used in visible-light photometry covers most but not all of the visible spectrum – more than the G pixels in a DSLR's Bayer matrix, but not as much as R, G, and B together. The AAVSO is developing methods to combine the R, G, and B pixels in appropriate proportions to simulate a *V* filter. An alternative is simply to use the G pixels alone; they are twice as numerous as R and B, and the G spectral response is about the same whether or not the camera has been modified for extended red response. In the long run, "DSLR G" may become a new photometric standard.

An even cruder approach that works surprisingly well is to make measurements on the un-deBayered raw image from the camera, after doing dark-frame subtraction in the camera (long exposure noise reduction). In effect, you are measuring $R + 2G + B$ because there are two Gs in every group of four pixels, and the overall response is reasonably similar to photometric *V*.

18.6.2 Aperture Photometry

Measurements are made by comparing the pixels in a circle (the *aperture*) with a doughnut-shaped region around them (the *annulus*, Figure 18.7). The annulus provides a way to subtract sky background (which may even include nebulosity or the galaxy containing a supernova), giving a linear measurement of the brightness of the star.

One quick way to do aperture photometry is to use the Information window of *MaxIm DL* (including editions that do not include the Photometry tool). Figure 18.8 shows how this is done. You can also read the intensity of each pixel (to make sure the maximum value is never reached), the star image size (FWHM,

Figure 18.7. For photometric measurement, brightness of a star image is compared to an annulus, which does not contain stars. Image must be linear and not reach maximum brightness.

full-width half-maximum), and the signal-to-noise ratio (SNR, the higher the better). The accuracy of the results depends on how well you center each star in the aperture and whether there are background stars in the annulus.

18.6.3 Photometry Software

Nonetheless, the Information window, or its equivalent in other software packages, is not really the right tool for serious photometry. Accuracy depends on how accurately you hit the stars with the mouse, the analysis algorithm is not the most sophisticated, and analyzing large sets of observations is tedious.

What you want is software that will process multiple images automatically once you've identified the stars in one image, finding the centroid of each star image automatically. In advanced editions of *MaxIm DL*, the Photometry tool does this, and it can calibrate your images (with dark frames, flats, and flat darks) as you go. Results can be plotted as light curves and exported as CSV data files. Detailed step-by-step instructions come with the software.

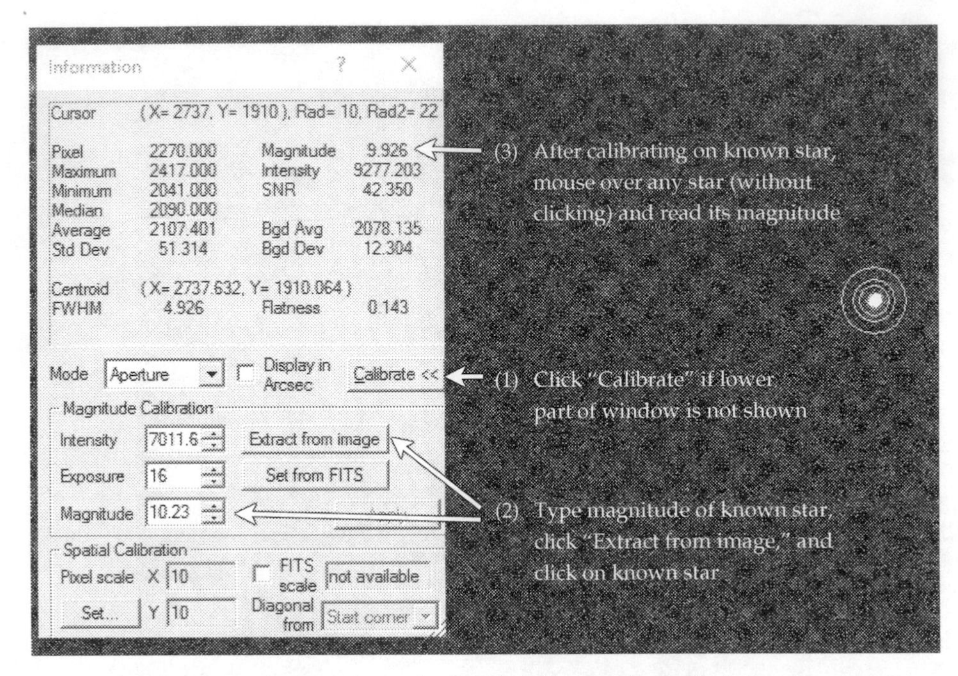

Figure 18.8. Aperture photometry can be performed manually in the Information window of *MaxIm DL*, including editions that do not include the Photometry tool. Adjust size of aperture by right-clicking on it.

Very similar functionality is provided by the freeware package *Muniwin* (http://c-munipack.sourceforge.net), by David Motl and colleagues at Masaryk University, Brno, Czech Republic. It is available compiled for Windows and as source code for UNIX-like operating systems. It, too, includes good step-by-step instructions.

Like a real photometer but unlike *MaxIm DL*, *Muniwin* outputs only differential magnitudes (the difference between the variable star and the comparison star), not actual magnitudes. A handy feature of *Muniwin* is that it tries several aperture sizes and lets you select the one that gives the best signal-to-noise ratio. A hint for *Muniwin* users: If you work with defocused star images, set their approximate size (FWHM) under *Project, Edit project settings,* so that the star matcher will recognize them.

Development in this area is proceeding rapidly, and I urge you to check the AAVSO's web site for updated recommendations.

18.6.4 Example: Light Curve of EH Librae

Figure 18.9 shows an example of what can be done. EH Librae is a short-period, low-amplitude variable star with a period of about two hours. I took 24 15-second exposures of it at five-minute intervals using a 300-mm $f/4$ lens at $f/5$

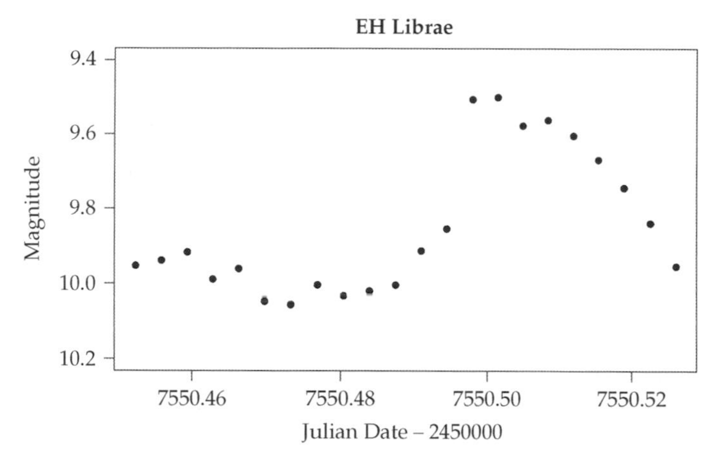

Figure 18.9. Light curve based on a series of 15-second exposures with a 300-mm lens at $f/5$, defocused, and a Canon 60Da. Reference star was HIP 73276, magnitude 10.23.

on a Canon 60Da body set to ISO 400 and Celestron AVX mount. The camera was deliberately out of focus to give star images about 10 pixels in diameter, and I verified, using the Information window in *MaxIm DL*, that they did not reach or approach maximum brightness. Dark-frame subtraction was done in the camera.

Then I used *MaxIm DL* to deBayer all the raw image files using its default settings for the Canon 60Da, convert all the images to monochrome, and process them with the Photometry tool. As a comparison star I used HIP 73276, taking its visual magnitude to be 10.23 as given in *Simbad*. Although *MaxIm DL* can plot light curves, I exported the data in CSV format and drew a more customized graph with the *R* statistical software package. For good measure, I analyzed the same data with *Muniwin* and got very much the same results. Both software packages helpfully converted the camera's timestamps into Julian dates.

18.7 Asteroid or Nova?

Unexpected stars that intrude into your deep-sky images are likely to be asteroids (minor planets). If you know what asteroid you're looking for, you can download its orbital data using *Stellarium* or other sky-charting software and plot its position. But what if you've found an apparent asteroid and don't know which one it is?

In that case, the service to use is *MPChecker*, available by Web from the Minor Planet Center (www.minorplanetcenter.net) under "Observers." Give it a right ascension, declination, and date, and it will tell you all the known asteroids that are within a specified distance of that point, down to magnitude 18 or fainter.

There was a time when amateurs could easily discover asteroids — a brief golden age from about 1995 to 2010 when amateur CCD cameras could reach

17th magnitude or better but most of the asteroids within their reach had not yet been discovered. Nowadays, goverment-funded surveys for near-earth objects (NEOs) catch practically all the asteroids, so that opportunity is gone, but we might still need to know whether an unexpected star is an asteroid or a nova. Compounding the problem, unlike the observatories of the film-and-plate era, we use exposures so short that the motion of an asteroid in a single exposure will not leave a streak. Thus the need for *MPChecker*.

The Minor Planet Center also provides ephemerides of asteroids and comets in the traditional format, computed for your location and date. Another treasure-trove of ephemerides is CalSky, www.calsky.com, which performs a wide range of calculations and predictions of interest to astronomical observers.

18.8 Research Literature On Line

Finally, the world's greatest astronomy library is now free on the Web. At http://adsabs.harvard.edu is the online library of the SAO/NASA Astrophysics Data System (ADS). Besides a searchable database of research papers, it also contains full-text scans of major astronomy journals going back to the early 1800s, observatory publications, and selected historical books. Recent issues of journals are not included – if they were given away free, the journals would have no source of income – but abstracts of recently published journal articles often include *arXiv e-prints,* which are preprints provided in digital form by the author.

ADS is, of course, where *Simbad* usually sends you when you look up papers about celestial objects. It is also enjoyable to browse. No longer must amateurs rely on third-hand accounts of the discovery of nebulae and galaxies – we can read the discoverers' accounts for ourselves, and perhaps even investigate long-standing puzzles.

If you're looking for a long-standing puzzle, go and read about Baxendell's Unphotographable Nebula (NGC 7088, not in *Simbad* because it is thought not to exist). The story starts with Baxendell's article in *Monthly Notices of the Royal Astronomical Society,* **41**, p. 48 (1881), with follow-up half a century later by J. G. Hagen in the same journal (**90**, pp. 331–333, 1930) and a possibly final wrap-up (in German, with a useful map) by W. Strohmeier and A. Güttler in *Astronomische Nachrichten,* **280**, pp. 254–258 (1950). Strohmeier and Güttler think the nebula is an optical illusion – a lack of bright stars in part of the field makes the sky background seem brighter there. But has the mystery really been solved? A DSLR with extended red response is the ideal instrument for investigating further. While you're at it, look for recent research on "galactic cirrus" or "integrated flux nebulosity" (IFN), the thin nebulosity that is now known to extend all over our galaxy.

Part V
Appendices

Appendix A
Digital Processing of Film Images

You can, of course, process film images with the same software that you use for DSLR images. Noise takes the form of random grain only; there is no fixed-pattern noise. Stacking multiple images builds contrast and reduces grain.

First you have to get the film images (slides or negatives) into digital form. There are many methods. The best is a film scanner with a resolution of at least 2400 dpi (about 100 pixels/mm). That is enough to do justice to almost any film image, even though it corresponds to less than 9 megapixels for the full frame. Some scanners can detect and electronically remove dust and scratches, which are distinguishable from the dyes in color film because they are opaque to infrared light.

Professional-quality film scanners have been abundant since the 1990s (actually before the heyday of digital cameras), and I use a vintage Nikon Coolscan III LS-30, which gives excellent results. Unfortunately, drivers for classic scanners are not necessarily available for the newest operating systems. The software package *VueScan* (www.hamrick.com) supports scanners old and new without drivers, and I rely on it. *Vuescan* gives you an exceptional level of control over the scanner, including the ability to scan repeatedly and combine the results for a lower-noise image.

Many flatbed scanners also scan film. My results with them have been mixed. Some flatbed scanners rival film scanners, provided the glass is kept scrupulously clean; others obviously acquire a rather blurred image and then "fake it" by applying a strong sharpening filter. That is not as good as scanning the image faithfully in the first place.

You can use your DSLR to digitize film images. Any slide duplicator attachment that fits a film SLR will also work with a DSLR, except that it may not cover the whole slide because the DSLR sensor is smaller than a film frame. The alternative is to use the DSLR with a macro lens and light box, and simply photograph the slide or negative (Figure A.1). To get the camera perpendicular to

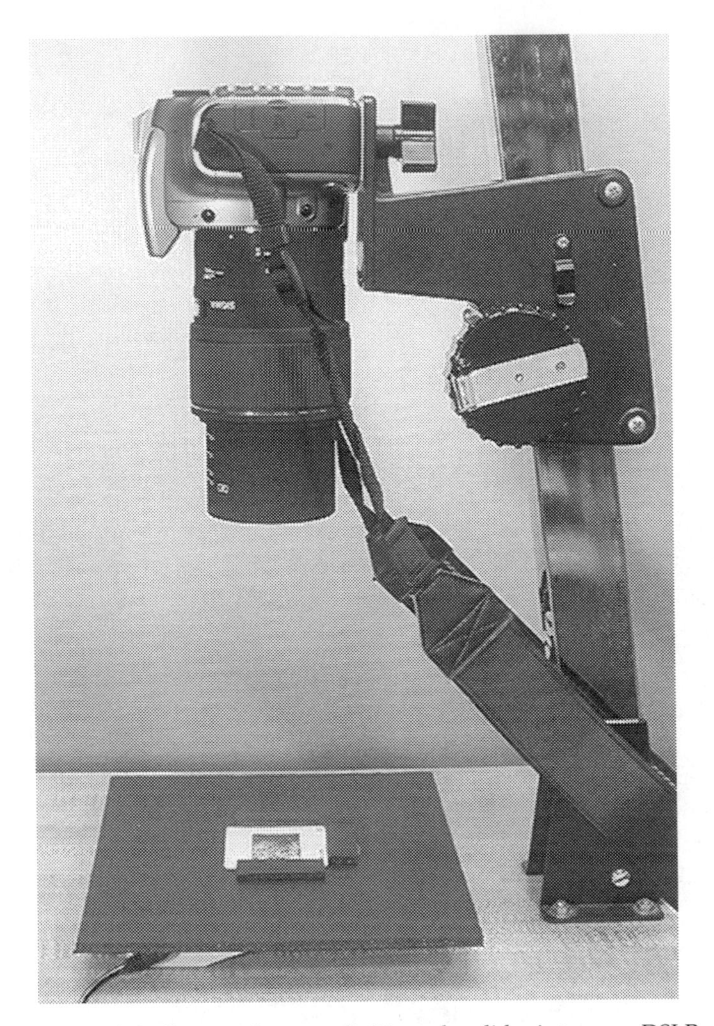

Figure A.1. One good way to digitize color slides is to use a DSLR with a macro lens and light box. Work in a darkened room.

the slide, put a mirror where the slide is going to be, and aim the camera at its own reflection.

Once the images are digitized, proceed as with DSLR or CCD images, except that gamma correction is not needed. Figure A.2 shows an example. Three Ektachrome slides were scanned. Each was scanned twice, once normally and once with the slide reversed end to end, to reduce the effect of any fixed-pattern noise that might exist in the scanner. That gave a total of six images, three of which were upside down, and not all were exactly the same size because of *VueScan*'s automatic cropping to margins. I used *PixInsight* to crop them to the same size and turn the inverted images right side up, then stacked them with *DeepSkyStacker* and went back to *PixInsight* to flatten the background, adjust contrast, and reduce noise.

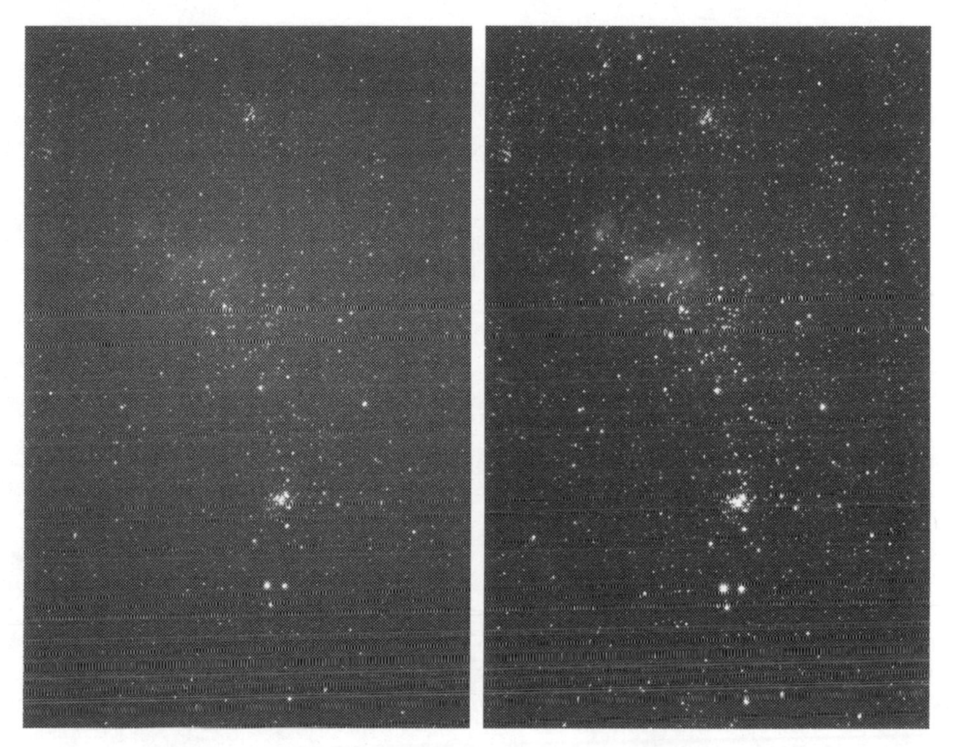

Figure A.2. Effect of stacking and processing film images. *Left:* Nebula IC 4628 in Scorpius; central part of single 7-minute exposure on Kodak Elite Chrome 200 film with Nikon F3 and 300-mm $f/4$ lens. *Right:* Result of stacking three exposures with *DeepSkyStacker* and processing in *PixInsight.*

Appendix B
Exposure Tables

The following tables give suggested exposure times for a wide variety of celestial objects. Your own trial and error should always take precedence over published exposures, but these tables will give you a good indication in advance of what settings might work.

B.1 Sun

	m″	ISO	*f*/4	*f*/8	*f*/16	*f*/32
Sun, uneclipsed or partial eclipse, 6.0 D filter		400	1/2000 s	1/500 s	1/125 s	1/30 s
Sun, uneclipsed or partial eclipse, 5.0 D filter		400	—	1/4000 s	1/1000 s	1/250 s
Sun, total eclipse, prominences	4.8	1600	1/4000 s	1/1000 s	1/250 s	1/60 s
Sun, total eclipse, inner corona (3° field)	7.5	1600	1/500 s	1/125 s	1/30 s	1/8 s
Sun, total eclipse, outer corona (10° field)	9.0	1600	1/125 s	1/30 s	1/8 s	1/2 s

B.2 Moon

	m″	ISO	*f*/4	*f*/8	*f*/16	*f*/32
Moon, thin crescent	6.7	400	1/250 s	1/60 s	1/15 s	1/4 s
Moon, wide crescent	5.9	400	1/500 s	1/125 s	1/30 s	1/8 s
Moon, quarter phase	5.2	400	1/1000 s	1/250 s	1/60 s	1/15 s
Moon, gibbous	4.4	400	1/2000 s	1/500 s	1/125 s	1/30 s
Moon, full	3.4	400	1/4000 s	1/1000 s	1/250 s	1/60 s

continued over

Moon, dimly lit features on terminator	5.9	1600	1/2000 s	1/500 s	1/125 s	1/30 s
Moon, brightly lit features on terminator	5.2	1600	1/4000 s	1/1000 s	1/250 s	1/60 s
Moon, earthshine on unlit portion	13.5	1600	1/2 s	3 s	—	—
Moon, eclipse, partial	4.4	400	1/2000 s	1/500 s	1/125 s	1/30 s
Moon, eclipse, umbra + penumbra	10.5	1600	1/30 s	1/8 s	1/2 s	—
Moon, eclipse, total, relatively light	12.2	1600	1/4 s	1 s	3 s	—
Moon, eclipse, total, relatively dark	14.8	1600	2 s	8 s	35 s	—

B.3 Planets

	m''	ISO	*f*/4	*f*/8	*f*/16	*f*/32
Mercury	4.0	400	1/2000 s	1/500 s	1/125 s	1/30 s
Venus	1.5	400	—	—	1/2000 s	1/500 s
Mars	4.0	400	1/2000 s	1/500 s	1/125 s	1/30 s
Jupiter	5.3	400	1/1000 s	1/250 s	1/60 s	1/15 s
Mercury	4.0	1600		1/2000 s	1/500 s	1/125 s
Venus	1.5	1600	—	—	—	1/2000 s
Mars	4.0	1600	—	1/2000 s	1/500 s	1/125 s
Jupiter	5.3	1600		1/1000 s	1/250 s	1/60 s
Saturn	6.7	1600	1/1000 s	1/250 s	1/60 s	1/15 s
Uranus	8.6	1600	1/125 s	1/30 s	1/8 s	1/2 s
Neptune	9.3	1600	1/60 s	1/15 s	1/4 s	1 s

B.4 Deep-sky Objects

No ISO setting is specified for deep-sky objects because deep-sky work is so different from daytime photography; pictures are normally underexposed (by daytime standards) and strongly stretched in processing. The ISO value used in the calculations was 1600, but some cameras perform best at settings as low as 200.

	m''	*f*/4	*f*/8	*f*/16	*f*/32
General deep-sky imaging	18 (typ.)	40 s	3 min	12 min	—
Nebulae, brightest cores (M42, M17, M8)	16	6 s	25 s	2 min	7 min
Nebulae, faint (California, Horsehead)	20	4 min	15 min	—	—
Galaxies, bright cores	18	40 s	3 min	12 min	—
Galaxies, outer regions	20	4 min	15 min	—	—

B.5 How Exposures are Calculated

The surface brightness of celestial objects is measured as magnitude per square arc-second (m''), where 1 m'' is equivalent to the light of a star of magnitude 1.0 spread out over one square arc-second of sky, and just as with star magnitudes, higher numbers indicate fainter objects. Of course, most objects do not have the same brightness over their entire surface, so averages are used. Exposures are calculated by the formula:

$$t = \frac{f^2}{I \times 2.512^{(9-m'')}},$$

where t is exposure time, f is the f-ratio, I is the ISO speed, 2.512 is the base of the magnitude system, and 9 accounts for the difference between the magnitude scale and the brightness scale used with ISO film speeds.

The exposures are rounded to standard shutter speeds and are inherently approximate; even twice or half as much exposure makes only a small difference. In particular, deep-sky work is so unlike daytime photography that the ISO speed does not have its usual meaning; image brightness is established during postprocessing, and the camera's job is simply to capture photons. In that situation, an ISO value of 1600 is a good starting point for calculations even with cameras that work better at a lower setting, still capturing the same number of photons.

Instead of m'', you may sometimes see brightnesses of faint objects given as magnitude per square arc-minute (m'). Those are 8.89 magnitudes brighter (lower), so that, for example, $10\ m' = 18.89\ m''$.

Of course, with digital sensors, unlike film, no adjustment for reciprocity failure is needed. Another difference between digital and film photography is that sky fog limits are more generous; the sky background does not have to be dark because it will be subtracted out.

Index

Printed in the United States
By Bookmasters